Drug Carriers Production Processes for Innovative Human Applications

Drug Carriers Production Processes for Innovative Human Applications

Guest Editor

Paolo Trucillo

MDPI

Basel • Beijing • Wuhan • Barcelona • Belgrade • Novi Sad • Cluj • Manchester

Guest Editor
Paolo Trucillo
Department of Chemical,
Material and Industrial
Production Engineering
University of Naples Federico II
Naples
Italy

Editorial Office
MDPI AG
Grosspeteranlage 5
4052 Basel, Switzerland

This is a reprint of the Special Issue, published open access by the journal *Processes* (ISSN 2227-9717), freely accessible at: https://www.mdpi.com/journal/processes/special_issues/Drug_Carriers_Human_Applications.

For citation purposes, cite each article independently as indicated on the article page online and as indicated below:

Lastname, A.A.; Lastname, B.B. Article Title. *Journal Name* **Year**, *Volume Number*, Page Range.

ISBN 978-3-7258-3987-2 (Hbk)
ISBN 978-3-7258-3988-9 (PDF)
https://doi.org/10.3390/books978-3-7258-3988-9

Cover image courtesy of Paolo Trucillo

Contents

About the Editor

Paolo Trucillo

Paolo Trucillo is currently an Assistant Professor in the Department of Chemical, Materials, and Industrial Production Engineering at the University of Naples Federico II, Italy. His scientific activity focuses on the design of sustainable materials, the development of drug delivery systems, and the physicochemical behavior of polymer-based structures. He collaborates with the FoamLab research group, where he is involved in the design of advanced polymeric carriers for controlled drug release, particularly through the use of bio-based materials and physical foaming processes. His recent work has focused on modeling the release kinetics of active compounds within biodegradable matrices, with applications in pharmacology and tissue engineering. He has also coordinated interdisciplinary studies on the environmental impact of polymer production and the substitution of fossil-based plastics with renewable alternatives. He has authored several peer-reviewed articles and served as the Guest Editor of several Special Issues for *Processes*. His work is driven by a strong interest in combining materials science with biomedical applications to develop efficient and eco-sustainable solutions for the healthcare industry.

Preface

This Special Issue reprint focuses on the advancement of drug delivery technologies, with particular emphasis on nanocarriers, controlled release systems, and their applications in pharmacology and biomedical engineering. The collection features original research and reviews contributed by scientists working at the intersection of pharmaceutical sciences, medical research, and materials engineering.

The aim of this Special Issue is to highlight recent efforts in designing innovative therapeutic systems that improve bioavailability, target specificity, and treatment efficacy. The papers explore a wide range of delivery platforms, such as hydrogels, liposomes, transdermal systems, nanoemulsions, and polymeric carriers, highlighting both experimental and modeling approaches. The topics span from antibacterial and anticancer applications to wound healing and chronic disease management, reflecting the urgent demand for effective and personalized therapies.

This reprint is intended for an audience of researchers, clinicians, engineers, and students engaged in drug formulation, biomaterial development, and therapeutic innovation. We hope that it will serve as both a valuable resource and a source of inspiration for future work in this rapidly evolving field.

We extend our sincere thanks to all contributing authors for their high-quality submissions and to the reviewers for their critical evaluations and constructive feedback, which helped ensure the scientific rigor and relevance of the articles included in this Special Issue.

Paolo Trucillo
Guest Editor

processes MDPI

Editorial

Advances in Drug Delivery System Designs and Therapeutics

Paolo Trucillo

Dipartimento di Ingegneria Chimica, dei Materiali e della Produzione Industriale, University of Naples Federico II, Piazzale Vincenzo Tecchio, 80, 80125 Naples, Italy; paolo.trucillo@unina.it

The field of drug delivery has experienced significant advancements in the design of novel carrier systems, alternative administration routes, and innovative therapeutic strategies. This Editorial provides an overview of the recent contributions to drug delivery research, including transdermal patches, liposome-based carriers, hydrogels, and innovative therapeutic formulations.

Innovations such as opioid transdermal patches, nanoemulsions for antimicrobial applications, and functionalized nanodiamonds for cancer therapy highlight the shift toward personalized medicine and improved therapeutic outcomes [1–3]. In this context, the mathematical modeling of drug release kinetics has played a fundamental role in enhancing formulation designs and ensuring predictable pharmacokinetics. Additionally, 3D-printed drug delivery systems demonstrate the potential for patient-specific treatments through controlled release profiles. Advances in glucose monitoring technology further emphasize the intersection of drug delivery with digital health solutions and remote sensing. The reviewed studies underscore the growing impact of nanotechnology and bioengineered materials in controlled drug delivery, highlighting a promising path toward effective and patient-centered therapies. Future research should focus on integrating artificial intelligence, computational modeling, and sustainable biomaterials to refine drug delivery capabilities. These developments are currently shaping the future of pharmaceutical sciences, offering more targeted, efficient, and safe drug administration methods.

Transdermal patches emerged as an effective method for administering opioids to patients with chronic pain [4–6]. On this topic, De Marco's study highlights the use of fentanyl and buprenorphine in patch formulations, emphasizing the benefits of sustained drug release and reduced gastrointestinal side effects [7]. The research discusses the different features of reservoir and matrix patch designs and suggests that patient-specific factors, such as age and site of application, should guide the choice of formulation.

Silva explores the development of hydrogels [8–12] incorporating cellulose nanocrystals that are loaded with vitamins, aimed at addressing vitamin D deficiency and associated health conditions [13]. These hydrogels demonstrated high biocompatibility and efficient controlled release properties, making them suitable for applications in dermatology and systemic dietary supplements.

Liposomes have been widely studied for their well-recognized potential in targeted drug delivery [14,15]. Ilić-Stojanović et al. provide a comprehensive review of patented liposome formulations, highlighting their advantages in enhancing drug stability and bioavailability [16]. The study underscores the role of intellectual property in protecting and supporting innovation in pharmaceutical development.

A critical aspect of drug delivery research is the mathematical modeling of release kinetics. Trucillo presents a review of widely used empirical, semi-empirical, and theoretical mathematical models to describe drug release phenomena, examining the physical,

Received: 21 March 2025
Accepted: 7 April 2025
Published: 13 April 2025

Citation: Trucillo, P. Advances in Drug Delivery System Designs and Therapeutics. *Processes* **2025**, *13*, 1176. https://doi.org/10.3390/pr13041176

chemical, and empirical meaning of each variable involved [17]. These models help to predict drug behavior in various formulations and optimize carrier design.

Being involved in drug delivery kinetic determination and prediction, Trucillo also provides an insightful description of his personal experience with discrete and continuous glucose monitoring systems for type 1 diabetes management [18]. This specific study highlights the advantages of automated insulin delivery (semi-automated artificial pancreas) in reducing glycemic variability and improving patients' quality of life.

Sultan et al. introduce a novel palm-oil-based α-mangostin nanoemulsion for use as an antimicrobial endodontic irrigant [19]. These findings demonstrate efficacy against biofilms of *Enterococcus faecalis*, *Staphylococcus epidermidis*, and *Candida albicans*, demonstrating the potential of this formulation as an alternative to conventional irrigation solutions in root canal therapy.

Agbo et al. present a redesigned quinine formulation using nanotechnology for intranasal administration, offering a promising alternative to oral quinine delivery [20]. The study shows that nanosuspensions significantly enhance drug permeation through nasal mucosa, potentially improving patient adherence and reducing gastrointestinal side effects.

Khalifa et al. investigated the development of a tea-tree-oil-based nanoemulgel loaded with azithromycin for enhanced antibacterial activity [21]. The paper demonstrates that nanoemulgels improve drug permeation and transdermal flux, providing a viable option for treating bacterial skin infections.

Zhu et al. review the role of antioxidants and drug carriers in promoting skin wound repair [22]. The study discusses the benefits of nanofibers and hydrogels in enhancing the protection and subsequent bioavailability of antioxidants, leading to improved wound healing outcomes.

Acuña-Aguilar et al. explore the application of functionalized nanodiamonds loaded with metformin in breast and ovarian cancer treatment [23]. Their study indicates enhanced cytotoxic effects and improved metformin delivery efficiency, highlighting the potential of nanodiamonds as promising carriers for anticancer agents.

Djilali et al. present a comprehensive stability analysis of haloperidol, focusing on its degradation under various stress conditions, including oxidation, hydrolysis, photolysis, and thermal exposure [24]. Using advanced chromatographic and thermal analytical techniques, the study identifies critical degradation pathways and highlights the importance of optimized formulation strategies to ensure pharmaceutical stability. The findings provide valuable insights into the chemical integrity of haloperidol, which is widely used for treating psychiatric disorders, thus influencing storage recommendations and formulation improvements.

Nascimento et al. developed 3D-printed filaments with pH-dependent solubility for controlled drug release applications [25]. By utilizing Fused Deposition Modeling (FDM), the study demonstrates the feasibility of producing polymeric filaments that release drugs in a pH-specific receiving medium, aiming at achieving targeted gastrointestinal therapies. The research emphasizes the potential of 3D printing in personalized medicine, enabling the fabrication of patient-specific dosage forms with tunable drug release profiles.

The studies reviewed in this Editorial underscore the dynamic nature of drug delivery research and its impact on medical therapies. From personalized pain management to advanced cancer treatment strategies, these contributions reflect a collective effort to enhance drug efficacy, patient compliance, and treatment outcomes. Future research should continue to integrate nanotechnology, computational modeling, and sustainable materials to further refine drug delivery systems, thus improving human knowledge on these topics.

Conflicts of Interest: The author declares that this editorial was written in total absence of any commercial or financial relationship that could constitute potential conflict of interest.

References

1. Allen, T.M.; Cullis, P.R. Liposomal Drug Delivery Systems: From Concept to Clinical Applications. *Adv. Drug Deliv. Rev.* **2013**, *65*, 36–48. [CrossRef] [PubMed]
2. Mitchell, M.J.; Billingsley, M.M.; Haley, R.M.; Wechsler, M.E.; Peppas, N.A.; Langer, R. Engineering Precision Nanoparticles for Drug Delivery. *Nat. Rev. Drug Discov.* **2021**, *20*, 101–124. [CrossRef]
3. Park, K. Nanotechnology: What it can do for drug delivery. *J. Control. Release Off. J. Control. Release Soc.* **2008**, *120*, 1. [CrossRef] [PubMed]
4. Prausnitz, M.R.; Langer, R. Transdermal Drug Delivery. *Nat. Biotechnol.* **2008**, *26*, 1261–1268. [CrossRef] [PubMed]
5. Escobar-Chávez, J.J.; Díaz-Torres, R.; Rodríguez-Cruz, I.M.; Domínguez-Delgado, C.L.; Morales, R.S.; Ángeles-Anguiano, E.; Melgoza-Contreras, L.M. Nanocarriers for transdermal drug delivery. *Res. Rep. Transdermal Drug Deliv.* **2012**, *2012*, 3–17. [CrossRef]
6. Basu, S.; Biswas, P.; Anto, M.; Singh, N.; Mukherjee, K. Nanomaterial-enabled drug transport systems: A comprehensive exploration of current developments and future avenues in therapeutic delivery. *3 Biotech* **2024**, *14*, 289. [CrossRef]
7. De Marco, I. Transdermal Patches Containing Opioids in the Treatment of Patients with Chronic Pain. *Processes* **2023**, *11*, 2673. [CrossRef]
8. Li, J.; Mooney, D.J. Designing Hydrogels for Controlled Drug Delivery. *Nat. Rev. Mater.* **2016**, *1*, 16071. [CrossRef]
9. Jeong, W.Y.; Kwon, M.; Choi, H.E.; Kim, K.S. Recent advances in transdermal drug delivery systems: A review. *Biomater. Res.* **2021**, *25*, 1–15. [CrossRef]
10. El-Sherbiny, I.M.; Yacoub, M.H. Hydrogel Scaffolds for Tissue Engineering: Progress and Challenges. *Glob. Cardiol. Sci. Pract.* **2013**, *2013*, 38. [CrossRef]
11. Zhang, A.; Jung, K.; Li, A.; Liu, J.; Boyer, C. Recent advances in stimuli-responsive polymer systems for remotely controlled drug release. *Prog. Polym. Sci.* **2019**, *99*, 101164. [CrossRef]
12. Zhang, Y.; Yu, T.; Peng, L.; Sun, Q.; Wei, Y.; Han, B. Advancements in hydrogel-based drug sustained release systems for bone tissue engineering. *Front. Pharmacol.* **2020**, *11*, 622. [CrossRef] [PubMed]
13. Silva, N.D.C.; Silva, C.J.T.; Gonçalves, M.P.; Borsagli, F.G. Carboxymethyl-Cellulose-Based Hydrogels for Controlled Drug Delivery. *Processes* **2024**, *12*, 1437. [CrossRef]
14. Yang, C.; Merlin, D. Lipid-based drug delivery nanoplatforms for colorectal cancer therapy. *Nanomaterials* **2020**, *10*, 1424. [CrossRef] [PubMed]
15. Tang, L.; Winkeljann, B.; Feng, S.; Song, J.; Liu, Y. Recent advances in superlubricity of liposomes for biomedical applications. *Colloids Surf. B Biointerfaces* **2022**, *218*, 112764. [CrossRef]
16. Ilić-Stojanović, S.; Cakić, S.; Nikolić, N.; Petrović, S. Innovative and Patented Liposome-Based Drug Carriers. *Processes* **2024**, *12*, 1970. [CrossRef]
17. Trucillo, P. Drug Carriers: A Review on the Most Used Mathematical Models for Drug Release. *Processes* **2022**, *10*, 1094. [CrossRef]
18. Trucillo, P. Discrete and Continuous Glucose Monitoring Systems. *Processes* **2022**, *10*, 2706. [CrossRef]
19. Sultan, O.S.; Kantilal, H.K.A.; Phaik, K.S.; Choudhury, H.; Davamani, F. Formulation and Characterization of a Novel Palm-Oil-Based α-Mangostin Nano-Emulsion (PO-AMNE) as an Antimicrobial Endodontic Irrigant: An In Vitro Study. *Processes* **2023**, *11*, 798. [CrossRef]
20. Agbo, C.P.; Ugwuanyi, T.C.; Eze, O.C.; Onugwu, A.L.; Echezona, A.C.; Nwagwu, C.S.; Uzondu, S.W.; Ogbonna, J.D.; Ugorji, L.O.; Nnamani, P.O.; et al. Quinine: Redesigned and Rerouted. *Processes* **2023**, *11*, 1811. [CrossRef]
21. Khalifa, N.E.; Abdallah, M.H.; Elghamry, H.A.; Khojali, W.M.; Khafagy, E.S.; El-Sayed, E.H.; Shawky, S. Development of Tea Tree Oil-Based Nanoemulgel for Enhancing Antibacterial Activity. *Processes* **2023**, *11*, 1836. [CrossRef]
22. Zhu, X.; Yuan, W.; Li, Z.; Lin, Y.; Li, W.; Ji, L.; Wang, D.; Zhang, H.; Wang, Y. Progress of research on antioxidants and carriers for skin wound repair. *Processes* **2023**, *11*, 2069. [CrossRef]
23. Acuña-Aguilar, L.E.; Conejo-Dávila, A.S.; Miki-Yoshida, M.; Hernández-de la Cruz, O.N.; Sánchez-Sánchez, G.; López-Camarillo, C.; Salas-Leiva, J.S.; Zaragoza-Contreras, E.A.; Reyes-Martinez, R.; Orrantia-Borunda, E. Improved Cytotoxic Capacity of Functionalized Nanodiamonds with Metformin. *Processes* **2023**, *11*, 2616. [CrossRef]

24. Djilali, K.; Maachi, R.; Boutoumi, H.; Lekmine, S.; Mesbah, Z.A.; Attia, S.M.; Nasrallah, N.; Bouallouche, R.; Zhang, J.; Tahraoui, H.; et al. Comprehensive Stability Analysis of Haloperidol: Insights from Advanced Chromatographic and Thermal Analysis. *Processes* **2025**, *13*, 904. [CrossRef]
25. Nascimento, G.M.; Cardoso, P.H.N.; da Silva, E.M.E.; Tavares, G.F.; Olivier, N.C.; Faia, P.M.; Araùjo, E.S.; Silva, F.S. FDM 3D Printing Filaments with pH-Dependent Solubility: Preparation, Characterization and In Vitro Release Kinetics. *Processes* **2024**, *12*, 2916. [CrossRef]

processes

MDPI

Article

Comprehensive Stability Analysis of Haloperidol: Insights from Advanced Chromatographic and Thermal Analysis

Khadidja Djilali [1], Rachida Maachi [1], Hocine Boutoumi [2], Sabrina Lekmine [3], Zohra Ait Mesbah [4], Sabry M. Attia [5], Noureddine Nasrallah [1], Rachida Bouallouche [1], Jie Zhang [6], Hichem Tahraoui [7,8] and Abdeltif Amrane [8,*]

1 Laboratory of Reaction Engineering, Department of Mechanical and Process Engineering, University of Science and Technology Houari Boumediene, Algiers-Bab Ezzouar 16111, Algeria; khadidjadjilali@gmail.com (K.D.); racha.maachi@gmail.com (R.M.); nas_nour@yahoo.fr (N.N.); r.bouallouche@mesrs.dz (R.B.)
2 Laboratoire de Génie Chimique, Département de Génie des Procédés, Faculté de Technologie, Université Saad Dahlab, Blida 1, BP270, Blida 09000, Algeria; ybentoumi@ymail.com
3 Biotechnology, Water, Environment and Health Laboratory, Abbes Laghrour University, Khenchela 40000, Algeria; sabrina.lekmin.400@gmail.com
4 Laboratory of Energy Processes and Nanotechnology, Blida University 1 rue de Soumaa, BP 270, Blida 09000, Algeria; zohraait.mesbah@gmail.com
5 Department of Pharmacology and Toxicology, College of Pharmacy, King Saud University, Riyadh 11451, Saudi Arabia; attiasm@ksu.edu.sa
6 School of Engineering, Merz Court, Newcastle University, Newcastle upon Tyne NE1 7RU, UK; jie.zhang@newcastle.ac.uk
7 Laboratoire de Biomatériaux et Phénomènes de Transport (LBMPT), Université de Médéa, University of Médéa, Nouveau Pôle Urbain, Médéa 26000, Algeria; hichemm.tahraouii@gmail.com
8 Ecole Nationale Supérieure de Chimie de Rennes (ENSCR), CNRS, ISCR–UMR6226, University de Rennes, F-35000 Rennes, France
* Correspondence: abdeltif.amrane@univ-rennes.fr

Abstract: In this study, we thoroughly investigated the stability of haloperidol using a comprehensive set of chromatographic and thermal analyses. Various stress conditions were examined, including exposure to oxidizing agents (such as hydrogen peroxide), dry heat, photolytic conditions, and acid and alkaline hydrolysis. Significant degradation was observed in acidic and alkaline environments, leading to the formation of degradation by-products, specifically DPA, DPB, DPC, and DPD for acidic and basic conditions. In contrast, haloperidol demonstrated robust stability under photolytic, oxidative, and dry-heat conditions. For the analysis of the drug and its degradation products, a C-18 column was employed, coupled with a mobile phase consisting of methanol and a phosphate buffer (pH = 9.8) in a 90:10 (v/v) ratio. The analytical method was rigorously validated according to ICH Q2 (R1) guidelines, ensuring its accuracy and reliability. This method exhibited excellent linearity within a concentration range of 1 to 50 µg/mL, with an R^2 of 0.999. Additionally, this method is applicable to commercial formulations, without the need for prior extraction. LC-MS/MS analysis revealed distinct m/z values and fragmentation spectra corresponding to the degradation products, including an impurity not documented in the European Pharmacopoeia monograph for the drug. Three additional degradation products were identified based on m/z values and base fragments. Thermal analyses, including thermogravimetric analysis (TGA), differential scanning calorimetry (DSC), and differential thermal analysis (DTA), provided further evidence of the active ingredient's thermal stability, with a melting temperature of approximately 150 °C. These results collectively offer valuable insights into the degradation behavior of haloperidol, providing critical implications for its pharmaceutical quality and integrity under various environmental conditions.

Academic Editors: Alina Pyka-Pająk and Paolo Trucillo

Received: 30 December 2024
Revised: 3 March 2025
Accepted: 17 March 2025
Published: 19 March 2025

Citation: Djilali, K.; Maachi, R.; Boutoumi, H.; Lekmine, S.; Ait Mesbah, Z.; Attia, S.M.; Nasrallah, N.; Bouallouche, R.; Zhang, J.; Tahraoui, H.; et al. Comprehensive Stability Analysis of Haloperidol: Insights from Advanced Chromatographic and Thermal Analysis. *Processes* **2025**, *13*, 904. https://doi.org/10.3390/pr13030904

Keywords: haloperidol; validation; stability; stress conditions; HPLC; LC-MS/MS; TGA/DSC/DTA

1. Introduction

Ensuring the accurate composition of drugs is a paramount concern in the pharmaceutical industry [1,2]. The reliability and accuracy of the active pharmaceutical ingredients (APIs) declared by manufacturers are crucial factors in determining the efficacy and safety of medical treatments [3–5]. Various analytical techniques are utilized to verify the qualitative and quantitative compositions of drugs. These methods play a vital role in quality control during drug production, the evaluation of generic alternatives, and the identification of counterfeit products [6,7]. Among the preferred techniques is high-performance liquid chromatography (HPLC) coupled with UV-Vis detection. Additionally, more advanced methods, such as liquid chromatography combined with mass spectrometry (LC-MS), are employed to precisely identify active pharmaceutical ingredients. Furthermore, thermal analysis techniques like thermogravimetric analysis (TGA) can be used for comprehensive characterization [8,9].

This study focuses on haloperidol, a molecule belonging to the class of basic butyrophenones, commonly used as a neuroleptic agent. Haloperidol is prescribed for the treatment of acute and chronic psychoses and psychotic disorders, such as schizophrenia, manic states, and delusions [10,11]. Widely used in many countries, it is considered equally effective compared to other low-potency psychotropic drugs, including chlorpromazine and thioridazine [12]. Chemically known as 4-[4-(4-chlorophenyl)-4-hydroxy-1-piperidinyl]-1-(4-fluorophenyl)-1-butanone, haloperidol is a white, crystalline powder with a molecular weight of 375.9 g/mol. It has a significant pKa value, typically reported to be around 8.3, which corresponds to the phenolic hydroxyl group. This group plays a crucial role in determining the solubility and ionic form of haloperidol, factors that are essential for its absorption and bioavailability in the body. The phenolic hydroxyl group, consisting of a hydroxyl group (-OH) attached to a benzene ring, is directly connected to the main aromatic ring of the molecule. The chemical structure of haloperidol, highlighting the phenol hydroxyl group responsible for its pKa value, is illustrated in Figure 1 [13]. Although haloperidol exhibits low solubility in water [14,15], it can dissolve in the presence of lactic acid or tartaric acid [14,16–18].

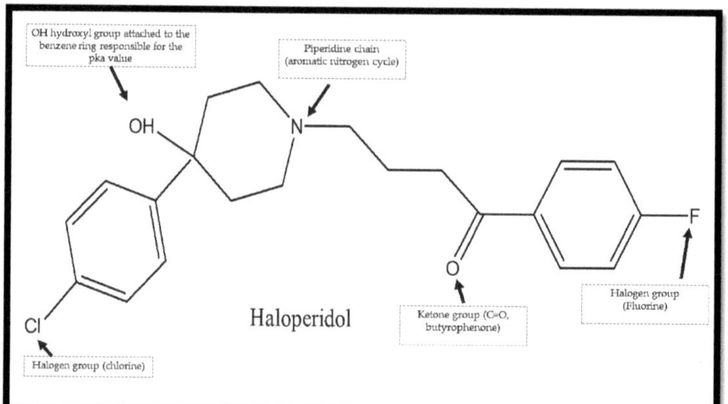

Figure 1. Chemical structure of haloperidol.

Haloperidol is available in tablet, oral, and injectable forms, so it is crucial to ensure the long-term stability of these formulations [16]. Numerous studies have focused on the determination of haloperidol in pharmaceutical forms, particularly in solid formulations. The most commonly used analytical methods include HPLC, UV-visible spectrophotometry, and derivative spectrophotometry [5,15,19,20], often combined with extraction or complexation steps [21–30]. One research team prepared a haloperidol solution with a concentration of 2 mg/mL using purified water, adding a small excess of lactic acid to ensure stability. The solution was then protected from exposure to sunlight or daylight to maintain its integrity [17,31]. The results demonstrated that haloperidol is sensitive to light, as the solutions stored in clear glass vials showed discoloration within hours and developed a greyish precipitate over several weeks [16]. In contrast, the solutions stored in amber glass vials remained unchanged, with no discoloration or precipitation observed even after eighteen months. Further investigations in 1981 confirmed that haloperidol solutions containing lactic acid at pH 3 remained stable for up to 5 years at room temperature, 2 years at 40 °C, and 6 months at 60 °C. However, exposure to natural light caused cloudiness, discoloration, and a reduction in haloperidol content [18]. In 1983, a new HPLC method was developed to identify haloperidol and detect its degradation products, revealing its instability under high temperatures, light exposure, and varying pH conditions [32,33]. By 1988, researchers had examined various physicochemical properties of haloperidol, particularly regarding the effects of oxygen and light. Their findings over a two-year stability period indicated that oxygen had minimal impact on haloperidol solutions stored in amber glass [34]. In 2002, an HPLC method for haloperidol determination was introduced, enabling effective separation of its degradation products. This method involved the use of an octadecylsilane stationary phase column under isocratic conditions, with degradation experiments conducted using hydrochloric acid, sodium hydroxide, and hydrogen peroxide, leading to the identification of major degradation products [35]. Additionally, in 2003, a rapid and sensitive HPLC technique was developed and validated for the simultaneous determination of haloperidol and its degradation products (Cis-PFBoxide, hydroxy-PFB, 2H-PFB, and trans-PFBoxide) using a porous graphitic carbon (PGC) column [36]. The extensive body of literature highlights haloperidol's susceptibility to hydrolysis, oxidation, photolysis, and heat, emphasizing the need for careful handling to protect the drug from these degrading factors, whether in liquid or solid form. Various chromatographic techniques, such as UPLC, RP-HPLC, capillary electrophoresis, and thin-layer chromatography, have been employed to evaluate the stability of haloperidol and its formulations. However, there remains a significant gap in fully characterizing and identifying its degradation products [37,38].

The inconsistencies in findings regarding storage conditions, packaging, and duration underscore the need for a more in-depth investigation into the stability of haloperidol. However, little research has been conducted on its quantitative analysis. In this study, we aimed to explore the degradation behavior of haloperidol under hydrolytic, oxidative, thermal, and photolytic conditions while also characterizing the resulting degradation products using a chromatographic approach combined with mass spectrometry. There is a pressing need to develop a rapid and straightforward HPLC method for haloperidol analysis. Although numerous studies have documented HPLC and LC-MS/MS methods for various drugs, there is a lack of stability-indicating methods specifically designed for haloperidol. Recent recommendations highlight the importance of developing HPLC methods for active pharmaceutical ingredients that ensure effective separation between the drug and its degradation products; established methodologies are available for this purpose [39,40]. Additionally, a recent study provides a critical review of the use of HPLC-MS in this context [41]. Therefore, the primary objective of this study was to introduce a new and rigorously validated stability-indicating HPLC method for the quantitative determination of haloperidol, addressing degradation under acidic, basic, oxidative, thermal, and light-induced conditions. The chemical structures of the degradation products formed under stress conditions were thoroughly investigated using LC-MS/MS, while thermal analysis via TGA/DTA/DSC was employed to confirm the thermal stability of haloperidol. The insights gained from the thermal stability studies and understanding the thermal decomposition pathways are essential for advancing our theoretical knowledge of haloperidol's chemical properties. Furthermore, these findings could have practical applications in optimizing the production, processing, and storage of haloperidol. Notably, no prior research has explored the thermal stability and decomposition of haloperidol. This study aims to address this gap by investigating the thermal decomposition mechanisms of haloperidol under nitrogen atmospheres using thermogravimetric analysis (TGA) and differential scanning calorimetry (DSC). Another goal of this research was to utilize the proposed HPLC method to monitor the stability of haloperidol under various stress conditions. It is important to note that forced degradation studies lack standardized protocols [42]. In accordance with ICH guidelines, we conducted a comprehensive assessment of haloperidol's forced degradation under diverse hydrolytic, oxidative, photolytic, and thermal conditions [43].

2. Materials and Methods

2.1. Instruments and Equipment

For this research, we employed a Hewlett Packard 1100 series chromatograph (manufactured in the USA), which included a G1312A binary pump, a G1322A degasser, a G1315A UV detector, and a G1316A column furnace with a manual injector (Agilent Technologies, USA). The HPLC system was controlled, and data acquisition and chromatogram generation were managed using ChemStation software version B.04.03, Agilent Technologies, Santa Clara, CA, USA. Additionally, a Shimadzu 8040 UPLC-ESI-MS-MS system (produced in Kyoto, Japan) was utilized, featuring ultra-high sensitivity along with UFMS technology and equipped with a Nexera XR LC-20AD binary pump Kyoto, JP. Thermogravimetric analysis (TGA) was conducted using a thermal analyzer (Model SDT Q600, TA Instruments, New Castle, DE, USA) for all three experiments. The analysis covered a temperature range from 20 °C to 500 °C, with a heating rate of 5 °C/min, using aluminum crucibles containing 10 mg of the sample. Nitrogen was used as purge gas, and the resulting data were processed using the TA Instruments Universal Analysis software.

2.2. Materials

Haloperidol was kindly supplied as a gift sample by the GROUPE SAN-TE-ALGERIE laboratory in Algeria, with a purity of 99%.

HPLC-grade methanol, monobasic phosphate, lactic acid, and sodium hydroxide were procured from SIGMA-ALDRICH (USA). Hydrochloric acid was sourced from BIOCHEM (Germany), while hydrogen peroxide was obtained from SCHARLAU (Spain). Purified water and methylparaben were supplied by PUBCHEM (Germany).

2.3. Methods

Lactic acid was chosen as the solubilizing solvent for haloperidol based on bibliographic data indicating its low solubility in water (0.1 mg/mL) and methanol (approximately 16.7 mg/mL at 25 °C) as well as its favorable solubility in ethanol, methylene chloride [14,15,19], and lactic acid [18].

2.3.1. Chromatographic Conditions

Preliminary tests were conducted to fine-tune critical parameters and enhance the reproducibility of retention times. The composition of the mobile phase was explored by testing different ratios (80:20, 90:10, and 95:5) of methanol to the phosphate buffer to achieve optimal separation of degradation products. In addition to pH 9.8, more moderate pH values (7.4 and 8.0) were also evaluated, allowing for the confirmation that pH 9.8 provided superior peak resolution and effective separation. A flow rate of 1 mL/min was chosen after comparing 0.8 mL/min, 1 mL/min, and 1.2 mL/min, as this rate offered the best compromise between analysis time and separation efficiency.

The chromatographic conditions employed in the analysis included a stationary phase comprising a C18 column (250 × 4.6 mm, 5 μm) and a mobile phase. The composition of the mobile phase was precisely defined in the software to ensure accurate control of the mixture. It consisted of methanol and potassium phosphate buffer (pH 9.8) in a 90:10% v/v ratio. The potassium phosphate buffer solution was prepared following the protocol established in our earlier study [31], where 100 mg of monobasic phosphate was dissolved in 1000 mL of ultrapure water, adjusted to pH 9.8 using 1M NaOH, and filtered through a 0.45 μm syringe filter. Consequently, the mobile phase comprised 90% methanol and 10% potassium phosphate buffer by volume. For every 100 mL of mobile phase, 90 mL corresponded to methanol, and 10 mL corresponded to potassium phosphate buffer. This specific ratio was meticulously chosen to optimize the separation of sample components. The analysis was performed at room temperature with a flow rate of 1 mL/min, and eluent detection was conducted at 248 nm.

2.3.2. Forced Degradation Studies

Forced degradation studies were carried out on haloperidol with the aim of achieving a degradation rate of 5–20%. Acidic and alkaline hydrolysis were performed at 70 °C under controlled pH conditions for 7 days. Oxidative degradation was induced by exposing haloperidol to 0.3% and 3% v/v H_2O_2 at 60 °C in the absence of light for 7 days. Photolytic stress was applied by exposing haloperidol to UV light using a UV lamp (Ultraviolet light UVA 24 W (Philips PL/L, Amsterdam), 270 nm, 20 mW/cm^2) for 48 h. Additionally, dry-heat studies were conducted by subjecting haloperidol to temperatures of 60 °C and 80 °C for 15 days in a solid-state hot-air oven. After exposure, the drug samples were cooled and analyzed to ensure a final drug concentration of 20 μg/mL had been reached. Degradation was evaluated based on a reduction in the peak area of haloperidol and/or the appearance of additional peaks.

2.3.3. Analysis of Haloperidol via HPLC and Validation of the Method

Analytical methods are frequently validated in accordance with the ICH Q2 (R1) guidelines [44]. In this research, a haloperidol solution at a concentration of 100 μg/mL was analyzed using HPLC, with 1% lactic acid serving as the dissolution solvent and the mobile phase acting as the blank, measured at a wavelength of 248 nm. The linearity of the calibration curve was assessed across a concentration range of 1 μg/mL to 50 μg/mL using haloperidol standard solutions. These standards were prepared by diluting the 100 μg/mL haloperidol stock solution with methanol as the diluent. The peak areas of the standard solutions, covering concentrations from 1 μg/mL to 50 μg/mL, were measured at a maximum wavelength (λmax) of 248 nm. The haloperidol quantification method was validated according to ICH guidelines [44,45], focusing on parameters such as linearity, precision (evaluated through repeatability and intermediate precision), and accuracy.

The linearity of the calibration curve was confirmed by a very high correlation coefficient (R = 0.999) between absorbance and standard concentrations. Repeatability was assessed by determining intra-day variation for three concentrations (1, 5, and 25 μg/mL), with three repetitions for each concentration on the same day. Intermediate precision was evaluated by inter-day variation for the same three concentrations (1, 5, and 25 μg/mL), with two repetitions for each concentration analyzed over three different days.

Accuracy, which reflects the systematic error of an assay method, was evaluated at three different concentration levels (1, 5, and 25 μg/mL) once linearity was established [31,46]. Variations in the organic concentration percentage, flow rate, and wavelength of the system were intentionally introduced under the ideal chromatographic conditions described above, with the parameters recorded. The effectiveness of the proposed method was confirmed by successfully separating haloperidol and its degradation products, with no peaks observed in blank assays during the retention time of haloperidol and its degradation products.

2.3.4. Application of the Method to the Analysis of a Finished Product

The approach described in this research was utilized to examine samples of a 0.2% oral solution of haloperidol (ISOPERIDOL®) obtained from pharmacies. For the preparation of the injection sample solution, 0.5 mL of the final product was mixed with 25 mL of methanol. This mixture was then filtered using a 0.22 μm filter and subsequently analyzed using HPLC.

2.3.5. Characterization of Degradation Products Using LC-MS/MS

Samples exposed to stress conditions (1 N HCl, 1 N NaOH for 7 days, and 48 h of UV light irradiation) were analyzed using LC-MS/MS. Positive electrospray ionization (ESI) was conducted with the following parameters: a CID gas pressure of 230 KPs, a conversion dynode voltage of −6.00 kV, an interface temperature maintained at 350 °C, a desolvation line (DL) temperature of 250 °C, a nebulizing gas flow rate of 3.00 L/min, a thermal block temperature of 400 °C, and a drying gas flow rate of 15.00 L/min. The analysis was performed using a mass spectrometer. Sample separation was achieved via HPLC with a mobile phase suitable for mass spectrometry, consisting of solvent A (20% water and 0.1% formic acid) and solvent B (80% methanol). The LC-MS/MS data obtained were subsequently used for structural identification.

2.3.6. Analysis of Haloperidol via TGA/DTG/DSC

The thermal stability of haloperidol was evaluated using TGA over a temperature range of 20 °C to 500° C. The experiments were performed using a thermal analyzer (model SDT Q 600-TA, New Castle, DE, USA). Samples were heated from 20 °C to 500 °C at a rate of 5 °C/min in aluminum crucibles containing 10 mg of the sample. Nitrogen was used as the purge gas, and the data were analyzed using TA Instruments' Universal Analysis software. During the heating process, any endothermic or exothermic reactions were monitored. The variations relative to the reference were recorded and translated into a specific thermograph corresponding to the thermal process.

3. Results and Discussion

3.1. Calibration Curve and Linearity

The linearity of the calibration curve was confirmed by the high correlation coefficient (R = 0.999) between absorbance and the standard concentrations.

3.2. Chromatographic Conditions for Analysis

To ensure accurate analysis of haloperidol and its degradation products, a series of experiments were conducted to identify the most suitable mobile phase. The goal was to achieve optimal retention of haloperidol while effectively separating it from its degradation products. Various mobile phase combinations were evaluated, each aimed at delivering the desired chromatographic performance. Under the selected conditions, haloperidol was adequately retained, with a peak retention time of 3.3 ± 0.05 min.

The system's suitability was confirmed to be acceptable under these conditions. The optimized chromatographic method successfully separated excipients and degradation products from the haloperidol peak, a process that is essential for ensuring precise quantification of haloperidol without interference from other sample components. Additionally, all the eluents were efficiently detected at the optimal wavelength of 248 nm, allowing for accurate identification and quantification of haloperidol and its related compounds. Figure 2 illustrates the chromatogram obtained using these optimized conditions, highlighting the method's effectiveness in accurately analyzing haloperidol samples while maintaining system suitability and peak resolution [35,47–50].

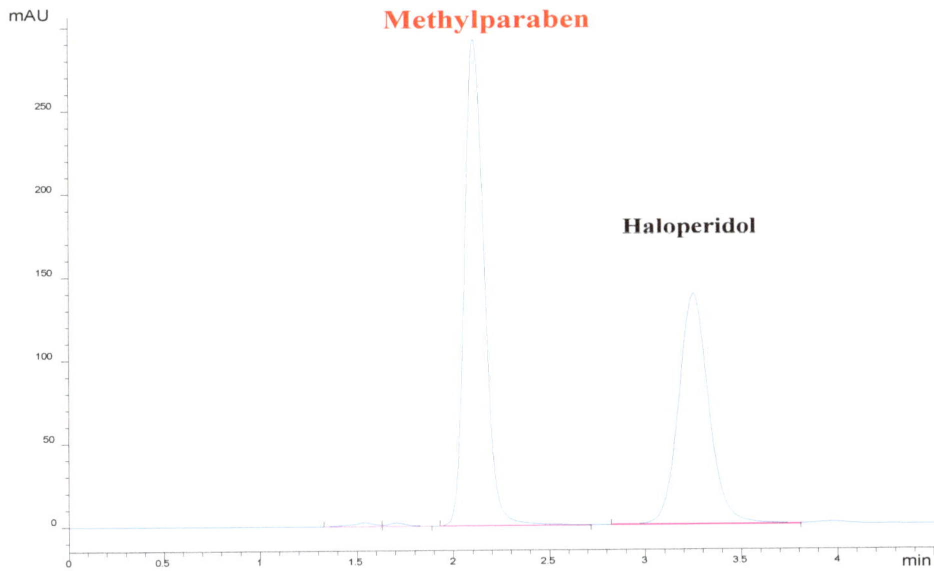

Figure 2. Chromatogram of haloperidol standard solution (50 µg/mL).

3.3. Forced Degradation Studies

Understanding the degradation pathways of haloperidol is crucial for evaluating its stability and developing effective formulation strategies. In this study, forced degradation experiments were performed to assess its behavior under hydrolytic, oxidative, thermal, and photolytic stress conditions (Table 1).

Haloperidol underwent hydrolytic degradation in both acidic and alkaline environments, revealing its sensitivity to changes in pH. When exposed to oxidative stress via hydrogen peroxide, degradation products were formed, confirming its vulnerability to oxidation. Thermal degradation was observed at higher temperatures (60 °C and 80 °C), indicating its instability under heat. Furthermore, exposure to UV light caused photolytic degradation, underscoring the importance of protecting this drug from light in formulations. These results provide valuable insights into the primary degradation pathways of haloperidol, offering essential information for enhancing its stability and guiding the development of appropriate pharmaceutical formulations and packaging.

Table 1. Forced-degradation results for haloperidol.

Forced Degradation Conditions	Percentage Degradation (%)	Degraded Concentration (µg/mL)
0.1 N HCl, 60 °C 7 days	16	3.2
1.0 N HCl, 60 °C, 7 days	25	5
0.1 N NaOH, 60 °C, 7 days	16	3.2
1.0 N NaOH, 60 °C, 7 days	17	3.4
H_2O_2 0.3%, 25 °C, 7 days	0.00	0.00
H_2O_2 0.3%, 60 °C, 7 days	0.00	0.00
H_2O_2 3.0%, 25 °C, 7 days	0.00	0.00
H_2O_2 3.0%, 60 °C, 7 days	0.00	0.00
Haloperidol powder, 60 °C, 15 days	0.00	0.00
Haloperidol powder, 80 °C, 15 days	0.00	0.00
Haloperidol solution, 60 °C, 15 days	10.00	2.00
Haloperidol solution, 80 °C, 15 days	17.00	3.40
Haloperidol powder, UV light, 48 h	0.00	0.00
Haloperidol solution, UV light, 48 h	13	2.6

3.4. Method Validation

The calibration curve exhibited excellent linearity across the concentration range of 1.0–50 µg/mL, with a correlation coefficient (R^2) of 0.999. The relative standard deviation (RSD) was consistently below 2%, demonstrating the method's precision (Table S1, Figure 3A). Furthermore, the residual plot (Figure 3B) displayed a random distribution around the zero line, confirming the absence of systematic errors and aligning with ICH guidelines. This streamlined presentation focuses on the key findings, as requested by the reviewer, omitting repetitive methodological details.

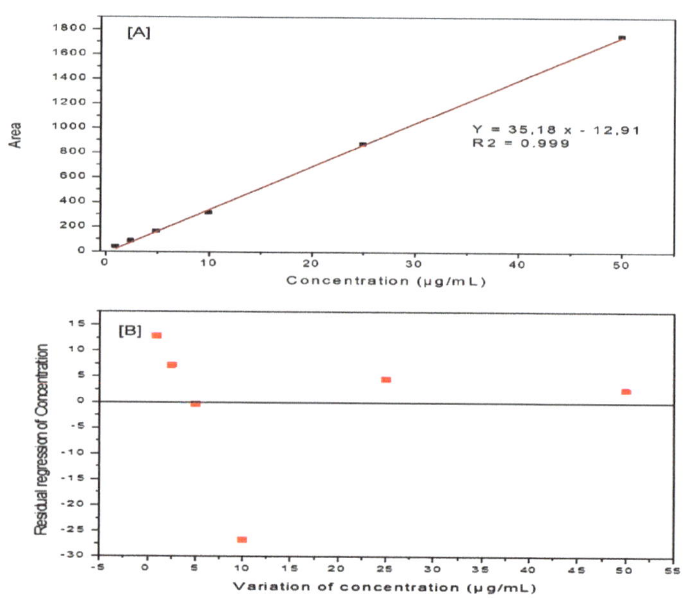

Figure 3. (**A**) Linearity of the method. (**B**) Residual graph of the linearity of the method.

The validation results for the proposed HPLC method for haloperidol analysis are summarized in Table S2. These results are consistent with those reported in the literature [37] and meet the criteria outlined in the ICH guidelines [44]. Consequently, the method can be considered reliable for the determination of haloperidol in liquid dosage forms [19,47,51].

3.5. Application of the Method to the Analysis of Haloperidol in a Finished Product (ISOPERIDOL) (2 mg/mL Solution)

When the sample solution, prepared via the simple dilution of the finished product, was injected, two clear chromatographic peaks were observed: one at approximately 3.30 min, corresponding to haloperidol, and another at 2.11 min, representing methylparaben, as shown in Figure 4. These compounds were efficiently eluted and separated, with a resolution greater than 2.

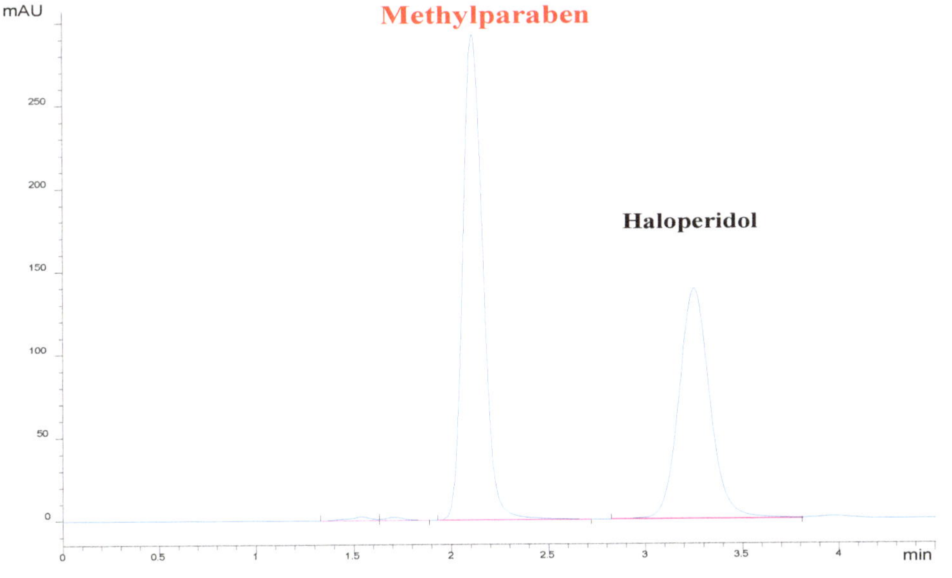

Figure 4. Chromatogram of the sample (0.2% haloperidol oral solution).

The accurate quantification of haloperidol in the final product is crucial to ensure the product's quality and adherence to regulatory standards. The reported concentration of (2 ± 0.54) mg/mL, derived from three replicate measurements, demonstrates the high precision of the analytical method used. The percentage estimate of 100.5% indicates that the product complies with the strict requirements of the British Pharmacopoeia, which typically mandates that the concentration of active pharmaceutical ingredients should fall within the range of 95% to 100%. This level of accuracy and compliance highlights the reliability of the analytical method and ensures the efficacy and safety of the finished product for its intended use [19,37].

The HPLC method presented in this study is a dependable approach for evaluating the content of haloperidol in oral solutions. Its specificity and robustness make it highly suitable for quality control applications in pharmaceutical laboratories. A significant advantage of this method is its simple mobile-phase composition, which contrasts with the more complex formulations often described in the literature and pharmacopeias. This streamlined approach improves analytical efficiency and laboratory practicality, making it a valuable and effective alternative for pharmaceutical quality control. In summary, this method combines simplicity and reliability, proving to be highly useful in quality assurance protocols.

3.6. Characterization of Degradation Products Using LC-MS/MS

LC-MS/MS analysis in ESI+ mode was employed to characterize the degradation products of haloperidol by detecting their molecular mass (m/z) values and fragmentation patterns. This approach facilitated the identification of key degradation products, as summarized in Table 2, and provided valuable insights into their structural elucidation. The results, depicted in Figure S1, enhance our understanding of haloperidol's degradation pathways, aiding in the optimization of its stability for pharmaceutical formulations.

Table 2. Potential degradation products of haloperidol.

Name	Molecular Formula	Structural Form and Chemical Name	Molecular Weight (g/mol)
Haloperidol	$C_{21}H_{23}ClFNO_2$	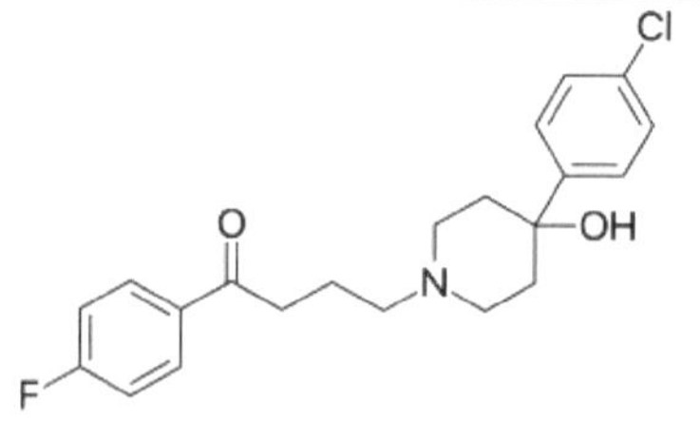	376
Degradation Product I	$C_{21}H_{25}Cl_2FNO_2$	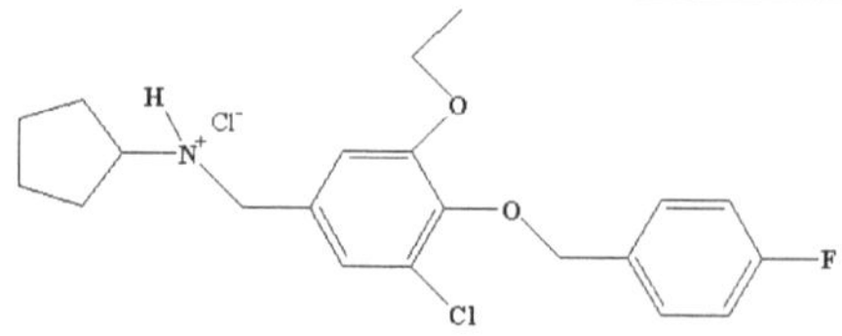	411.5
Degradation Product II	$C_{26}H_{32}ClFN_2O_3$	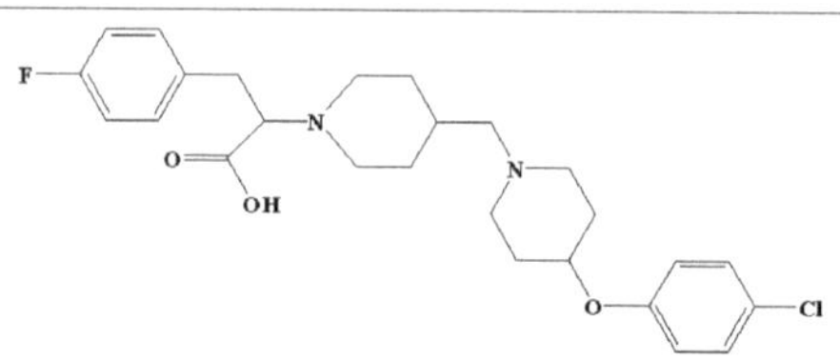	476
Degradation Product III	$C_{26}H_{33}Cl_2FN_2O_3$	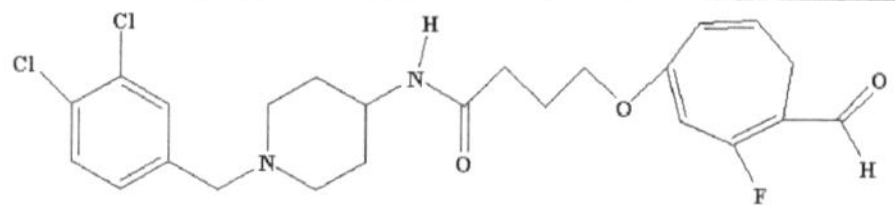	511.5
Degradation Product IV	$C_{21}H_{23}Cl\ FNO_3$	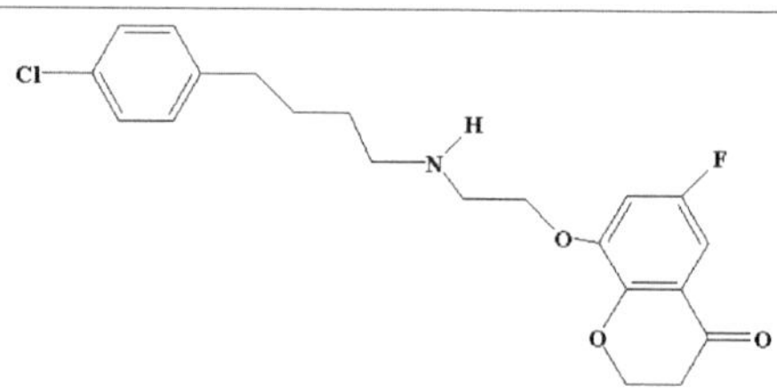	392

The results of the haloperidol analysis conducted using LC-MS/MS are presented in Figure 5. The m/z value for haloperidol in the reported methods was found to be 376.

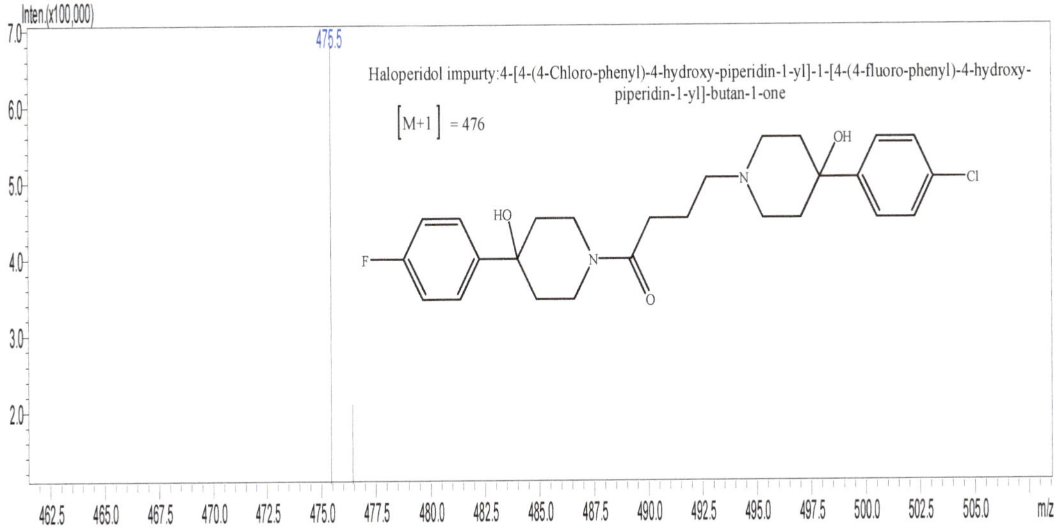

Figure 5. Mass spectrum of haloperidol.

3.6.1. Acidic Stress: HCl 1N

The mass spectra of the haloperidol solution under acidic conditions, presented in Figures S2–S4, provide critical insights into its degradation behavior. By analyzing the mass-fragmentation patterns, degradation signals were identified, confirming the formation of degradation products. Under acidic stress, the degradation product [A] was detected with a significant signal at $m/z = 411$. Additionally, a key fragment with a mass-to-charge ratio of $m/z = 362.5$ was observed, further corroborating the presence of this degradation product.

In the case of degradation product [B], illustrated in Figure S3, multiple fragments were observed in the mass spectra. A notable signal corresponding to this product was identified at $m/z= 476$. It is hypothesized that degradation product [B] may originate from a synthetic impurity. Importantly, this impurity has not been previously reported or documented in the literature. The detection of this previously unidentified impurity highlights the necessity of comprehensive characterization and analysis in pharmaceutical research. Understanding the nature and sources of such impurities is essential for ensuring the quality, safety, and efficacy of pharmaceutical formulations.

The mass spectra of degradation product [C], shown in Figure S4, revealed several fragments, including a prominent signal at $m/z = 511.5$. A fundamental fragment with a mass-to-charge ratio of $m/z = 453.6$ was also observed. These findings contribute to a deeper understanding of the degradation pathways of haloperidol under acidic conditions. Identifying and characterizing degradation products such as [A] and [C] are vital for evaluating the stability and integrity of haloperidol formulations.

3.6.2. Basic Stress NaOH 1N

The mass spectra of the haloperidol solution exposed to basic conditions, as shown in Figures S5–S7, offer important insights into its degradation behavior. These spectra enable the identification of degradation signals through their mass-fragmentation patterns, characterized by their mass-to-charge ratios (m/z). For degradation caused by basic conditions, the resulting degradation product [D] exhibited several fragments. A significant signal corresponding to this degradation product was detected at $m/z = 392$. Additionally, a fundamental fragment with a mass-to-charge ratio of $m/z = 363.1$ was observed alongside the main degradation product.

Degradation product [B], observed under basic stress conditions, displayed multiple fragments in the mass-spectra analysis. Specifically, the signal corresponding to this degradation product was identified at $m/z = 476$.

Similarly, another degradation product, degradation product [B], that formed under basic stress conditions showed multiple fragments in the mass spectra. The signal associated with this degradation product was detected at $m/z = 511.5$, and a fundamental fragment was observed at $m/z = 453.6$.

This detailed analysis provides valuable insights into the degradation pathways of haloperidol under basic conditions, underscoring the importance of monitoring degradation products to ensure the stability and efficacy of pharmaceutical formulations.

3.6.3. Photolytic Stress

The mass spectra of the haloperidol degradation product formed under photolytic conditions are shown in Figure S8. Degradation signals were identified based on their mass-fragmentation patterns (m/z). Specifically, degradation product [B], resulting from photolytic stress, displayed multiple fragments, with the primary signal detected at $m/z = 476$. This analysis enhances our understanding of the photolytic degradation pathways of haloperidol and underscores the importance of ensuring proper storage conditions to preserve its stability.

The active ingredient, haloperidol, remained stable under UV light radiation. The chromatogram shows two peaks: the main peak corresponds to the predominant active ingredient, haloperidol, while the second peak represents impurity [B]. This impurity, detected in all the stability tests under HCl and NaOH conditions, is not documented in the literature and is likely a byproduct formed during the synthesis of haloperidol. Based on the LC-MS results, four degradation products were identified under various hydrolytic stress conditions using the forced degradation protocol. One degradation product, labeled [C], was common to both acidic and basic stress conditions. Under basic stress, haloperidol degrades and forms *cis*-4-[4-(4-chlorophenyl)-4-hydroxy-1-piperidinyl]-1-(4-fluorophenyl)-1-butanone N-oxide (Product [D]). However, in the literature, product [D] is described as an oxidative degradation product [32,52]. Similar findings regarding the formation of N-oxide derivatives under such conditions have been reported for other nitrogen-containing heterocyclic compounds [18,53].

The separation of degradation products from the haloperidol peak and its potential impurities demonstrates the stability-indicating capability of the method [54]. The proposed new stability-indicating method was validated for specificity, precision, accuracy, limit of quantification, limit of detection, linearity, robustness, and ruggedness, following International Conference on Harmonization (ICH) guidelines.

The validated RP-HPLC method effectively detected most degradation products formed during the forced degradation studies, confirming its reliability and robustness for stability-indicating purposes.

3.7. *Haloperidol Analysis via TGA/DTG/DSC*

The thermal stability of haloperidol was evaluated under conditions ranging from 20 to 500 °C with thermo-analytical data collected at a heating rate of 5 °C/min, as illustrated in Figure 6.

The thermal analysis results for haloperidol confirm its stability up to 230 °C, as evidenced by the TGA, which shows no mass loss within this range. The DSC thermogram reveals a melting point at 149.5 °C, indicating the transition from solid to liquid form without decomposition [38]. Beyond 230 °C, haloperidol undergoes thermal degradation through a single-step mechanism, with complete decomposition occurring at 341.5 °C, leaving a residual mass of approximately 1% [18,55]. The DTG curve supports this behavior, displaying a distinct decomposition phase between 230 °C and 341.5 °C. The DSC thermogram also shows an endothermic peak at 149.5 °C, corresponding to the melting process, with a heat of fusion of 151.69 J/g [13,56]. These findings confirm the thermal stability of haloperidol up to 230 °C and provide critical insights into its degradation pathway under heat stress.

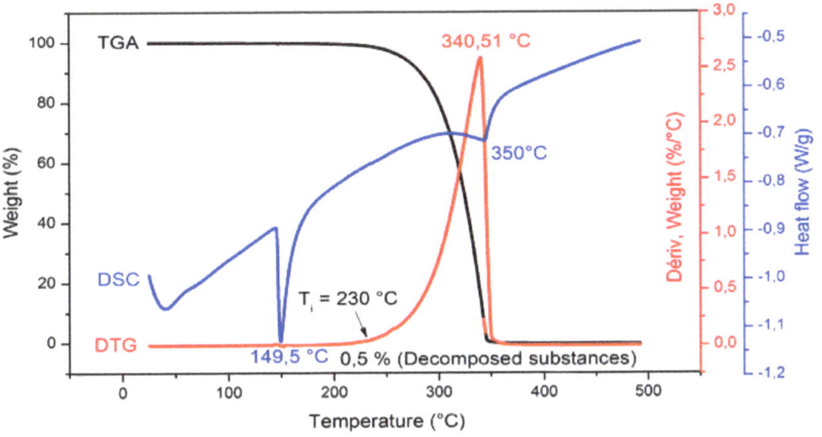

Figure 6. Thermal analysis curves for haloperidol (heating rate: 5 °C/min).

3.8. Comparative Analysis of the Stability of Haloperidol and the Analytical Methods

The comparison table (Table 3) highlights the significant differences in the analytical methods used for haloperidol quantification. Different stationary phases, including C18, CN, monolithic silica, and carbon columns, influence the retention time and sensitivity of the analyses. C18-based columns are most commonly used due to their efficiency in retaining haloperidol while ensuring stable peak separation. Non-porous and monolithic silica columns, while allowing rapid analysis, can be more expensive and require specialized equipment.

The composition of the mobile phase plays a crucial role in the efficiency of the separation. In most methods, a combination of a phosphate buffer and organic solvents such as methanol, acetonitrile, or tetrahydrofuran is used. The pH of the mobile phase significantly affects the retention time and peak resolution. Lower pH values (2.0–3.0) improve separation but may shorten column lifetime, while neutral to high pH values (6.5–9.8) improve stability and reduce interference from degradation products. Some methods involve the use of gradient elution, which provides superior peak resolution but results in longer analysis times compared to isocratic elution.

Sensitivity varies widely among methods, as demonstrated by the limits of detection (LOD) and quantitation (LOQ). The most sensitive approaches, such as those using nonporous silica and monolithic silica columns, achieve detection limits in the nanogram-per-milliliter range. In contrast, traditional methods using C18 columns typically provide LOQs in the microgram-per-milliliter range. In contrast, the LC-MS/MS in our study is distinguished by an exceptionally low LOD of 1 ng/mL, making it particularly suitable for the detection of trace amounts of degradation products.

Table 3. Table comparing results between the current study and the literature.

Methodology Used	Stationary Phase	Mobile Phase	λ (nm)	Retention Time (min)	LOD	LOQ	Pharmaceutical Form	Strengths	Limitations	Reference
HPLC, UV-Vis, spectrofluorimetry	Various columns	Various solvents	Variable	Variable	Variable	Variable	Solid and liquid forms	Covers multiple analytical techniques	Limited stability studies	[37]
HPLC	C18 (250 × 4.6 mm, 5 μm)	Methanol/tetrabutyl ammonium sulfate (55:45)	254 nm	7	0.90 μg/mL	2.75 μg/mL	Oral solution	Good sensitivity	Longer retention time	[47]
HPLC	ODS-A (150 × 4.6 mm, 3 μm)	TBA sulfate/acetonitrile/isopropanol (gradient)	230 nm	37.9	0.137 μg/mL	0.458 μg/mL	Injection	Highly sensitive	Very long analysis time	[57]
HPLC	ODS (33 × 4.6 mm, 1.5 μm)	Phosphate buffer/TEA/acetonitrile (77:23:10)	220 nm	1.3	1 ng/mL	---	Tablet	Ultra-fast, high sensitivity	LOQ has not been determined	[58]
HPLC	RP-18 (25 cm × 4.6 mm)	Phosphate buffer/acetonitrile/tetrahydrofuran/TEA (63:34:3:0.1)	246 nm	9.3	15 ng/mL	50 ng/mL	Tablet	High accuracy	Longer analysis time	[35]
HPLC	C18 (250 × 4.6 mm, 5 μm)	Methanol/acetonitrile (50:50)	244 nm	2.23	0.40 μg/mL	1.20 μg/mL	Tablet	Good sensitivity	Stability not studied	[48]
HPLC	Monolithic silica (100 × 4.6 mm)	Phosphate buffer/acetonitrile (70:30)	230 nm	4.26	1 ng/mL	3 ng/mL	Injection	Very sensitive	High column cost	[59]
HPLC	Carbon (100 × 4.6 mm, 7 μm)	Tetrahydrofuran/water/trichloroacetic acid (55:45)	254 nm	4.25	0.1 μg/mL	---	Tablet	Good separation	LOQ has not been determined	[36]
HPLC	XDB C18 (50 × 4.6 mm, 1.8 μm)	Organic phase/phosphate buffer/acetonitrile (gradient)	230 nm	3.77	1.16 μg/mL	3.86 μg/mL	---	Good sensitivity	More complex method	[49]
HPLC	CN (30 cm × 3.9 mm)	Tetrahydrofuran/water/phosphoric acid (40:60)	254 nm	5.4	---	---	---	Standard method	LOD and LOQ not specified	[32]
HPLC	C18 (250 × 4.6 mm, 5 μm)	Methanol/phosphate buffer/TEA (50:50:0.2)	254 nm	---	---	---	Solution	Good accuracy	Incomplete information	[60]
HPLC	C8 (150 × 4.6 mm, 5 μm)	Acetonitrile/tetramethylammonium perchlorate (pH 2.8)	230 nm	6.4	---	---	Solution	Good separation	LOD and LOQ not specified	[61]
HPLC (green chromatography)	C18 (250 × 4.6 mm, 5 μm)	Methanol/phosphate buffer (pH 9.8) (90:10)	248 nm	<4	0.40 μg/mL	1.20 μg/mL	Solution and powder	Eco-friendly, validated fast method	Lacks impurity characterization	[31]
HPLC, LC-MS/MS, TGA/DSC/DTA	C18 (250 × 4.6 mm, 5 μm)	Methanol/phosphate buffer (pH 9.8) (90:10)	248 nm	4.2	1 ng/mL	3 ng/mL	All forms	Comprehensive stability study	Needs large-scale validation	In this study
HPLC, LC-MS/MS, TGA/DSC/DTA	C18 (250 × 4.6 mm, 5 μm)	Methanol/phosphate buffer (pH 9.8) (90:10)	248 nm	4.2	1 ng/mL	3 ng/mL	All forms	Comprehensive stability study	Needs large-scale validation	In this study

The pharmaceutical form analyzed also influences the choice of method. Tablets and oral solutions require robust methods with high accuracy, while injectable formulations require ultra-sensitive detection to identify analytes at low concentrations. The green chromatography approach developed by Djilali et al. (2025) offers a rapid and environmentally friendly alternative, making it ideal for routine quality control applications [31]. On the other hand, traditional HPLC methods remain reliable but may lack the advanced ability to differentiate impurities and degradation byproducts.

Among the methods compared, this study demonstrates a distinctive advantage in integrating HPLC with LC-MS/MS and thermal analysis techniques (TGA/DSC/DTA). This comprehensive approach allows for a more in-depth study of haloperidol's stability under stress conditions. While green chromatography is preferred for routine analyses due to its speed and durability, the methodology used in this study excels in characterizing degradation products and ensuring pharmaceutical integrity.

4. Discussions

The primary innovation of this research lies in the thorough characterization of haloperidol's stability under various stress conditions, employing both chromatographic and thermal analyses. This study expands on existing knowledge by identifying specific conditions under which haloperidol undergoes significant degradation, particularly emphasizing its stability under dry-heat and photolysis conditions, while also noting notable degradation under acidic and alkaline environments. Additionally, the use of advanced analytical techniques such as LC-MS/MS enabled the precise identification of degradation products, including some not listed in the European Pharmacopoeia monograph for the respective medicinal product. This detailed characterization of degradation products enhances our understanding of the potential degradation pathways of haloperidol, essential for ensuring its quality and stability as an active ingredient in pharmaceutical formulations. In summary, this study significantly contributes to our understanding of haloperidol's stability, highlighting its behavior under various environmental stresses and providing a precise analytical methodology for assessing its stability in pharmaceutical formulations.

A novel RP-HPLC method was developed for the analysis of haloperidol, utilizing a mobile phase composed of methanol and a phosphate buffer (90:10 ratio, pH = 9.8) with isocratic elution. The method was rigorously validated in accordance with industry regulatory standards, confirming its high linearity, accuracy, sensitivity, and specificity. The calibration curve for haloperidol in a standard solution exhibited excellent linearity across a concentration range of 1–50 μg/mL, with a correlation coefficient of 0.999 and a mean relative standard deviation (RSD) below 2%. This method was successfully applied to determine the dosage (2 mg/mL) of haloperidol in commercial liquid samples without the need for extraction or pre-treatment steps.

This study outlines the development and validation of an HPLC method designed to accurately quantify haloperidol in bulk formulations while accounting for potential degradation products. This stability-indicating isocratic HPLC method was carefully developed to evaluate haloperidol content under various stress conditions, including exposure to acidic, basic, oxidative, thermal, and UV light environments. Using this method, we effectively separated the drug from its degradation products. Notably, haloperidol in powder form demonstrated resilience against dry-heat and photolytic stress, while its solution counterpart was susceptible to degradation under hydrolytic and photolytic conditions, though it remained stable under oxidative stress. Based on these findings, the industrial manufacturing process for haloperidol solution could be revised by eliminating the nitrogen bubbling stage and potentially using buffer solutions to stabilize the pH.

Additionally, we propose optimizing the dry pharmaceutical formulation, which could offer greater stability given the thermal stability of haloperidol powder.

Linearity testing confirmed a reliable range of 1 to 50 μg/mL for haloperidol. Validation of the HPLC method demonstrated high specificity, linearity, precision, and accuracy. Mass spectrometry facilitated the characterization of degradation products, revealing different compounds depending on the applied stressors. Acidic stress led to the formation of DPA, DPB, and DPC, while basic stress resulted in DPB, DPC, and DPD. Photolytic stress introduced a new compound, DPB, identified as an impurity in haloperidol synthesis. Further analysis using TGA/DTG/DSC confirmed haloperidol's stability under thermal stress, with a melting temperature of approximately 150 °C. Although the primary goal of the forced-degradation study was to assess haloperidol's stability rather than identify specific degradation products, it was evident that pH significantly influenced haloperidol's stability in solutions. The simplicity of the proposed HPLC method allows for the straightforward determination of haloperidol content in pharmaceutical oral solutions at a concentration of 0.2%, requiring only simple dilution without prior sample extraction or treatment. Furthermore, this method shows promise for conducting stability studies on haloperidol.

Supplementary Materials: The following supporting information can be downloaded at: https://www.mdpi.com/article/10.3390/pr13030904/s1, Figure S1: Possible degradation products of haloperidol. Figure S2. Mass spectrum of haloperidol degradation product [A] (haloperidol hydrochloride) and its fragmentation products. Figure S3. Mass spectrum of haloperidol degradation product [B] (Haloperidol impurity) and its fragmentation products. Figure S4. Mass spectrum of haloperidol degradation product [C] (haloperidol hydrochloride) and its fragmentation products. Figure S5. Mass spectrum of haloperidol degradation product [D] (haloperidol-N-Oxide) and its fragmentation products. Figure S6. mass spectrum of haloperidol impurity [B]., Figure S7. Mass spectrum of haloperidol degradation product [C] (haloperidol hydrochloride) and its fragmentation products. Figure S8. Mass spectrum of haloperidol degradation product [B] (haloperidol impurity) and its fragmentation products. Table S1. Results of the method linearity study. Table S2. Results of validation method.

Author Contributions: Conceptualization, K.D., R.M., H.B., S.L., Z.A.M., S.M.A., N.N., R.B., J.Z., H.T. and A.A.; data curation, K.D., S.M.A., H.T. and A.A.; formal analysis, K.D., R.M., H.B., S.L., Z.A.M., S.M.A., N.N., R.B., J.Z., H.T. and A.A.; funding acquisition, S.M.A.; investigation, K.D., R.M., Z.A.M., S.M.A., N.N., R.B., J.Z., H.T. and A.A.; methodology, K.D., R.M., H.B., S.L., Z.A.M., S.M.A., N.N., R.B., J.Z., H.T. and A.A.; project administration, R.M., N.N., R.B., J.Z., H.T. and A.A.; resources, K.D., R.M., H.B., S.L., Z.A.M., S.M.A., N.N., R.B., J.Z., H.T. and A.A.; software, K.D., R.M., S.M.A., N.N., J.Z., H.T. and A.A.; supervision, R.M., N.N., J.Z., H.T. and A.A.; validation, K.D., R.M., H.B., S.L., Z.A.M., S.M.A., N.N., R.B., J.Z., H.T. and A.A.; visualization, K.D., R.M., H.B., S.L., Z.A.M., S.M.A., N.N., R.B., J.Z., H.T. and A.A.; writing—original draft, K.D.; writing—review and editing, R.M., H.B., S.L., Z.A.M., S.M.A., N.N., R.B., J.Z., H.T. and A.A. All authors have read and agreed to the published version of the manuscript.

Funding: King Saud University, RSPD2025R748

Data Availability Statement: The data that support the findings of this study are available within the article.

Acknowledgments: The authors wish to extend their sincere gratitude to the Center for Scientific and Technical Research in Physical and Chemical Analysis (CRAPCC) in Algeria for allowing to use instruments for this work. The authors also extend thanks to the University of Saad Dahlab Blida 1 (USDB), and the University of Science and Technology Houari Boumediene, Algiers-Bab Ezzouar, Algeria, for providing the facilities and resources used in this research work. The authors acknowledge and extend their appreciation to the Researchers Supporting Project Number (RSPD2025R748), King Saud University, Riyadh, Saudi Arabia for funding this study.

Conflicts of Interest: The authors declare that there are no known competing financial or personal interests that could have appeared to influence this research work.

References

1. Pushpakom, S.; Iorio, F.; Eyers, P.A.; Escott, K.J.; Hopper, S.; Wells, A.; Doig, A.; Guilliams, T.; Latimer, J.; McNamee, C. Drug Repurposing: Progress, Challenges and Recommendations. *Nat. Rev. Drug Discov.* **2019**, *18*, 41–58. [CrossRef]
2. Grangeia, H.B.; Silva, C.; Simões, S.P.; Reis, M.S. Quality by Design in Pharmaceutical Manufacturing: A Systematic Review of Current Status, Challenges and Future Perspectives. *Eur. J. Pharm. Biopharm.* **2020**, *147*, 19–37. [CrossRef]
3. Murugan, T.; Singh Jai, W. (Eds.) *Cybersecurity and Data Science Innovations for Sustainable Development of HEICC: Healthcare, Education, Industry, Cities, and Communities*; CRC Press: Boca Raton, FL, USA, 2025.
4. Lang, J.; Li, L.; Quan, Y.; Tan, R.; Zhao, J.; Li, M.; Zeng, J.; Chen, S.; Wang, T.; Li, Y. LC-MS-Based Metabolomics Reveals the Mechanism of Anti-Gouty Arthritis Effect of Wuwei Shexiang Pill. *Front. Pharmacol.* **2023**, *14*, 1213602. [CrossRef]
5. Kanjal, M.I.; Muneer, M.; Ullah, S.; Sabir, S.; Boudraa, R.; Amrane, A.; Mouni, L. UV Radiation-Induced Degradation of Moxifloxacin: Toxicity Evaluation and Conditions Optimization. *Euro-Mediterr. J. Environ. Integr.* **2025**, 1–13. [CrossRef]
6. Shi, S.; Li, K.; Peng, J.; Li, J.; Luo, L.; Liu, M.; Chen, Y.; Xiang, Z.; Xiong, P.; Liu, L. Chemical Characterization of Extracts of Leaves of *Kadsua coccinea* (Lem.) AC Sm. by UHPLC-Q-Exactive Orbitrap Mass Spectrometry and Assessment of Their Antioxidant and Anti-Inflammatory Activities. *Biomed. Pharmacother.* **2022**, *149*, 112828. [CrossRef]
7. Guo, Y.; Han, Z.; Zhang, J.; Lu, Y.; Li, C.; Liu, G. Development of a High-Speed and Ultrasensitive UV/Vis-CM for Detecting Total Triterpenes in Traditional Chinese Medicine and Its Application. *Heliyon* **2024**, *10*, e32239. [CrossRef]
8. Craig, D.; Reading, M. *Thermal Analysis of Pharmaceuticals*; CRC Press: Boca Raton, FL, USA, 2006.
9. Ahuja, S.; Dong, M. *Handbook of Pharmaceutical Analysis by HPLC*; Elsevier: Amsterdam, The Netherlands, 2005; ISBN 0-08-045518-2.
10. Missa, J.-N. Peut-on Parler de Militantisme Psychopharmacologique Chez Les Pionniers de La Psychiatrie Biologique (1952–1960)? *Sud/Nord* **2010**, *25*, 105–120. [CrossRef]
11. Granger, B.; Albu, S. The Haloperidol Story. *Ann. Clin. Psychiatry* **2005**, *17*, 137–140. [CrossRef]
12. Tardy, M.; Huhn, M.; Kissling, W.; Engel, R.R.; Leucht, S. Haloperidol versus Low-potency First-generation Antipsychotic Drugs for Schizophrenia. *Cochrane Database Syst. Rev.* **2014**. [CrossRef]
13. Xu, Q.A.; Madden, T.L. *Analytical Methods for Therapeutic Drug Monitoring and Toxicology*; John Wiley & Sons: New York, NY, USA, 2011; ISBN 0-470-92279-6.
14. Européenne, P. EUROPÉENNE, Pharmacopée. 6ème Edition, Conseil d'Europe. 2008. Available online: https://www.decitre.fr/livres/pharmacopee-europeenne-supplement-6-6-9789287165084.html?srsltid=AfmBOorD91kigqt2-0ItuGvWfXxT1PTY38_EGPiUcMDgMoR5rd1a2ifx (accessed on 16 March 2025).
15. Pharmacopeia, U. *The United States Pharmacopeia, USP 41/The National Formulary*; United States Pharmacopeial Convention: Rockville, MD, USA, 2018.
16. Demoen, P.J. Properties and Analysis of Haloperidol and Its Dosage Forms. *J. Pharm. Sci.* **1961**, *50*, 350–353. [CrossRef]
17. Djilali, K.; Maachi, R.; Mesbah, Z.A.; Nasrallah, N.; Touzout, N.; Tahraoui, H.; Zhang, J.; Amrane, A. Breaking Barriers in Pharmaceutical Analysis: Streamlined UV Spectrometric Quantification and Stability Profiling of Haloperidol and Methylparaben in Liquid Formulations. *Anal. Biochem.* **2024**, *695*, 115632. [CrossRef]
18. Janicki, C.A.; Ko, C.Y. Haloperidol. In *Analytical Profiles of Drug Substances*; Elsevier: Amsterdam, The Netherlands, 1981; Volume 9, pp. 341–369, ISBN 0099-5428.
19. Pharmacopoeia, B.H. *The British Pharmacopoeia Commission Secretariat of the Medicines and Healthcare Products Regulatory Agency (MHRA)*; TSO (The Stationery Office): Edinburgh, UK, 2019.
20. Boudraa, R.; Talantikite-Touati, D.; Souici, A.; Djermoune, A.; Saidani, A.; Fendi, K.; Amrane, A.; Bollinger, J.-C.; Tran, H.N.; Mouni, L. Breaking New Grounds: Solid-State Synthesis of TiO_2–La_2O_3–CuO Nanocomposites for Degrading Brilliant Green Dye under Visible Light. *J. Clean. Prod.* **2024**, *481*, 144126. [CrossRef]
21. Ouanes, S.; Kallel, M.; Trabelsi, H.; Safta, F.; Bouzouita, K. Zero-Crossing Derivative Spectrophotometry for the Determination of Haloperidol in Presence of Parabens. *J. Pharm. Biomed. Anal.* **1998**, *17*, 361–364. [CrossRef]
22. Rahman, N.; Khatoon, A.; Rahman, H. Studies on the Development of Spectrophotometric Method for the Determination of Haloperidol in Pharmaceutical Preparations. *Quím. Nova* **2012**, *35*, 392–397. [CrossRef]
23. Yasir, M.; Sara, U. Development and Validation of UV Spectrophotometric Method for the Estimation of Haloperidol. *Br. J. Pharm. Res.* **2014**, *4*, 1407–1415. [CrossRef]
24. Sambamurty Raju, S.; Raju, P.; Ashok Babu, R.; Anjani Devi, A. Validated UV and Visible Spectrophotometric Methods for the Quantification of Haloperidol in Pharmaceutical Dosage Forms. *Actapharmica* **2015**, *2*, 107–111.
25. Mennickent, S.; Pino, L.; Vega, M.; Godoy, C.G.; de Diego, M. Quantitative Determination of Haloperidol in Tablets by High Performance Thin-layer Chromatography. *J. Sep. Sci.* **2007**, *30*, 772–777. [CrossRef]

26. Mennickent, S.; Pino, L.; Vega, M.; de Diego, M. Chemical Stability of Haloperidol Injection by High Performance Thin-layer Chromatography. *J. Sep. Sci.* **2008**, *31*, 201–206. [CrossRef]
27. Rahman, N.; Sameen, S.; Kashif, M. Spectroscopic Study on the Interaction of Haloperidol and 2,4-Dinitrophenylhydrazine and Its Application for the Quantification in Drug Formulations. *Anal. Chem. Lett.* **2016**, *6*, 874–885. [CrossRef]
28. Wate, S.; Borkar, A. Simultaneous Spectrophotometric Estimation of Haloperidol and Trihexyphenidyl in Tablets. *Indian J. Pharm. Sci.* **2010**, *72*, 265.
29. Saidani, A.; Boudraa, R.; Fendi, K.; Benouadah, L.; Benabbas, A.; Djermoune, A.; Salvestrini, S.; Bollinger, J.-C.; Alayyaf, A.A.; Mouni, L. Effect of Calcination Temperature on the Photocatalytic Activity of Precipitated ZnO Nanoparticles for the Degradation of Rhodamine B Under Different Light Sources. *Water* **2024**, *17*, 32. [CrossRef]
30. Zhuang, Q.; Li, X.; Lian, X.; Hu, H.; Wang, N.; Wu, J.; Miao, K.; Feng, G.; Luo, X. Catalysis Enhancement of Co_3O_4 through the Epitaxial Growth of Inert ZnO in Peroxymonosulfate Activation: The Catalytic Mechanism of Surface Hydroxyls in Singlet Oxygen Generation. *Cryst. Growth Des.* **2024**, *25*, 319–329. [CrossRef]
31. Djilali, K.; Maachi, R.; Danish, M.; Lekmine, S.; Hadjadj, M.; Ait Mesbah, Z.; Benslama, O.; Tahraoui, H.; Ola, M.S.; Ali, A. A Novel Mobile Phase for Green Chromatographic Determination of Haloperidol: Application to Commercial Pharmaceutical Products and Forced Degradation Studies. *Processes* **2025**, *13*, 260. [CrossRef]
32. Panaggio, A.; Greene, D. High Pressure Liquid Chromatographic Determination of Haloperidol Stability. *Drug Dev. Ind. Pharm.* **1983**, *9*, 485–492. [CrossRef]
33. Boudraa, R.; Talantikite-Touati, D.; Djermoune, A.; Souici, A.; Kebir, M.; Merzeg, F.A.; Amrane, A.; Bollinger, J.-C.; Mouni, L. Comprehensive Characterization and Unprecedented Photocatalytic Efficacy of TiO_2-CuO-La_2O_3 and TiO_2-CuO-Bi_2O_3 Nanocomposites: A Novel Approach to Environmental Remediation. *Mater. Sci. Eng. B* **2025**, *312*, 117863. [CrossRef]
34. Öl¸er, M.; Hakyemez, G. Investigations of Some Physicochemical Properties of Haloperidol Which May Affect Its Activity. *J. Clin. Pharm. Ther.* **1988**, *13*, 341–349. [CrossRef]
35. Trabelsi, H.; Bouabdallah, S.; Bouzouita, K.; Safta, F. Determination and Degradation Study of Haloperidol by High Performance Liquid Chromatography. *J. Pharm. Biomed. Anal.* **2002**, *29*, 649–657. [CrossRef]
36. Monser, L.; Trabelsi, H. A Rapid LC Method for the Determination of Haloperidol and Its Degradation Products in Pharmaceuticals Using a Porous Graphitic Carbon Column. *J. Liq. Chromatogr. Relat. Technol.* **2003**, *26*, 261–271. [CrossRef]
37. Kumar, R.S.; Nalini, C. Analytical Determinations of Haloperidol and Its Combinations in Pharmaceutical Dosage Forms and Biological Matrices. *J. Liq. Chromatogr. Relat. Technol.* **2021**, *44*, 33–51. [CrossRef]
38. Pawar, A.; Pandita, N. Statistically Designed, Targeted Profile UPLC Method Development for Assay and Purity of Haloperidol in Haloperidol Drug Substance and Haloperidol 1 Mg Tablets. *Chromatographia* **2020**, *83*, 725–737. [CrossRef]
39. Guideline, I.H.T. Impurities in new drug substances Q3A (R2). Proceedings of the International Conference on Harmonization of Technical Requirements for Registration of Pharmaceuticals for Human Use, Geneva, Switzerland. 2006, Volume 25. Available online: https://database.ich.org/sites/default/files/Q3A(R2)%20Guideline.pdf (accessed on 16 March 2025).
40. Sesharamsingh, B.; Kumar, J.S.; Viswanath, I.K.; Rao, G.S.; Anna, V.R. Development of Stability-Indicating HPLC Method for Quantification of Pharmacopeia Impurities of Zuclopenthixol and Characterization of Its Stress Degradation Products by LCMS/MS. *Future J. Pharm. Sci.* **2023**, *9*, 101. [CrossRef]
41. Narayanam, M.; Handa, T.; Sharma, P.; Jhajra, S.; Muthe, P.K.; Dappili, P.K.; Shah, R.P.; Singh, S. Critical Practical Aspects in the Application of Liquid Chromatography–Mass Spectrometric Studies for the Characterization of Impurities and Degradation Products. *J. Pharm. Biomed. Anal.* **2014**, *87*, 191–217. [CrossRef]
42. Baertschi, S.W.; Alsante, K.M.; Reed, R.A. *Pharmaceutical Stress Testing: Predicting Drug Degradation*; CRC Press: Boca Raton, FL, USA, 2016; ISBN 1-4398-0180-0.
43. Guideline, I. Stability Testing of New Drug Substances and Products. *Q1A R2 Curr. Step.* **2003**, *4*, 1–24.
44. Guideline, I.H.T. Validation of Analytical Procedures: Text and Methodology. *Q2 R1* **2005**, *1*, 5.
45. Vaidya, S.J.; Sanghai, D.B. Forced Degradation Study on Valbenazine and Application of Validated Stability-Indicating HPLC Method. *J. Reatt. Ther. Dev. Divers.* **2023**, *6*, 1743–1752.
46. Alladio, E.; Amante, E.; Bozzolino, C.; Seganti, F.; Salomone, A.; Vincenti, M.; Desharnais, B. Effective Validation of Chromatographic Analytical Methods: The Illustrative Case of Androgenic Steroids. *Talanta* **2020**, *215*, 120867. [CrossRef]
47. Gadhavi, R.; Patel, J. Development and Validation of Stability Indicating Assay Method of Haloperidol in Oral Solution. *JPSBR* **2014**, *4*, 319–329.
48. Jain, A.K.; Dubey, B.; Khare, S.; Joshi, A.; Ahirwar, M.; Jain, P. Comparison of RP-HPLC and UV Spectrophotometric Methods for Estimation of Haloperidol in Pure and Pharmaceutical Formulation. *J. Drug Deliv. Ther.* **2018**, *8*, 277–282. [CrossRef]
49. Petkovska, R.; Dimitrovska, A. Use of Chemometrics for Development and Validation of an RP-HPLC Method for Simultaneous Determination of Haloperidol and Related Compounds. *Acta Pharm.* **2008**, *58*, 243–256. [CrossRef]
50. Erukulla, K.K.; Renjitham, S.S. Bio-Analytical Method Development and Validation for Estimation of Zaltoprofen in Human Plasma by Reverse Phase-HPLC Method. *Curr. Pharm. Anal.* **2021**, *17*, 774–781. [CrossRef]

51. Babu, S.K.; Rao, M.V.; Babu, S.P.; Chakka, M. Chemometric Assisted Development and Validation of a Stability-Indicating Lc Method for Determination of Related Substances in Haloperidol Decanoate Injection. *Indian J. Pharm. Educ. Res.* **2021**, *55*, 904–915. [CrossRef]
52. Djilali, K.; Maachi, R.; Tahraoui, H.; Mesbah, Z.A.; Amrane, A. Advancing Thermal Stability Analysis of Haloperidol: Integrative Approaches and Optimization Strategies for Enhanced Pharmaceutical Formulations. *J. Mol. Struct.* **2024**, *1315*, 138870. [CrossRef]
53. Fernandes, F.; Dias-Teixeira, M.; Delerue-Matos, C.; Grosso, C. Critical Review of Lipid-Based Nanoparticles as Carriers of Neuroprotective Drugs and Extracts. *Nanomaterials* **2021**, *11*, 563. [CrossRef]
54. Alam, P.; Shakeel, F.; Alshehri, S.; Alhaiti, A.; Alqarni, M.H.; Foudah, A.I.; Aljarba, T.M. Analysis of the Greenness to Determine Risperidone in Commercial Products Using a Green Stability-Indicating HPTLC Method. *Arab. J. Chem.* **2024**, *17*, 105693. [CrossRef]
55. Soradech, S.; Kengkwasingh, P.; Williams, A.C.; Khutoryanskiy, V.V. Synthesis and Evaluation of Poly (3-Hydroxypropyl Ethylene-Imine) and Its Blends with Chitosan Forming Novel Elastic Films for Delivery of Haloperidol. *Pharmaceutics* **2022**, *14*, 2671. [CrossRef]
56. Alhamhoom, Y.; Kumaraswamy, T.; Kumar, A.; Nanjappa, S.H.; Prakash, S.S.; Rahamathulla, M.; Thajudeen, K.Y.; Ahmed, M.M.; Shivanandappa, T.B. Formulation and Evaluation of pH-Modulated Amorphous Solid Dispersion-Based Orodispersible Tablets of Cefdinir. *Pharmaceutics* **2024**, *16*, 866. [CrossRef]
57. Goud, E.S.; Reddy, V.K.; Krishnadevaraya, S. Development and validation of a reverse-phase liquid chromatographic method for assay and related substances of haloperidol for 50 mg/mL and 100 mg/mL. *Int. J. Pharm. Sci.* **2013**, *5*, 389–396.
58. Xu, X.; Stewart, J. HPLC analysis of haloperidol and its related compound in the drug substance and a tablet dosage form using a non-porous silica octadecylsilane column. *J. Liq. Chromatogr. Relat. Technol.* **1999**, *22*, 1857–1866. [CrossRef]
59. Ali, I.; Aboul-Enein, H.Y. Fast Determination of Haloperidol in Pharmaceutical Preparations Using HPLC with a Monolithic Silica Column. *J. Liq. Chromatogr. Relat. Technol.* **2005**, *28*, 3169–3179. [CrossRef]
60. Driouich, R.; Trabelsi, H.; Bouzouita, K.A. Stability—Indicating assay for haloperidol syrup by high-performance liquid chromatography. *Chromatographia* **2001**, *53*, 629–634. [CrossRef]
61. Raggi, M.; Casamenti, G.; Mandrioli, R.; Sabbioni, C.; Volterra, V. A rapid LC method for the identification and determination of CNS drugs in pharmaceutical formulations. *J. Pharm. Biomed. Anal.* **2000**, *23*, 161–167. [CrossRef] [PubMed]

processes

MDPI

Article

FDM 3D Printing Filaments with pH-Dependent Solubility: Preparation, Characterization and In Vitro Release Kinetics

Gustavo M. Nascimento [1], Pedro H. N. Cardoso [2], Eumara M. E. da Silva [2], Ginetton F. Tavares [3], Nelson C. Olivier [4], Pedro M. Faia [5], Evando S. Araújo [2,*] and Fabrício S. Silva [1,6]

[1] Postgraduate Program in Biotechnology, Department of Biological Sciences, State University of Feira de Santana (UEFS), Feira de Santana 44036-900, Brazil; 22225018@discente.uefs.br (G.M.N.); fabricio.souzasilva@univasf.edu.br (F.S.S.)

[2] Research Group on Electrospinning and Nanotechnology Applications (GPEA-Nano), Department of Materials Science, Federal University of São Francisco Valley, Juazeiro 48902-300, Brazil; pedro.cardoso@discente.univasf.edu.br (P.H.N.C.); eumara.mayra@discente.univasf.edu.br (E.M.E.d.S.)

[3] Research and Extension Center, Laboratory of Fuels and Materials (NPE/LACOM), Department of Chemistry, Federal University of Paraíba, Campus I, João Pessoa 58051-900, Brazil; ginetton@reitoria.ufpb.br

[4] Department of Materials Science, Federal University of São Francisco Valley, Juazeiro 48902-300, Brazil; nelson.cardenas@univasf.edu.br

[5] Electrical and Computer Engineering Department—CEMMPRE, FCTUC, Polo 2, University of Coimbra, 3030-290 Coimbra, Portugal; faia@deec.uc.pt

[6] Laboratory of Experimental Pharmacology, Department of Pharmacy, Federal University of Vale do São Francisco (UNIVASF), Petrolina 56304-917, Brazil

* Correspondence: evando.araujo@univasf.edu.br; Tel.: +55-74-2102-7645

Abstract: The process of manufacturing drug delivery systems (DDSs) by fused deposition modeling (FDM) with 3D printing requires the availability of a polymeric filament containing the drug of interest. This filament is fused in the printer heating system and used to print polymer/drug volumetric parts. Polymers with pH-dependent solubility are widely known in the literature for their controlled release and drug dissolution-enhancing properties, biocompatibility, and variety of release profiles. Given these characteristics, the study of pH-responsive 3D printing filaments appears as a potential alternative for the development of new 3D printing functional materials for healthcare area applications. In this sense, this work aimed at the preparation and characterization of pH-dependent filaments of the Eudragit E 100 copolymer (E100) containing the model drug Amlodipine (Aml) for potential application in the manufacturing of DDSs by 3D printing. The E100/Aml filaments with two distinct drug concentrations were produced by hot-melt extrusion at 105 °C. The posterior chemical protonation treatment of the filaments for 60 min provided a significant improvement in their flexibility. Microstructural analysis (SEM, XRD, FTIR, and DLS) and thermal studies by DSC proved the feasibility of producing the filaments by hot-melt extrusion without the degradation of their constituent materials. The in vitro dissolution profiles of the E100/Aml samples were evaluated in simulated gastric and intestinal fluids. The facilitated solubility of the polymer in an acidic medium (pH = 1.2) was preserved in the filament form, with rapid and reproducible drug release from the polymer matrix. The saturation of the drug concentration in the medium occurred after 30 min of testing for E100/Aml models. A customized 3D part with geometry and fill control was also printed from E100/Aml filaments as proof of concept.

Keywords: materials characterization; pH-responsive polymer; drug delivery systems; filament fabrication

Citation: Nascimento, G.M.; Cardoso, P.H.N.; da Silva, E.M.E.; Tavares, G.F.; Olivier, N.C.; Faia, P.M.; Araújo, E.S.; Silva, F.S. FDM 3D Printing Filaments with pH-Dependent Solubility: Preparation, Characterization and In Vitro Release Kinetics. *Processes* **2024**, *12*, 2916. https://doi.org/10.3390/pr12122916

Academic Editor: Paolo Trucillo

Received: 21 October 2024
Revised: 10 December 2024
Accepted: 11 December 2024
Published: 19 December 2024

1. Introduction

The 3D printing techniques represent a set of technologies developed for the manufacture of three-dimensional objects. The parts production process involves the deposition of successive layers of the molten material of interest on the working surface of a 3D printer, based on the user's design, using software associated with the equipment [1,2]. Currently, selective laser sintering (SLS), stereolithography (SLA), and fused deposition modeling (FDM) are the most common methods used for the development of devices such as sensors and biosensors, supercapacitors, flexible circuits, biomedical implants, and functional and customized drug delivery devices [3–6]. Polymers are the first choice of materials used in these applications since they present a great diversity of synthetic and natural types with desirable mechanical and thermal characteristics and compatibility with other chemical materials [6,7].

FDM printing stands out for its versatility and diversity of fabricated materials for applications in various areas of science and technology. In the process of manufacturing a three-dimensional polymeric part using this technique, a cylindrical solid filament (generally based on thermoplastic polymers and/or their composites) with a millimeter in diameter is required. This filament is heated to a temperature higher than its glass transition point in the 3D printer extrusion nozzle, making it reach a fluid state. With the computer-assisted movement of the extruder nozzle and maintaining constant temperature and extrusion flow, it is possible to form overlaid layers (layers deposited sequentially on top of the previous one) of the material on the printer working surface until the 3D part is completely produced [6,8].

Currently, the development of functional and pharmacologically active FDM 3D printing filaments is on the rise, and an open field of research, both in academia and industry, has formed [9,10]. Synthetic polymers such as polyethylene glycol, polyethylene oxide, polyvinylalcohol, and polyvinylpyrrolidone, have been the preferred choice to produce these filaments, due to their biocompatibility, biodegradability, and low melting point in comparison to other polymers, including their interaction with drugs, and different release profiles based on their formation with other polymers and/or additives characteristics [11–13].

The use of FDM polymeric filaments proves to be an efficient and economical option for the three-dimensional printing of dosage forms [5] due to the possibility of producing pharmaceutical forms with different geometries, formulations with new functional materials, and production scalability in comparison with other processes and (or) protocols that have been adopted for this purpose, such as rotary manufacturing processes [14]. The development of controlled drug release devices, with improved drug bioavailability and solubility at a desired pH, is also achievable based on the interaction between the drug and the selective polymeric matrices [15,16].

In particular, the development of unconventional FDM polymer/drug filaments with pH-dependent solubility is justified by the constant search for new materials, methods, designs, and personalized formulations that provide the programmed action of the active ingredient, with greater precision regarding the aimed body region, and increased administration interval concerning user's life quality improvement [17]. Other aspects, such as specific permeability, porosity, hydrophobicity/hydrophilicity, and drug release kinetics, also justify the exploration of these filaments for application in the 3D printing of drug delivery systems (DDSs) [15].

Among the polymers with pH medium solubility dependency, it is possible to highlight the cationic block copolymer poly(dimethylaminoethyl methacrylate/butyl methacrylate/methyl methacrylate) in a 2:1:1 monomer ratio, which belongs to the Eudragit family of thermoplastic pH-soluble polymers (commercially known as EudragitE100) [18]. The E100 is soluble in acidic media (below pH 5.0), biocompatible, biodegradable, and non-genotoxic. It has no irritation and/or sensitization potential, and it is classified as safe for use in pharmaceutical formulations by the United States Department of Food and Drugs (with acceptable daily intakes of 20 mg/kg bw) [19].

In the case of E100, the hydration of dimethylamino groups and their molecules in an acidic environment (for example, in gastric fluid) causes a protonation reaction, which favors repulsion between these units. Subsequently, these units can interact with anions in the media to form salts, breaking the original molecule and, consequently, facilitating the polymer's solubility (in contrast to what happens in a basic environment, where its polymer chain tends to become immobilized, and the material functions as a device to protect the drug from contacting the external environment [20,21]).

This polymer is already used in the pharmaceutical industry as a coating material in conventional pharmaceutical forms for oral administration, such as capsules and tablets, masking the taste and odor of medicines [19]. It is also used in the production of micro/nanocapsules, where polymeric matrices act as a support for active ingredients to be released after the ingestion of capsules [22,23]. However, studies showing its use in the development of FDM 3D printing filaments for applications in drug delivery systems are scarce in the literature [21,24], largely due to the difficulty of overcoming their brittle and sticky aspect. Cardoso and his collaborators [21] developed FDM 3D printing filaments of this copolymer containing hesperidin, a flavonoid extracted from citrus fruits. In the work of Choudhury et al. [24], E100 FDM 3D printing filaments containing the berberine chloride drug were fabricated. The authors cited the difficulty of producing the filaments due to the brittle and sticky aspect of the polymer, but they overcame this problem by using an optimized combination of additives. The filaments were used to print custom capsules of the drug. In vitro release tests showed that the capsule models maintained the selective solubility of the polymer in simulated gastric fluid (pH = 1.2), for which the first contact of the drug with the medium took place at around 15 min, while the release of approximately 70% of the active ingredient occurred after 4 h. In addition, the printed parts proved to be matrices for slow drug release in the phosphate buffer (pH = 6.8), with capsule swelling for up to 12 h, with 90% of the drug released after 36 h. The authors concluded that the prototypes were potential candidates for the production of personalized pharmaceutical forms for the programmed release of active substances.

Aiming to significantly contribute to this field of research, this work proposes the preparation and characterization of pH-soluble FDM 3D printing filaments based on the combination of the E100 copolymer and of the model drug Amlodipine Besylate (indicated as a first-choice medication in the treatment of hypertension and angina pectoris), and free of chemical and graft additives.

Polymer/drug composite filaments with different concentrations of Amlodipine were prepared by the hot-melt extrusion process. Microstructural, thermal, and mechanical characterizations, combined with the drug release profiles in different media, were performed. The main objective of the present work was to demonstrate the feasibility of developing filaments containing both components (a support polymer and the drug of interest) and their ability to maintain their original characteristics even after the printing process and, consequently, their potential application in personalized drug delivery systems.

2. Materials and Methods

2.1. Materials

Poly(dimethylaminoethyl methacrylate/butyl methacrylate/methyl methacrylate) (E100, 2:1:1, Figure 1a), in granules, and pH-soluble (below 5.0) was purchased from Evonik Industries, Essen, Germany. The model drug Amlodipine Besylate (Aml, Figure 1b) ($C_{20}H_{25}ClN_2O_5C_6H_5SO_3H$, white powder, purity $\geq$ 98%, the molecular weight of 567.05, saturation solubility of 7.4 µg/mL (in water), LD50 oral-rat at 393 mg/kg with acute toxicity (Category 4, H302-Regulation (EC) 1272/2008 of the European Parliament and Council), and monopotassium phosphate salt (MKP, powder) were purchased from Sigma Aldrich, St. Louis, MO, USA. Sodium hydroxide was obtained from Êxodo Científica, Sumaré, Brazil. All reagents were used as received.

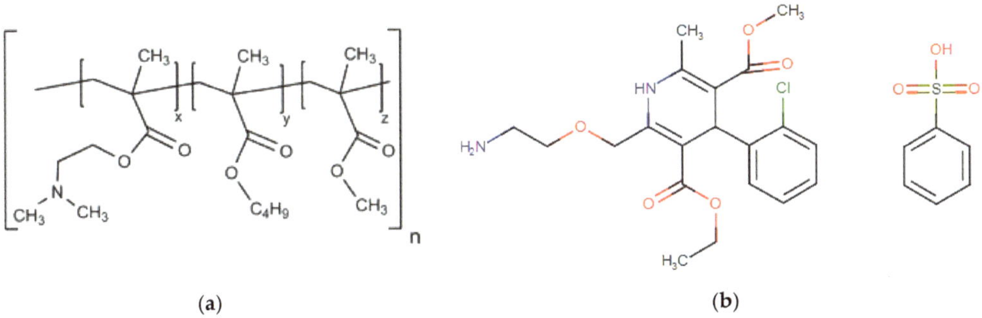

Figure 1. The 2D representation of the molecular structures of (**a**) Eudragit® E100 (E100) and (**b**) Amlodipine Besylate (Aml).

2.2. Sample Preparation

Initially, variable amounts of the model drug (100 and 200 mg) were mechanically mixed with 24 g of the polymer (samples were named E100/Aml(−) and E100/Aml(+), respectively) in a beaker until a homogeneous dispersion of the materials was obtained (Figure 2a). Samples of the unmixed polymer were also used for comparative purposes.

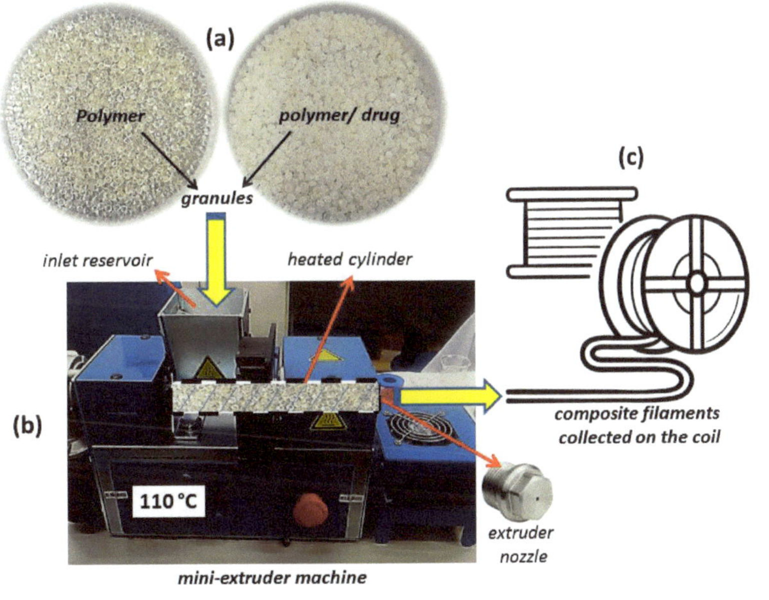

Figure 2. (**a**) Unmixed polymer and polymer/drug granules; (**b**) mini-extruder machine details; and (**c**) illustration of the filaments collected in a circular coil.

Then, each sample was taken to a Filmaq3D CV mini-extruder (Filmaq3D, Curitiba, Brazil) so that the composite filaments could be produced by the hot-melt extrusion process (Figure 2b). In this step, the materials were physically mixed and inserted into the extrusion machine reservoir, which was connected to a cylindrical galvanized metal tube. During the procedure, a screw inside the tube pushed the material (rotation speed of 10 rpm) to the other end of the cylinder (connected to the extrusion nozzle possessing a diameter of 1.75 mm). The cylindrical extruder tube was initially maintained at 110 °C for 1 h before starting the extrusion process. Then, the extrusion process was activated at 105 °C, with an extrusion rate of 30 cm·min^{-1}. Three independent samples of each filament model were produced and then collected in a circular coil (Figure 2c) for subsequently receiving chemical treatment (for 15, 30, 45, or 60 min) to increase their flexibility. The obtained materials were placed in a plastic reservoir (30 × 40 × 30 cm) to which a humid air flow terminal was initially connected (60% relative humidity concentration) and subsequently subjected to the vapor produced from a solution of 6.8 g of MKP in 0.2 M sodium hydroxide (1 L of deionized water, pH = 6.5). Finally, the samples were left to dry for 24 h at room temperature.

2.3. Bending Testing

The bending resistance of the filaments was evaluated in terms of the elastic modulus ($E = \Delta\sigma/\Delta\varepsilon$, Young's modulus) and rupture time (RT), both using data obtained from the elastic zones of the stress (strain) plots and utilizing an EMIC DL 10000 universal mechanical testing machine (INSTRON, Brazil). For the tests, the samples were fixed using two supports separated from each other by 4 cm, and a load cell speed (maximum normal force of 500 N) of 10 mm·min^{-1} was applied in accordance with the ASTM D790 standard [25]. Three samples of each filament configuration were considered when analyzing the results.

2.4. Microstructural Properties

2.4.1. Scanning Electron Microscopy (SEM)

Scanning electron microscopy (SEM) images were obtained on a Hitachi TM-1000 SEM device (Hitachi, Ltd., Chiyoda, Japan) with an accelerating voltage of 20 kV. Samples were previously covered with a gold layer applied by a Q150R Quorum device (Quorum Technologies, Lewes, UK) using a 5 nm·min^{-1} rate for 5 min.

2.4.2. X-Ray Diffraction (XRD)

X-ray diffraction patterns of the drug and filament samples were achieved on a Miniflex Rigaku device (Rigaku Corporation, Osaka, Japan), Kα radiation, with a scanning rate of 0.02°·s^{-1} and diffraction angle range of 5–50°.

2.4.3. Fourier-Transform Infrared (FTIR) Spectroscopy

Fourier-transform infrared (FTIR) spectroscopy was performed on a Shimadzu Prestige 21 device (Shimadzu Corporation, Kyoto, Japan). FTIR spectra were established in the wavenumber range of 500 to 3500 cm^{-1} (90 scans and a resolution of 16 cm^{-1}). The samples for testing were prepared from a mixture of 10 mg of the filament and 1 g of KBr, which was soaked and pressed into the pellets.

2.4.4. Dynamic Light Scattering (DLS) Measurements

The mean hydrodynamic size of the particle collection (Z-average size) was obtained by the dynamic light scattering (DLS) technique (in triplicate) on a Malvern Zetasizer Nano Analyzer (Malvern Panalytical Ltd., Malvern, UK). The z-average value is a central and stable measurement parameter for particle size quality control analysis, as defined in the ISO13321 standard [26] (particle size analysis–photon correlation spectroscopy).

2.5. Thermal Analysis

The thermal behavior of the samples was studied by Differential Scanning Calorimetry (DSC). A Shimadzu DSC-60 Plus device (Shimadzu Corporation, Japan) was used for the experiments. Aluminum mini pans with lids (2.5 mm in diameter, Shimadzu Corporation) were used to dispose of 2 mg of each type of filament. DSC tests were conducted (i) with one heating cycle in the temperature range of 30 to 285 °C (heating rate of 10 °C.min^{-1}) and (ii) with two continuous heating and cooling cycles (in the range of 30 to 150 °C, using the same heating rate as above).

2.6. Statistical Analysis

ImageJ software (version 13.0.6.) was used to estimate the drug particle size and diameter of each filament type from the SEM micrographs. Data were presented as the mean $\pm$ standard deviation, calculated from thirty observations ($n = 30$). The hypothesis test for the difference between means with unknown population deviation was performed as an inferential statistics test, with a significance level (α) of 5% (the hypothesis of statistically equal means is accepted for p-values greater than α).

2.7. In Vitro Release Kinetics

The in vitro kinetics of the Amlodipine drug's release from the polymeric filament-shaped matrices was analyzed based on the results of the temporal variation in the drug concentration in gastric (pH = 1.2) and intestinal (pH = 7.2) simulated fluids. These acidic and basic solutions were prepared according to the procedures established in the United States Pharmacopeia [27,28]. The gastric fluid was prepared from a solution of 3 g of NaCl dissolved in 1.5 L of deionized water. The pH of the solution was adjusted to 1.2 $\pm$ 0.1 using dilute hydrochloric acid. The intestinal fluid was prepared from a solution of 10 g of potassium phosphate dissolved in 1 L of deionized water. The pH of the solution was altered to 7.2 $\pm$ 0.1 using 1 N sodium hydroxide. For testing, 30 mg of each filament sample was immersed in 100 mL of each type of simulated fluid at 36.5 °C (arranged in a 250 mL beaker) while being kept under constant stirring (rotation speed of 100 rpm) [27,28]. The concentration of Amlodipine in each fluid medium as a function of time was determined from the drug calibration curve (absorbance versus concentration), established at its maximum absorption peak (λ = 243 nm). Three independent samples of each E100/Aml system were evaluated. For the tests, 2 mL aliquots of each solution containing the filaments were analyzed at fixed time intervals in a spectrophotometer Kasvi K-37 (Kasvi, Pinhais, Brazil). The equivalent absorbance intensities were converted into relative drug concentration, $(C_t - C_0)/C_0$, where C_0 and C_t are the initial and time-dependent concentrations, respectively. The drug release kinetics in the simulated media were analyzed for a 165 min period.

2.8. FDM 3D Printing

A commercial direct drive 3D printer (model Kywoo3D Tycoon was used to print the test part. In this printer, the filament is directed into a PTFE tube with an internal diameter of 2.0 mm in contact with the hot end of the extruder nozzle. Thus, the filament takes a short path (about 10–15 cm) to the hot end (compared to Bowden-type printers), minimizing the clogging of the printing system due to possible filament breaks and, consequently, considerable material loss. A three-dimensional dosage form model of Amlodipine (cylindrical geometry, with 6 mm in base diameter and 12 mm in height) was designed in a CAD environment. Thus, the part was printed using the Repetier Host slicer software v. 2.2.4, with a grid-type infill pattern (24 layers, 20% inner density) and an extrusion temperature of 145 °C.

3. Results

3.1. Structural and Morphological Characterization

Samples of the E100/Aml composite filaments are shown in Figure 2. The amount of material used in the preparation stage resulted in a production of approximately 3.8 m of continuous filament for each E100/Aml model (Figure 3a) without significant material loss during the hot-melt extrusion (HME) process.

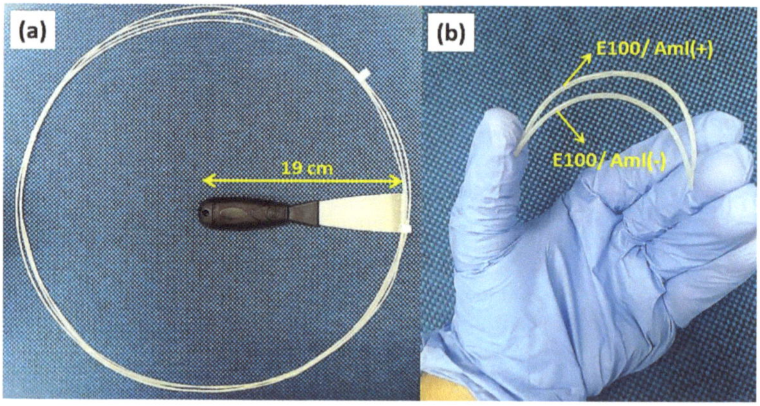

Figure 3. (**a**) Sample of E100/Aml continuous filament resulting from the extrusion process at 105 °C. (**b**) Photo image of small segments of E100/Aml filaments under bending after the chemical treatment.

It is important to highlight that the heating cylinder of the mini-extruder machine (Figure 2), which received the material for extrusion, is composed of two heating electrical resistors and that the entire system, including the extruder, is covered by a thermal blanket (15 cm in length, from the extruder nozzle). With this arrangement, the temperature gradient along the cylinder tends to decrease considerably compared to other extrusion systems (where heating occurs only at the end of the cylinder, close to the extruder nozzle). This directly results in a more homogeneous heat conduction through the material over a greater range of the tube length [29].

Thus, it was possible to initially stabilize the temperature of the cylindrical extruder tube at 110 °C for 1 h before starting extruding the E100/Aml mixtures using a lower temperature of 105 °C. Improvements noted in the filament production process included the reduction in the sticky aspect of the E100 (acrylic copolymer) inside the tube, which would be observed at higher extrusion temperatures [30] and the guarantee of its fluidity and the prevention of the thermal expansion of the filament at the exit of the extruder nozzle, which would negatively influence the expected diameter [31,32].

Then, the polymeric filaments were subjected to the chemical process of protonation [21]. Protonation is the action of adding a proton to an atom, molecule, or ion and differs from hydrogenation, which consists of a process by which hydrogen atoms are added to unsaturated fats and oils (unsaturated molecules are those that contain double bonds, while saturated ones contain only single bonds, and, consequently, these double bonds have the potential to accept hydrogen atoms and, thus, become hydrogenated); additionally, during protonation, a change in the charge of the protonated species takes place (and mass and chemical properties are altered), while in the hydrogenation process, it is unaffected.

In this final phase of filament preparation, they are exposed to a vaporized protonated solution (pH = 6.5), allowing the dimethylamino groups of their constituent molecules to receive H^+ protons from the solution. This causes the molecules to repel each other due to electrical repulsion from these neighboring protonated groups. This effect occurs radially, from the surface to the bulk of the material and, as a consequence, the filaments acquire greater flexibility when no chemical treatment is applied (Figure 3b).

In this sense, the improvement in the filaments' flexibility was quantified based on the analysis of the elastic modulus (E) variations as a function of the sample's exposure time, t, to chemical treatment (Figure 4).

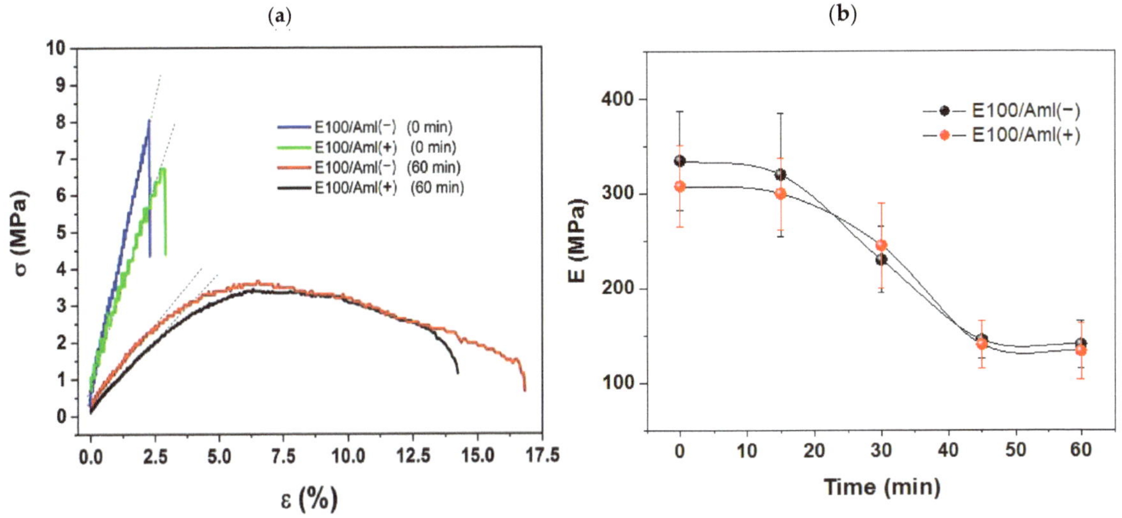

Figure 4. (**a**) The σ versus ε curves for E100/Aml(−) and E100/Aml(+) filament samples without (0 min) and with 60 min of chemical treatment; (**b**) elastic modulus (E) of the E100/Aml filaments as a function of the chemical reaction time.

E100/Aml(−) and E100/Aml(+) filaments without chemical treatment displayed purely elastic behavior (characterized by their respective σ vs. ε linear curves (0 min), Figure 4a), with high stiffness and brittleness compared to the normal stress applied during the bending tests. These characteristics were confirmed by the estimated average values of the elastic modulus, E, of approximately 335 and 308 MPa, and of the rupture time, RT, at around 43.2 and 48.4 s, for the E100/Aml(−) and E100 /Aml(+) filaments, respectively. On the other hand, the samples that received the chemical treatment for 60 min had their elastic behavior reduced, with the emergence of a plastic deformation phase (Figure 4a) and a longer time for the parts to rupture.

With the increasing exposure time, from 15 to 60 min, for the polymeric matrix to the protonation reaction, a progressive decrease in the E values was observed. The inflection point of the E vs. t curves (Figure 4b) was observed at about 30 min of exposure time, with E = 230 MPa and RT = 75.8 s for the E100/Aml(−) filaments and E = 245 MPa and RT = 69.6 s for the E100/Aml(+) filaments. This represents a significant variation in these parameters when compared with the initial mechanical properties displayed by the filaments. The lower values of E were found for protonation reaction times in the interval between 45 and 60 min. For the last measured reaction time, the values of E = 140 MPa and E = 133 MPa (approximately 58% and 57% of reduction in relation to the initial E values), and RT = 104.5 s and RT = 95.2 s (with a 141% and 98% increase for samples with low- and high-drug concentrations, respectively) support the claim that there is an improvement in flexibility for E100 filaments for the proposed use.

SEM micrographs of the Amlodipine Besylate (Aml) particles and of the flexible E100, E100/Aml(−), and E100/Aml(+) filaments are shown in Figure 5. Amlodipine particles (Figure 5a) are characterized by a microplate-type geometry and have irregular dimensions [33,34]. From the analysis of the SEM micrographs, using the ImageJ software, it was possible to verify two distinct particle size populations, centered mostly at 23.45 ± 5.23 µm and at 2.15 ± 0.74 µm. This micrometer-scale size distribution of the Amlodipine particles is one of the main contributing factors to its low solubility in water [33,35].

Figure 5. SEM images of the (**a**) Amlodipine (Aml) particles; (**b**) E100, (**c**) E100/Aml(−), and (**d**) E100/Aml(+) filaments.

Drug-free E100 filaments (E100 sample, Figure 5b) were also produced and analyzed for comparison purposes with the E100/Aml(−) (Figure 5c) and E100/Aml(+) (Figure 5d) filaments. They all exhibited a surface without apparent structural defects (no bubbles, no cracks, and/or abrupt variations in diameter resulting from heating are visibly observable). The absence of these geometric discontinuities is a determining factor for the usability of these filaments in conventional 3D printers [36,37].

In addition, the pristine E100 filament presented an average sample diameter of 1.79 ± 2.01 mm ($n = 30$), while the estimated diameters of the E100/Aml(−) and E100/Aml(+) ones were 1.81 ± 2.08 mm ($n = 30$) and 1.80 ± 1.25 mm ($n = 30$), respectively. The statistical evaluation demonstrated that the average diameter of the filaments does not present significant variations (p-value $> \alpha$ for all sample comparisons). This indicates that the presence of the drug in the bulk of the filament does not cause evident changes in the volume/unit length relationship and, consequently, in the surface area of the samples in contact with the dissolution media for the study of drug release kinetics.

It is important to highlight that a low-mass concentration of the drug was used in the mixtures to prepare the filaments, which corroborates the invariability of the filament diameters. In other words, the control of these parameters shows that the drug concentration variability, which can be released from the polymer matrix in the gastrointestinal fluids, can be directly related to specific parameters, such as the pH-dependent solubility of the copolymer, and to the quantity and dispersion of the drug mass in the polymeric matrix [9].

Figure 6 highlights the XRD spectra obtained for the pristine Aml and for E100, E100/Aml(−), and E100/Aml(+) filament samples. The Aml spectrum confirmed the crystalline nature of the drug, associated with the presence of its characteristic diffraction peaks in the range of the 2θ angles studied (at 5.9°, 10.4°, 11.7°, 13.1°, 14.5°, 18.1°, 19.3°, 20.2°, 23.4°, and 24.4°) [38,39].

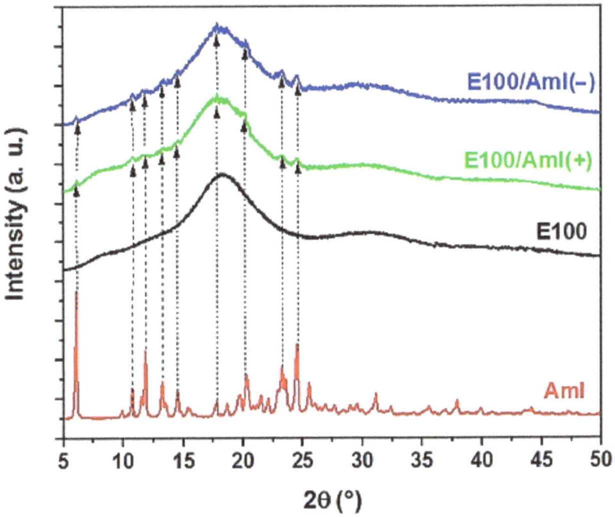

Figure 6. Characteristic X-ray diffraction patterns of the Amlodipin (Aml) drug and E100, E100/Aml(−), and E100/Aml(+) filament samples.

The X-ray spectrum of the E100 type filaments showed a broad peak, with the vertical displacement of the baseline, in a wide 2θ range between 10 and 25°, as a signature of an amorphous solid matrix. This result is in agreement with the XRD data of the polymer in its original form [40].

The XRD spectra of the E100/Aml filaments showed similar characteristics, as displayed in Figure 6. The diffractograms of the E100/Aml(−) and E100/Aml(+) filament samples showed a broad peak in the same 2θ angle region as observed for the pure E100 filament, which could be attributed to the predominance of the polymeric phase in the polymer/drug sample. In addition, in this 2θ range, although the Aml peaks were not as explicit as in the spectrum of the pure drug sample, it was possible to distinguish small peaks over the trace, indicating the presence of the Aml crystalline phase. This is probably due to the small amount of drug in the samples and/or the decrease in the degree of crystallinity of the drug due to its interaction with the polymeric matrix.

The molecular vibrations of the filament samples were studied by FTIR spectroscopy (Figure 7). The spectrum of pristine Amlodipine (as provided by the manufacturer) exhibited all of its characteristic vibrational bands, as follows: 3305 cm^{-1} for the molecular vibration of valence electrons of the N-H bond; 3159 cm^{-1} for valence electrons of the C-H bond (aromatic); 2982 cm^{-1} for valence electrons of the C-H bond (aliphatic); 1695 cm^{-1} for valence electrons of the C = O bond (aliphatic); 1675 cm^{-1} for valence electrons of the C = C bond (aliphatic); 1614 cm^{-1} for valence electrons of the C = C bond (aromatic); 1497 cm^{-1} for the out-of-plane folding of N-H bonds; 1304 cm^{-1} for the coplanar folding of pyridine; 1267 cm^{-1} for symmetric stretching of the C-O bond; 1207 cm^{-1} for asymmetric stretching of the S = O bond; 1090 cm^{-1} for symmetric stretching of the S = O bond; 1015 cm^{-1} for symmetrical stretching of the C-O-C bond; 758 cm^{-1} for the out-of-plane folding of the C-H bond; 730 cm^{-1} for the vibration of valence electrons of the C-S bond (aliphatic); 690 cm^{-1} for the coplanar folding of the C-S bond; and 565 cm^{-1} for the out-of-plane folding of the C-N bond [33,38,41–46].

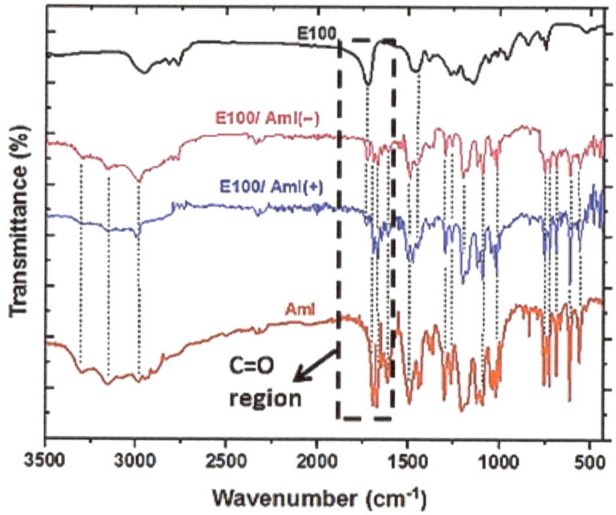

Figure 7. FTIR spectra of the pristine Aml, E100, and E100/Aml filaments. The details show the region of the polymer and drug C=O vibrational bands.

The FTIR spectrum of the E100 filament sample showed the typical bands of the copolymer in its original form, centered at 2956 cm^{-1} (with asymmetric stretching of the C-H bond), 2875 cm^{-1} (with symetric stretching of the C-H bond), 2820 cm^{-1} (with asymmetric stretching of the H_3C-N-CH_3 bond), 2772 cm^{-1} (with symmetric stretching of the H_3C-N-CH_3 bond), 1730 cm^{-1} (with symmetric stretching of the $C = O$ bond of the ester group), 1460 cm^{-1} (with asymmetric stretching of the H-C-H bond), 1388 cm^{-1} (with the coplanar folding of the C-H bonds), and 1150 cm^{-1} (with the asymmetric stretching of the C-O bond of the ester group) [41,47–49].

In addition, the FTIR spectra of the E100/Aml(+) and E100/Aml(−) filaments exhibited molecular vibration bands through the whole wavenumber range of interest (without significant differences). Whether using a high or low concentration of the drug dispersed in the polymer matrix, it is possible to discern the presence of a considerable number of vibrational bands for both constituent materials. In other words, the presence of most of the vibration bands for both the polymer and drug in the E100/Aml samples, located at the respective characteristic wavenumbers of the isolated substances, highlights the predominance of the physical interaction between the polymer and drug molecules. In fact, the molecular structures of these components (Figure 1) do not support significant chemical interactions for the formation of chemical complexes between these materials [50]. The most evident change in the FTIR spectrum of the E100/Aml filament samples is the intensity decrease in the $C = O$ carbonyl vibrational bands (at 1730 cm^{-1} for the polymer and at 1695 cm^{-1} for the drug) in relation to the spectrum of the pristine Amlodipine and the E100 filament. This characterizes polar interactions between the drug and the polymer, which are important to demonstrate the physical interaction concerning the constituent materials and the molecular stability of the drug after the HME process.

The thermal characteristics of the pristine Amlodipine drug (Aml), E100, E100/Aml(−), and E100/Aml(+) filaments were analyzed by DSC (Figure 8). The DSC spectrum of the Aml sample shows a narrow and well-defined peak at 210 °C, already known as its melting point (Figure 8a) [34,51]. This result is characteristic of a crystalline material, which was already validated by the XRD results. In addition, Aml degradation signals were not detected in the spectrum, demonstrating the thermal stability of the component in the analyzed temperature range.

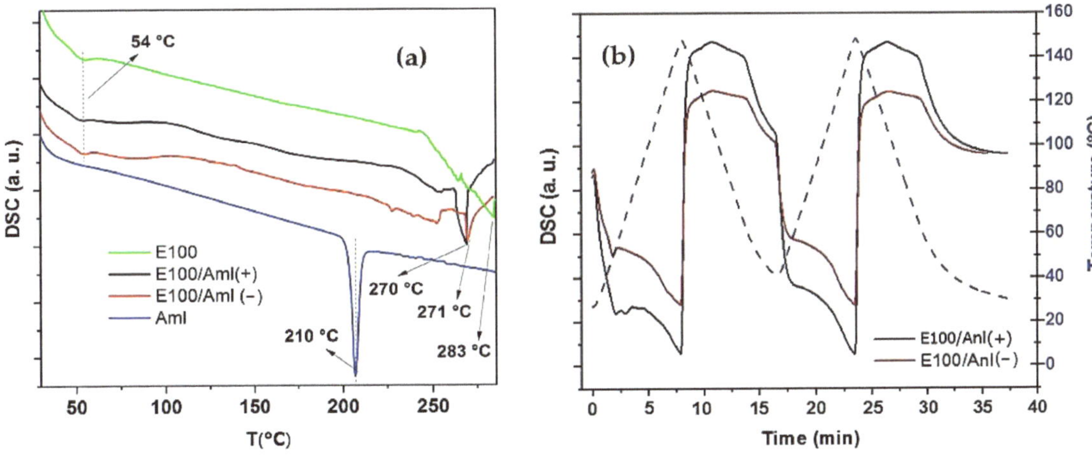

Figure 8. (**a**) DSC spectra (heating) of the Aml (pristine drug) and E100, E100/Aml(−), and E100/Aml(+) filaments; (**b**) DSC curves, with two continuous heating and cooling cycles for the E100/Aml(−) and E100/Aml(+) filaments (temperature range of 30–150 °C).

The DSC spectra of the E100 and polymer/drug filaments showed an endothermic signal at 54 °C due to the glass transition temperature of the E100 copolymer (Figure 8a) [30,52]. In these samples, the peaks observed at 283, 271, and 270°C are a signature of the degradation point of the E100 copolymer in E100, E100/Aml(−) and E100/Aml(+) filaments, respectively [52]. These data indicate that the extrusion process of the polymer/drug filaments at 105 °C (which exposes the materials thermally) is adequate without altering the properties of the drug and polymer molecules.

It is also possible to observe the absence of the drug melting point in the DSC spectra of the E100/Aml(−) and E100/Aml(+) filaments. Thus, as the presence of Amlodipine in these samples was confirmed by XRD and FTIR, the absence of its melting point in the DSC spectra suggests the better dispersion of the drug in the amorphous matrix. Specifically, a significant portion of the micrometric particles of Amlodipine (as confirmed by SEM analysis) tend to dissolve and decrease in size in the bulk of the filament, increasing the drug contact area [21]. This behavior is mentioned in the literature for solid dispersions of Amlodipine with other polymers, such as poly(vinyl pyrrolidone), chitosan, microcrystalline cellulose, and poly(lactic acid-co-glycolic acid), all aiming to facilitate the dissolution of the drug in the medium of interest [53–56].

The analysis of the hydrodynamic particle size of the Amlodipine drug obtained by DLS (Figure 9) in the acidic medium confirms this hypothesis. The drug particles, as supplied by the manufacturer, showed a heterogeneous particle size distribution, with three main populations at 175.5 ± 60.71 nm, 607.6 ± 203.2 nm, and 5380.0 ± 322.9 nm (with 31%, 23%, and 46% intensity, respectively). In addition, DLS data showed that the Z-average size for the drug particles was estimated to be 691.7 nm.

On the other hand, the Amlodipine particles that were released from the E100/Aml(−) and E100/Aml(+) matrices in the same medium presented two well-defined particle size populations, centered at 327.6 ± 173.2 nm (87% intensity) and 4626.0 ± 823.8 nm (13% intensity) for the E100/Aml(−) model, and centered at 321.5 ± 157.0 nm (93% intensity) and 4590.0 ± 838.8 nm (7% intensity) for the E100/Aml(+) model. In addition, their Z-mean calculated values were 332.4 nm and 284.8 nm for the filaments with low and high drug concentrations, respectively. Thus, it is possible to infer from the results that the interaction of the drug with the polymer promotes not only a significant reduction in the intensity of the micrometric particles of Amlodipine but also makes the majority of the population particles, around 90% of the intensity of the measurements, display at about 320 nm, and as a consequence, a reduction in the mean hydrodynamic size (Z-mean value) of the drug in the fluid medium occurs.

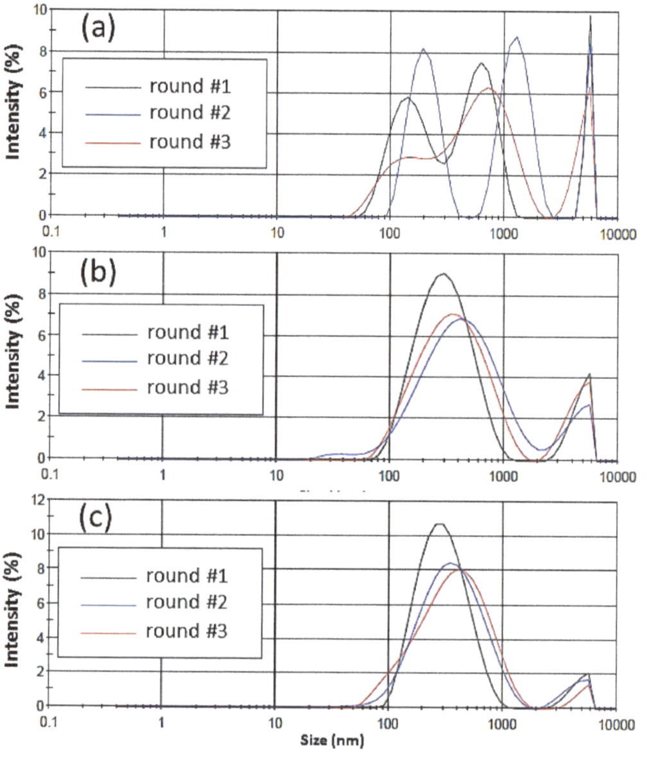

Figure 9. Hydrodynamic particle size in accordance with intensity of the (**a**) Amlodipine drug supplied by the manufacturer in an acidic medium and (**b**) Amlodipine drug released from the bulk of the E100/Aml(−) and (**c**) E100/Aml(+) filaments in an acidic medium. Three rounds of measurements were performed for each sample.

In fact, the pressure that disintegrates particles is significantly higher the smaller their size. Drugs characterized by a smaller particle size are known to have greater interfacial solubility due to a thinner diffusion layer and greater surface area, which is generally preferable for the oral administration of drugs [57]. The encapsulation of Amlodipine under these conditions in the pH-dependent matrix of the E100 polymer is another differentiating factor regarding the programmed release of the drug. It is known that the dispersion of drugs poorly soluble in water and with a low dissolution rate (such as Amlodipine) in amorphous solid solutions of the E100 copolymer can abruptly increase the drug dissolution [48,58]. A higher dissolution of Amlodipine can be expected from solid dispersion, probably due to the presence of amorphous Aml-particles incorporated into the E100 matrix. This indicates that the polymer can inhibit the recrystallization of the drug molecules [55].

In this sense, the choice of the E100 polymer for the production of 3D printing filaments can be justified, firstly, because it increases the surface area of action of the Amlodipine particles dispersed in the matrix and, secondly, since it provides 3D printing filaments of amorphous solid dispersions of the drug with much smaller particles than those supplied by the manufacturer, with a consequent improvement in its dissolution and bioactivity.

Figure 8b shows the DSC results of the E100/Aml filament samples subjected to two complete and continuous cycles of heating and cooling, in the range of 30–150 °C. It is important to highlight that the components of these samples (polymer and drug) had, previously, been subjected to a heating process at 105 °C in the heated tube of the extrusion machine. The thermal behavior study of the E100/Aml samples through two new cycles allowed the temperature gradient with which these filaments were subjected in conventional FDM 3D printers to be stimulated during the three-dimensional printing of the engineered solid pharmaceutical products, and/or on the possible reuse of the waste materials after FDM 3D in the production of new filaments (in both cases, considering a second and/or a third heating/cooling moment).

The results show that, in both cycles, the thermal signatures of the E100/Aml(−) and E100/Aml(+) samples are preserved in relation to the DSC spectrum obtained for the same samples after the first heating process (in the same temperature range). In the spectra, the original amorphous characteristic of the E100 copolymer is kept, and no degradation signal of the constituent materials is observed [21]. The data obtained are in agreement with the earlier reported results, which demonstrated that the initial mass loss and subsequent complete degradation of the drug [51] and polymer [30,59,60] only occurs at high temperatures (above 200 °C, and reaching up to 600 °C, to attain the total degradation of the drug).

3.2. Drug Release Evaluation

The drug release profiles from the polymeric matrix of the filaments in simulated gastric (pH = 1.2, acidic medium) and intestinal (pH = 7.2, basic medium) fluids are presented in Figure 10. It is possible to observe the kinetics of the average concentration of Amlodipine available in the acidic environment, which in the first few minutes of the trials displayed an exponential behavior for both E100/Aml(−) and E100/Aml(+) filaments. The low variability of the concentrations observed at each evaluated time indicates the homogeneous and reproducible distribution of the drug in the bulk of the filaments since the average concentration was obtained for independent samples of each E100/Aml system.

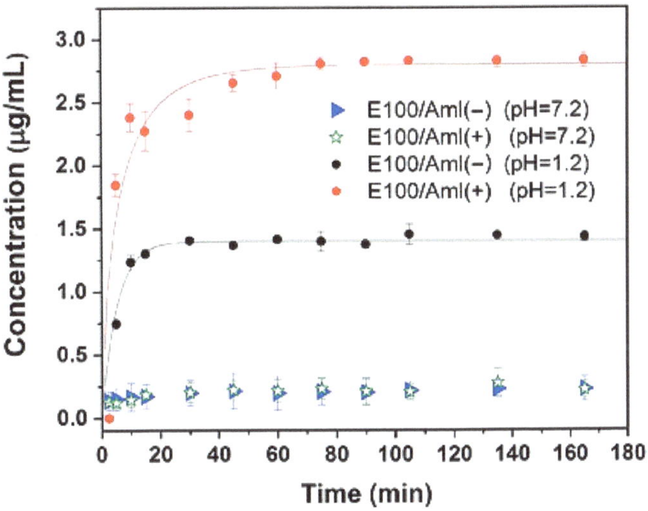

Figure 10. In vitro release kinetics of Amlodipine in acidic (pH = 1.2) and basic (pH = 7.2) media from the polymeric matrix of the E100/Aml(−) and E100/Aml(+) filaments.

The drug concentrations released into the medium as a function of time assumed an asymptotic behavior (saturation value) around 30 min after the tests were initiated for both hybrid filaments, achieving saturation concentrations of 1.38 µg/mL and 2.81 µg/mL for the E100/Aml(−) and E100/Aml(+) filaments, respectively.

These characteristics are typical of the E100 copolymer supplied by the manufacturer and confirm that the pH-dependent solubility of the polymer is maintained for usage as a pH-responsive 3D printing filament. When the Eudragit® E100 solid matrix containing Amlodipine is placed in contact with the simulated acid fluid, the superficial radial layers of the filament are hydrated, and the copolymer chains in this region are relaxed and then dismantled, releasing the drug into the medium via an erosion mechanism of the polymer matrix. This release mechanism is established rapidly from the surface to the core of the filament until all the previously encapsulated drugs are available in the solution. In addition, the rapid-release profile of Amlodipine in the acidic medium was successfully modeled using the Weibull function (with solid lines following the experimental data in Figure 8) for the case of drug diffusion from cylindrical matrices [61,62]:

$$\frac{C_t}{C_\infty} = 1 - \exp\left(-at^b\right). \quad (1)$$

In this expression, C_t is the available concentration of the drug at time t; C_∞ is the asymptotic concentration of the drug; a is the time scale factor; and b is the parameter representing the degree of kinetics release. The modeling allowed a value of $b = 1$ for both E100/Aml(−) and E100/Aml(+) samples to be estimated, which is quantitatively in accordance with the exponential kinetics of the drug release pattern found. The combination of these results indicates that the E100/Aml filaments provide the rapid release of Amlodipine, with reproducibility and a typical diffusive characteristic.

On the other hand, a negligible concentration of the drug available in the basic medium was observed throughout the analyzed time interval, indicating that the E100 matrix acted as a barrier device to minimize the drug's contact with the external medium.

The application of E100/Aml filaments in FDM 3D printing was exemplified by printing a model part of one typical pharmaceutical form previously designed in CAD software (https://www.autodesk.com/hk/solutions/3d-cad-software, accessed on 20 October 2024) (Figure 11).

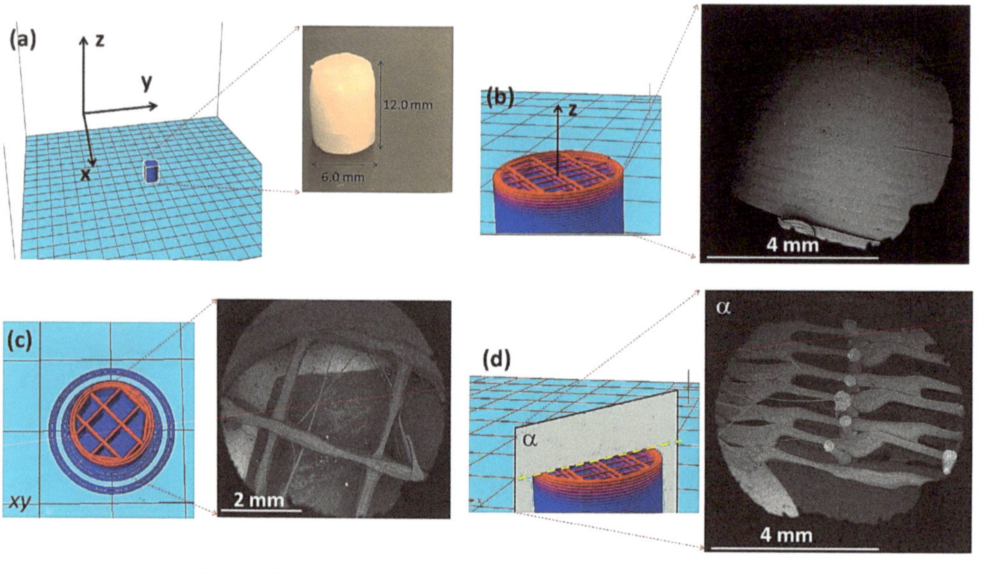

Figure 11. (**a**) CAD project of a 3D model part with a cylindrical shape and a photo image of the corresponding pharmaceutical form, produced from E100/Aml filaments; (**b**) CAD design and SEM image of the lateral surface of the part; (**c**) design of the part inner circular section and SEM image of the printed material; and (**d**) interior of the printed part seen from section α.

As mentioned in Section 2.8 (FDM 3D printing), the fabricated filaments were used with a commercial direct drive 3D printer to create a volumic shape as proof of concept. One part, with cylindrical geometry (Figure 11a), was previously subdivided into 24 layers (details in Figure 11b) using the slicer software. A grid type with internal filling was used (layers with parallel filling lines, perpendicular to the parallel filling lines of the subsequent layer—see details of the CAD project of the internal region of the part in Figure 11b,c), with a filling factor of 20%. The extrusion temperature of 145 °C was suitable so that the printing process could take place on the flat working surface of the printer. The total printing time of the part was 5 min and 24 s.

Figure 11b shows an SEM image of the lateral surface of the printed pharmaceutical form. It is possible to observe in detail the overlapping layers of E100/Aml material in the cylindrical geometry. SEM images in Figure 11c,d detail, in that order, the normal and parallel cross-sections (relative to the z-axis) of the three-dimensional printed part, demonstrating the high execution level of the printing procedure, even considering the small dimensions of the designed geometry. In this sense, given the characteristics observed for the polymer/drug filament, the filling level of the part obtained with them is an important factor to consider when determining the amount of drug available for parts with the same external dimensions.

As previously explained, although the importance of the E100 polymer in the pharmaceutical industry is recognized, a limited number of studies on the development of 3D printing filaments from E100 polymer are found in the literature [21,24]. The acrylic nature of the polymer, with brittle and sticky aspects, makes it difficult to obtain filaments with adequate flexibility for their workability in FDM 3D printers (given their extruded cylindrical geometric shape with millimeter diameters).

Cardoso and his collaborators [21] developed FDM 3D printing filaments of this copolymer containing hesperidin, which is a flavonoid extracted from citrus fruits. The authors highlighted the difficulties in optimizing filament extrusion process parameters. One of the actions involved was adapting the design of a commercial mini-extruder machine so that the mixture between the drug and polymer would be uniform in the heated reservoir of the device and so that the extruded filaments would not display surface defects and possess a constant diameter. The filaments were produced at a temperature of 130 °C and were then used to print cylindrical three-dimensional parts, 10 mm in diameter and 3 mm in height, with a grid-type infill pattern (60% infill), using a printing temperature equal to that of the filament production. The results of microstructural and thermal characterization and in vitro drug release tests demonstrated the feasibility of the materials for the production of customized pharmaceutical forms of the drug.

In the work by Choudhury et al. [24] E100 FDM 3D printing filaments containing the drug berberine chloride were fabricated and applied to print customized hollow capsules. The authors cited the difficulty of producing E100 filaments, but they overcame the problem by using an optimized combination of talc (20% *w*/*w*) and triethyl citrate (3.5% *w*/*w*) as additives. Microstructural and thermal analyses showed that the extrusion of the filaments at 100 °C and the printing process of the prototypes at 160 °C did not cause the degradation of either the polymer or drug, therefore preserving the original chemical characteristics of both materials in the printed parts.

Sadia et al. [63] used fused deposition modeling 3D printing to fabricate pH-soluble pharmaceutical tablets capable of the immediate release of various model drugs. They investigated the addition of non-melting fillers to the methacrylic matrix to facilitate FDM 3D printing and explored the impact of the filler nature, the compatibility with the gears of the 3D printer, and the polymer/filler ratio on the 3D printing process. The optimized filaments were based on Eudragit E and TCP (tribasic calcium phosphate). Following the two thermal fabrication processes, hot melt extrusion and fused deposition modeling 3D printing and drug contents of 94.22%, 88.53%, 96.51%, and 93.04% of 5-ASA, captopril, theophylline, and prednisolone, respectively, were added to the filaments. They concluded that the optimal range of the non-melting component of 20–50% within the filament composition enabled the fabrication of well-defined caplets. The level of drug incorporation was above 94% after the thermal fabrication steps, remaining intact in the tablets for the cases of theophylline, 5-ASA, and prednisolone, whilst a significant drop in captopril content was observed.

In comparison with previous results, the present study demonstrated that the production of E100 filaments containing the model drug Amlodipine was viable, significantly contributing to this field of study. The use of the chemical protonation process in the post-extrusion preparation of the filaments improved their flexibility for use in FDM 3D printers, eliminating the need for chemical additives and grafting. Potential challenges for production scalability and/or programmed modification in the release rates of three-dimensional drug parts produced from these filaments involved the adequate control of printing parameters such as extruder nozzle diameter, printing speed, geometry, and the infill rate of these parts.

4. Conclusions

The combination of the mechanical, microstructural, and thermal characterization results proved appropriate as the hot-melt extrusion process parameters were chosen for the production of flexible and uniform E100 filaments containing Amlodipine. The extrusion process was set at 105 °C, and one exemplifying three-dimensional parts was printed at 145 °C using the produced filament. The thermal stability of the polymer and the drug was preserved in the hybrid filaments after two heating and cooling cycles, and a smaller Z-average size of the drug particles released from the amorphous matrix was observed (with a reduction from 691.7 nm, as provided by the manufacturer to 332.4 nm and 284.8 nm 284.8–332.4 nm after their interaction with the polymer and release in the acidic medium). These observations were important to demonstrate the efficiency of the 3D printable Amlodipine delivery systems with improved action and the possibility of reusing the material to produce new filaments.

In vitro release tests showed that the pH-dependent characteristic of the polymer was preserved in the filaments, with rapid release of the drug occurring only in an acidic environment. In addition, the saturation of the accumulated drug concentration in the acidic medium occurred after 30 min of testing. The low variability of the mean drug concentration values in the acidic medium as a function of time, observed for independent filament samples, indicates a homogeneous dispersion of Amlodipine in the cylindrical polymeric matrix in accordance with the results of thermal and microstructural characterizations. The 3D printing of the amorphous solid solution of the drug (which was projected with customized surface and filling geometry) from these filaments, using a conventional FDM 3D printer, proved the potential of the technique for the development of new pharmaceutical forms of Amlodipine, with reproducibility, individualized therapy, controlled release, and improved drug dissolution.

After the satisfactory results of this initial study, future research will be continued by the authors, with the objective of studying variations in the drug release profile from copolymer matrices with different three-dimensional geometries of the same specific surface area (evaluating simultaneously the reproducibility of the printing process); establishing a correlation between the fill rate of a polymer/drug part and the concentration of the available drug; developing E100 filaments with other poorly water-soluble drugs and/or with broad-spectrum antimicrobial drugs for innovative applications in biotechnology.

Author Contributions: Conceptualization, E.S.A., G.M.N., P.H.N.C., G.F.T., F.S.S. and P.M.F.; methodology, E.S.A., G.M.N., P.H.N.C., G.F.T., N.C.O., F.S.S. and P.M.F.; writing—review and editing, E.S.A., G.M.N., P.H.N.C., G.F.T., P.M.F., F.S.S. and N.C.O.; data curation, E.S.A., G.M.N., P.H.N.C., G.F.T., P.M.F., N.C.O., F.S.S. and E.M.E.d.S.; writing—original draft preparation, E.S.A., G.M.N., P.H.N.C., P.M.F. and F.S.S.; visualization, E.S.A., G.M.N., F.S.S., P.H.N.C., G.F.T., P.M.F., N.C.O. and E.M.E.d.S.; project administration, E.S.A. and F.S.S. All authors have read and agreed to the published version of the manuscript.

Funding: This research was funded by Fundação de Amparo a Ciência e Tecnologia do Estado de Pernambuco—FACEPE, Brazil, APQ 1387-3.03/22 and IBPG-0849-3.03/20 projects; Conselho Nacional de Desenvolvimento Científico e Tecnológico—CNPq, Brazil; and Portuguese national funds through FCT—Fundação para a Ciência e a Tecnologia, Portugal, UIDB/00285/2020 and LA/P/0112/2020 projects.

Data Availability Statement: The original contributions presented in the study are included in the article, further inquiries can be directed to the corresponding author.

Conflicts of Interest: The authors declare no conflicts of interest.

References

1. Elhadad, A.A.; Rosa-Sainz, A.; Canete, R.; Peralta, E.; Begines, B.; Balbuena, M.; Alcudia, A.; Torres, Y. Applications and multidisciplinary perspective on 3D printing techniques: Recent developments and future trends. *Mater. Sci. Eng. R Rep.* **2023**, *156*, 100760. [CrossRef]
2. Gao, G.; Ahn, M.; Cho, W.-W.; Kim, B.-S.; Cho, D.-W. 3D Printing of Pharmaceutical Application: Drug Screening and Drug Delivery. *Pharmaceutics* **2021**, *13*, 1373. [CrossRef] [PubMed]
3. DePalma, K.; Walluk, M.R.; Murtaugh, A.; Hilton, J.; McConky, S.; Hilton, B. Assessment of 3D printing using fused deposition modeling and selective laser sintering for a circular economy. *J. Clean. Prod.* **2020**, *264*, 121567. [CrossRef]
4. Padmakumar, M. Additive manufacturing of tungsten carbide hardmetal parts by selective laser melting (SLM), selective laser sintering (SLS) and binder jet 3D printing (BJ3DP) techniques. *Lasers Manuf. Mater. Process* **2020**, *7*, 338–371.
5. Kafle, A.; Luis, E.; Silwal, R.; Pan, H.M.; Shrestha, P.L.; Bastola, A.K. 3D/4D Printing of Polymers: Fused Deposition Modelling (FDM), Selective Laser Sintering (SLS), and Stereolithography (SLA). *Polymers* **2021**, *13*, 3101. [CrossRef]
6. Cardoso, P.H.N.; Araújo, E.S. An Approach to 3D Printing Techniques, Polymer Materials, and Their Applications in the Production of Drug Delivery Systems. *Compounds* **2024**, *4*, 71–105. [CrossRef]
7. Reddy, N.; Ananthaprasad, M.G. Polymeric materials for three-dimensional printing. In *Additive Manufacturing*; Woodhead Publishing: Sawston, UK, 2021; pp. 233–274. [CrossRef]
8. Pistone, M.; Racaniello, G.F.; Arduino, I.; Laquintana, V.; Lopalco, A.; Cutrignelli, A.; Rizzi, R.; Franco, M.; Lopedota, A.; Denora, N. Direct cyclodextrin-based powder extrusion 3D printing for one-step production of the BCS class II model drug niclosamide. *Drug Deliv. Transl. Res.* **2022**, *12*, 1895–1910. [CrossRef]
9. Tan, D.K.; Maniruzzaman, M.; Nokhodchi, A. Advanced pharmaceutical applications of hot-melt extrusion coupled with fused deposition modelling (FDM) 3D printing for personalised drug delivery. *Pharmaceutics* **2018**, *10*, 203. [CrossRef]
10. Zhang, Z.; Feng, S.; Almotairy, A.; Bandari, S.; Repka, M.A. Development of multifunctional drug delivery system via hot-melt extrusion paired with fused deposition modeling 3D printing techniques. *Eur. J. Pharm. Biopharm.* **2023**, *183*, 102–111. [CrossRef]
11. Lu, M. Novel Excipients and Materials Used in FDM 3D Printing of Pharmaceutical Dosage Forms. In *3D and 4D Printing in Biomedical Applications: Process Engineering and Additive Manufacturing*; Mohammed Maniruzzaman, M., Ed.; Wiley: Hoboken, NJ, USA, 2019; pp. 211–237. [CrossRef]
12. Auriemma, G.; Tommasino, C.; Falcone, G.; Esposito, T.; Sardo, C.; Aquino, R.P. Additive Manufacturing Strategies for Personalized Drug Delivery Systems and Medical Devices: Fused Filament Fabrication and Semi Solid Extrusion. *Molecules* **2022**, *27*, 2784. [CrossRef]
13. Wei, C.; Solanki, N.G.; Vasoya, J.M.; Shah, A.V.; Serajuddin, A.T. Development of 3D printed tablets by fused deposition modeling using polyvinyl alcohol as polymeric matrix for rapid drug release. *J. Pharm. Sci.* **2020**, *109*, 1558–1572. [CrossRef]
14. Elkasabgy, N.A.; Mahmoud, A.A.; Maged, A. 3D printing: An appealing route for customized drug delivery systems. *Int. J. Pharm.* **2020**, *588*, 119732. [CrossRef]
15. Geraili, A.; Xing, M.; Mequanint, K. Design and fabrication of drug-delivery systems toward adjustable release profiles for personalized treatment. *View* **2021**, *2*, 20200126. [CrossRef]
16. Shojaie, F.; Ferrero, C.; Caraballo, I. Development of 3D-Printed Bicompartmental Devices by Dual-Nozzle Fused Deposition Modeling (FDM) for Colon-Specific Drug Delivery. *Pharmaceutics* **2023**, *15*, 2362. [CrossRef]
17. Mahmood, M.A. 3D Printing in Drug Delivery and Biomedical Applications: A State-of-the-Art Review. *Compounds* **2021**, *1*, 94–115. [CrossRef]
18. Sester, C.; Ofridam, F.; Lebaz, N.; Gagnière, E.; Mangin, D.; Elaissari, A. pH-Sensitive methacrylic acid–methyl methacrylate copolymer Eudragit L100 and dimethylaminoethyl methacrylate, butyl methacrylate, and methyl methacrylate tri-copolymer Eudragit E100. *Polym. Adv. Technol.* **2020**, *31*, 440–450. [CrossRef]
19. Patra, C.N.; Priya, R.; Swain, S.; Jena, G.K.; Panigrahi, K.C.; Ghose, D. Pharmaceutical significance of Eudragit: A review. *Future J. Pharm. Sci.* **2017**, *3*, 33–45. [CrossRef]
20. dos Santos, J.; da Silva, G.S.; Velho, M.C.; Beck, R.C.R. Eudragit®: A Versatile Family of Polymers for Hot Melt Extrusion and 3D Printing Processes in Pharmaceutics. *Pharmaceutics* **2021**, *13*, 1424. [CrossRef]
21. Cardoso, P.H.N.; Oliveira, C.Y.B.; Nunes, M.; Tavares, G.F.; Faia, P.M.; Araújo, E.S. Eudragit E100/Hesperidin 3D Printing Filaments: Preparation, Characterization, and In Vitro Release Studies. *Appl. Sci.* **2023**, *13*, 11558. [CrossRef]

22. Elizondo-Luevano, J.H.; Castro-Ríos, R.; Parra-Saldívar, R.; Larqué-García, H.; Garza-Tapia, M.; Melchor-Martínez, E.M.; Chávez-Montes, A. Influence of the polymer and solvent variables on the nanoencapsulation of the flavonoid quercetin: Preliminary study based on Eudragit® polymers. *Appl. Sci.* **2023**, *13*, 7816. [CrossRef]
23. Llera-Rojas, V.G.; Hernández-Salgado, M.; Quintanar-Guerrero, D.; Leyva-Gómez, G.; Mendoza-Elvira, S.; Villalobos-García, R. Comparative study of the release profiles of ibuprofen from polymeric nanocapsules and nanospheres. *J. Mex. Chem. Soc.* **2019**, *63*, 61–70. [CrossRef]
24. Choudhury, D.; Murty, U.S.; Banerjee, S. 3D printing and enteric coating of a hollow capsular device with controlled drug release characteristics prepared using extruded Eudragit® filaments. *Pharm. Dev. Technol.* **2021**, *26*, 1010–1020. [CrossRef]
25. *ASTM D790*; Standard Test Methods for Flexural Properties of Unreinforced and Reinforced Plastics and Electrical Insulating Material. ASTM International: West Conshohocken, PA, USA, 2001.
26. *ISO13321*; Particle Size Analysis—Photon Correlation Spectroscopy. ISO: Geneva, Switzerland, 1996.
27. United States Pharmacopeia. *Reagents, Simulated Gastric Fluid TS*; USP-NF: Rockville, MD, USA, 2023. [CrossRef]
28. United States Pharmacopeia. *Reagents, Simulated Intestinal Fluid TS*; USP-NF: Rockville, MD, USA, 2023. [CrossRef]
29. Vidakis, N.; Petousis, M.; Velidakis, E.; Mountakis, N.; Fischer-Griffiths, P.E.; Grammatikos, S.; Tzounis, L. Fused Filament Fabrication Three-Dimensional Printing Multi-Functional of Polylactic Acid/Carbon Black Nanocomposites. *C* **2021**, *7*, 52. [CrossRef]
30. Alhnan, M.A. Solid Dosage Form Production. Patent EP3191084B1, 28 November 2018.
31. Ai, J.R.; Peng, F.; Joo, P.; Vogt, B.D. Enhanced dimensional accuracy of material extrusion 3D-printed plastics through filament architecture. *ACS Appl. Polym. Mater.* **2021**, *3*, 2518–2528. [CrossRef]
32. Bandari, S.; Nyavanandi, D.; Dumpa, N.; Repka, M.A. Coupling hot melt extrusion and fused deposition modeling: Critical properties for successful performance. *Adv. Drug Deliv. Rev.* **2021**, *172*, 52–63. [CrossRef]
33. Pohar, A.; Likozar, B. Dissolution, nucleation, crystal growth, crystal aggregation, and particle breakage of amlodipine salts: Modeling crystallization kinetics and thermodynamic equilibrium, scale-up, and optimization. *Ind. Eng. Chem. Res.* **2014**, *53*, 10762–10774. [CrossRef]
34. Hadžidedić, Š.; Uzunović, A.; Šehić Jazić, S.; Kocova El-Arini, S. The impact of chirality on the development of robust and stable tablet formulation of (S-) amlodipine besylate. *Pharm. Dev. Technol.* **2014**, *19*, 930–941. [CrossRef]
35. Jalilov, A.K.; Khaidarov, V.R. Microscopic studies of the shape and size of particles of samples of amlodipine besylate substances from selected pharmaceutical manufacturers. *Am. J. Med. Sci. Pharm. Res.* **2023**, *5*, 16–24. [CrossRef]
36. Saleh, M.; Anwar, S.; AlFaify, A.Y.; Al-Ahmari, A.M.; Abd Elaty, E. Development of PLA/recycled-desized carbon fiber composites for 3D printing: Thermal, mechanical, and morphological analyses. *J. Mater. Res. Technol.* **2024**, *29*, 2768–2780. [CrossRef]
37. Gama, N.; Ferreira, A.; Barros-Timmons, A.; Evtuguin, D. Polyamide 6/modified pine bark particle composites for additive manufacturing. *J. Mater. Sci.* **2021**, *56*, 19093–19105. [CrossRef]
38. Zhao, Z.; Li, Y.; Wu, J.; Shi, Z.; Zhao, P.; Su, H.; Wang, Q.; Jin, L. Nanofiber orodispersible films based on carboxymethyl curdlan and PEO: New delivery system for amlodipine besylate. *Colloids Surf. A Physicochem. Eng. Asp.* **2022**, *635*, 128096. [CrossRef]
39. Pezik, E.; Gulsun, T.; Sahin, S.; Vural, I. Development and characterization of pullulan-based orally disintegrating films containing amlodipine besylate. *Eur. J. Pharm. Sci.* **2021**, *156*, 105597. [CrossRef] [PubMed]
40. Wang, X.; Zhu, Y.; Zhao, X.; Zhang, S.; Cao, M.; Wang, X.; Li, W. Development and characterization of an amorphous curcumin-Eudragit® E100 solid dispersions with improved solubility, stability, and pharmacokinetic properties. *Pharm. Dev. Technol.* **2022**, *27*, 965–974. [CrossRef] [PubMed]
41. Pavia, D.L.; Lampman, G.M.; Kriz, G.S.; Vyvyan, J.R. *Introduction to Spectroscopy*, 4th ed.; Cengage: Belmont, CA, USA, 2009; pp. 1–104.
42. Ahmed, Z.A.G.; Hussein-Al-Ali, S.H.; Ibrahim, I.A.A.; Haddad, M.K.; Ali, D.K.; Hussein, A.M.; Abu Sharar, A.A. Development and Evaluation of Amlodipine-Polymer Nanocomposites Using Response Surface Methodology. *Int. J. Polym. Sci.* **2022**, *2022*, 3427400. [CrossRef]
43. Kapor, A.; Nikolić, V.; Nikolić, L.; Stanković, M.; Cakić, M.; Stanojević, L.; Ilić, D. Inclusion complexes of amlodipine besylate and cyclodextrins. *Open Chem.* **2010**, *8*, 834–841. [CrossRef]
44. Nanda, A.; Sahoo, R.N.; Pramanik, A.; Mohapatra, R.; Pradhan, S.K.; Thirumurugan, A.; Das, D.; Mallick, S. Drug-in-mucoadhesive type film for ocular anti-inflammatory potential of amlodipine: Effect of sulphobutyl-ether-beta-cyclodextrin on permeation and molecular docking characterization. *Colloids Surf. B Biointerfaces* **2018**, *172*, 555–564. [CrossRef]
45. Hrichi, H.; Monser, L.; Adhoum, N. A novel electrochemical sensor based on electropolymerized molecularly imprinted poly (aniline-co-anthranilic acid) for sensitive detection of amlodipine. *J. Electroanal. Chem.* **2017**, *805*, 133–145. [CrossRef]
46. Beaman, C.W.; Lees, R.M.; Xu, L.H.; Billinghurst, B.E. FTIR synchrotron spectroscopy of lower modes of methyl-D3 mercaptan (CD3SH). *J. Mol. Spectrosc.* **2023**, *392*, 111739. [CrossRef]
47. Quinteros, D.A.; Manzo, R.H.; Allemandi, D.A. Interaction between Eudragit® E100 and anionic drugs: Addition of anionic polyelectrolytes and their influence on drug release performance. *J. Pharm. Sci.* **2011**, *100*, 4664–4673. [CrossRef]
48. Hanif, M.; Ameer, N.; Akram, H.; Mahmood, K.; Bano, S.; Qaiser, M.; Rahman, H.M.A. Raft-forming gastroretentive tablets incorporating solidly dispersed Curcumin-Eudragit E100; in vitro and in vivo approaches for treatment of gastric ulcer. *Polym. Bull.* **2023**, *80*, 9833–9851. [CrossRef]

49. Khizar, S.; Ahmad, N.M.; Ahmed, N.; Manzoor, S.; Elaissari, A. Encapsulation of doxorubicin in magnetic-polymer hybrid colloidal particles of Eudragit E100 and their hyperthermia and drug release studies. *Polym. Adv. Technol.* **2020**, *31*, 1732–1743. [CrossRef]
50. Carrascal, J.J.; Pinal, R.; Carvajal, T.; Pérez, L.D.; Baena, Y. Benzoic acid complexes with Eudragit E100®: New alternative antimicrobial preservatives. *Int. J. Pharm.* **2021**, *607*, 120991. [CrossRef] [PubMed]
51. Silva, A.C.M.; Gálico, D.A.; Guerra, R.B.; Perpétuo, G.L.; Legendre, A.O.; Rinaldo, D.; Bannach, G. Thermal stability and thermal decomposition of the antihypertensive drug amlodipine besylate. *J. Therm. Anal. Calorim.* **2015**, *120*, 889–892. [CrossRef]
52. Linares, V.; Yarce, C.J.; Echeverri, J.D.; Galeano, E.; Salamanca, C.H. Relationship between Degree of Polymeric Ionisation and Hydrolytic Degradation of Eudragitfi E Polymers under Extreme Acid Conditions. *Polym. J.* **2019**, *11*, 1010. [CrossRef]
53. Chourasiya, V.; Bohrey, S.; Pandey, A. Formulation, optimization, and characterization of amlodipine besylate loaded polymeric nanoparticles. *Polym. Polym. Compos.* **2021**, *29*, S1555–S1568. [CrossRef]
54. Ramirez-Tagle, R.; Salgado-Moran, G.; Mendez-Luna, D.; Correa-Basurto, J.; Villada, W.C.; Candia, L.G.; Mendoza-Huizar, L.H. Synthesis, chemical identification, drug release and docking studies of the amlodipine–chitosan nanobiopolymer composite. *J. Chil. Chem. Soc.* **2021**, *66*, 5063–5066. [CrossRef]
55. Ghobashy, M.M.; Alshangiti, D.M.; Alkhursani, S.A.; Al-Gahtany, S.A.; Shokr, F.S.; Madani, M. Improvement of In Vitro Dissolution of the Poor Water-Soluble Amlodipine Drug by Solid Dispersion with Irradiated Polyvinylpyrrolidone. *ACS Omega* **2020**, *5*, 21476–21487. [CrossRef]
56. Shaikh, S.A.; Shaikh, S.S.; Shahi, S.R.; Shookur, M.A.; Reddy, L.K.; Padalkar, A.N.; Thube, M. Formulation and evaluation of s-(-)-amlodipine besylate and nebivolol hydrochloride tablets. *J. Adv. Pharm. Technol. Res.* **2010**, *1*, 199–206. [CrossRef]
57. Sun, J.; Wang, F.; Sui, Y.; She, Z.; Zhai, W.; Wang, C.; Deng, Y. Effect of particle size on solubility, dissolution rate, and oral bioavailability: Evaluation using coenzyme Q_{10} as naked nanocrystals. *Int J Nanomed.* **2012**, *7*, 5733–5744. [CrossRef]
58. Lin, X.; Su, L.; Li, N.; Hu, Y.; Tang, G.; Liu, L.; Li, H.; Yang, Z. Understanding the mechanism of dissolution enhancement for poorly water-soluble drugs by solid dispersions containing Eudragit® E PO. *J. Drug Deliv. Sci. Technol.* **2018**, *48*, 328–337. [CrossRef]
59. Vlachou, M.; Kikionis, S.; Siamidi, A.; Kyriakou, S.; Tsotinis, A.; Ioannou, E.; Roussis, V. Development and Characterization of Eudragit®-Based Electrospun Nanofibrous Mats and Their Formulation into Nanofiber Tablets for the Modified Release of Furosemide. *Pharmaceutics* **2019**, *11*, 480. [CrossRef] [PubMed]
60. Vedha Hari, B.N.; Narayanan, N.; Dhevedaran, K. Efavirenz–eudragit E-100 nanoparticle-loaded aerosol foam for sus-tained release: In-vitro and ex-vivo evaluation. *Chem. Pap.* **2015**, *69*, 358–367. [CrossRef]
61. Kosmidis, K.; Argyrakis, P.; Macheras, P. A reappraisal of drug release laws using Monte Carlo simulations: The preva-lence of the Weibull function. *Pharm. Res.* **2003**, *20*, 988–995. [CrossRef] [PubMed]
62. Corsaro, C.; Neri, G.; Mezzasalma, A.M.; Fazio, E. Weibull Modeling of Controlled Drug Release from Ag-PMA Nanosystems. *Polymers* **2021**, *13*, 2897. [CrossRef]
63. Sadia, M.; Sośnicka, A.; Arafat, B.; Isreb, A.; Ahmed, W.; Kelarakis, A.; Alhnan, M.A. Adaptation of pharmaceutical excipients to FDM 3D printing for the fabrication of patient-tailored immediate release tablets. *Int. J. Pharm.* **2016**, *513*, 659–668. [CrossRef]

processes

MDPI

Article

Carboxymethyl-Cellulose-Based Hydrogels Incorporated with Cellulose Nanocrystals Loaded with Vitamin D for Controlled Drug Delivery

Nathália da Cunha Silva, Carla Jeany Teixeira Silva, Max Pereira Gonçalves and Fernanda G. L. Medeiros Borsagli *

Institute of Engineering, Science and Technology, Universidade Federal dos Vales do Jequitinhonha e Mucuri/UFVJM, Av. 01, 4050 Cidade Universitária, Janaúba 39440-039, MG, Brazil; nathalia.cunha@ufvjm.edu.br (N.d.C.S.); carla.silva@ufvjm.edu.br (C.J.T.S.); max.pereira@ufvjm.edu.br (M.P.G.)

* Correspondence: fernanda.borsagli@ufvjm.edu.br

Abstract: Currently, the development of innovative materials for the treatment of various diseases is highly interesting and effective. Additionally, in recent years, environmental changes, including the search for a sustainable world, have become the main goal behind developing sustainable and suitable materials. In this context, this research produced innovative hydrogels that incorporate cellulose nanocrystals and nanofibres from underutilised fibres from a semiarid region of Brazil; the hydrogels were loaded with vitamin D to evaluate controlled drug release for the treatment of diverse diseases. Spectroscopic (FTIR, Raman, UV–VIS), X-ray diffraction, zeta potential and morphology (SEM, TEM) analyses were used to characterise these hydrogels. In addition, biocompatibility was assessed using a resazurin assay, and the in vitro kinetic accumulative release of vitamin D was measured. The results showed that nanocrystals and nanofibres changed the structure and crystallinity of the hydrogels. In addition, the chemical groups of the hydrogels were red- and blueshifted in the FTIR spectra when the nanocrystals, nanofibres and vitamin D were incorporated. Moreover, the nanocrystals and nanofibres were homogeneously spread into the hydrogel when vitamin D was loaded into the hydrogel matrix. Furthermore, the cytotoxicity was greater than 90%. Additionally, the in vitro accumulative kinetic data of vitamin D release were robust (close to 40 ng·mL^{-1}), with equilibrium being reached in the first 30 min. These results confirm the potential of using these hydrogels as therapeutic biomaterials for diverse diseases and problems in humans, mainly in women, who are the most harmed by vitamin D deficiency.

Keywords: women's health; sustainability; social and economic increase; nanomaterial

Citation: Silva, N.d.C.; Silva, C.J.T.; Gonçalves, M.P.; Borsagli, F.G.L.M. Carboxymethyl-Cellulose-Based Hydrogels Incorporated with Cellulose Nanocrystals Loaded with Vitamin D for Controlled Drug Delivery. *Processes* **2024**, *12*, 1437. https://doi.org/10.3390/pr12071437

Academic Editors: Paolo Trucillo and Adina Musuc

Received: 17 May 2024
Revised: 14 June 2024
Accepted: 23 June 2024
Published: 9 July 2024

1. Introduction

Currently, climate change, the rapid growth of the population and animal production, energy consumption and various other transformations have garnered human research on diverse frontiers, here with the goal of searching for a more sustainable, economic and social world. Along these lines, the UN has promoted 17 goals for reaching the sustainable development of the planet. Among these 17 goals, it is important to highlight some that are linked to human health: zero hunger and sustainable agriculture, health and wellness, poor eradication, affordable and clean energy, industry, innovation and infrastructure, sustainable cities and communities, responsible production and consumption, decent work and economic growth [1]. Brazil can be characterised as a semiarid region, a poor region with sparse rain during the year and a climate that is difficult for agriculture. Additionally, there are unique plant yields that produce a large amount of fibre in the spring, namely *Ceiba speciosa*. *Ceiba speciosa* produces fibre that is underutilised and discharged into the environment every year; however, this fibre could be modified to produce an innovative nanomaterial that can have social and economic benefits [2]. Indeed, these nanomaterials are incredible materials for diverse areas, such as pharmaceuticals, energy, biomaterials and agriculture.

These nanomaterials have been highlighted for their use in pharmaceutical products, such as systems used to transport molecules of therapeutic interest. The nanoparticles that mediate this transport can be solid materials or stable colloidal suspensions, or they can be used to transport substances and control their release in cells and tissues because of their small size (10^{-9} m) [3]. In this context, different types of nanoparticles have been produced, including polymeric nanoparticles, which have attracted increasing attention because of their therapeutic potential and better stability in biological fluids, and magnetic nanoparticles; these nanoparticles have a wide range of functions, such as hyperthermia and disease diagnosis, and the application of these nanomaterials is immense, ranging from agriculture to medical applications [4].

Currently, investments in nanotechnology are enormous, and more than 60 countries possess national and/or private initiatives aimed at developing nanoscale materials, with investments exceeding USD 15 billion [5]. This nanotechnology has been applied in diverse sectors and has improved material properties, reduced costs and improved living standards. In this context, several studies have been conducted in academic and industrial contexts to implement this technology in everyday life [5]. Furthermore, because of its range of applications, nanotechnology presents a multidisciplinary approach because it involves physics, chemistry, biology, mathematics and engineering [5].

Based on this, investments in different sectors have become promising in nanotechnology, and interest in this class of materials has been exponentially increasing [6]. Considering this, over the past two decades, nanotechnology has played a crucial role in the medical sector, leading to significant improvements in diagnosis, monitoring, control, prevention and disease treatment. This intersection between nanotechnology and medicine has resulted in the rise of an innovative disciplinary field known as nanomedicine. The application of nanomedicine has had a significant impact on areas such as drug delivery, therapy, diagnosis and biomaterials [7].

One of the most promising areas of nanomedicine is controlled drug delivery. Nanoparticles can be designed to transport drugs in a targeted and controlled manner to specific locations in the body, such as tumours. This allows for a gradual and prolonged release of the medication, increasing its effectiveness and reducing its side effects. Furthermore, nanoparticles can be functionalised by targeting molecules that bind to specific receptors on target cells, thereby increasing treatment selectivity [8].

Nanoparticles loaded into hydrogel systems have great potential for minimising pharmacotherapeutic problems, increasing the therapeutic efficacy of chemotherapeutic agents and reducing their side effects and toxicity. Furthermore, these systems present high drug encapsulation efficiency, protecting them against degradation and reducing the possibility of allergies in patients because of their polymeric coating [9]. In drug release, these materials and technologies have been used to construct nanodelivery systems [10].

These hydrogels can be synthesised from a variety of polymers, including polyacrylates, polyacrylamides, alginates, collagen and gelatine. Each type of polymer has specific characteristics that determine the properties of the resulting hydrogel, such as its adsorption capacity, mechanical resistance and biocompatibility [9]. Because of their properties, hydrogels have been widely used in various applications, such as wound dressings, controlled drug release systems and supports for tissue engineering [11]. They are also applied in cosmetic products, such as moisturising creams and facial masks, because of their ability to retain moisture and promote skin hydration [12].

In summary, hydrogels are often used as wound dressings because of their ability to maintain a moist environment and promote healing. They can adsorb wound exudate, providing a sterile environment and promoting tissue regeneration [13]. They are also excellent materials for use in the controlled release of bioactive molecules, such as medicines or growth factors. The porous structure of hydrogels allows molecules to be incorporated into their matrix and released gradually over time, providing a sustained and controlled release [14]. Ghorbanzadeh et al. [15] developed hydrogels with vitamin D incorporated within as a way to treat psoriasis, a hard disease that affects the skin. Their research promoted a nanocarrier for vitamin D, a hydrophobic drug, using chitosan as a polymeric hydrogel matrix, with promising results for psoriasis treatment. In addition, these hydrogels have been used in biomedical implants such as contact lenses, intraocular devices and dermal fillers. Their ability to retain water and their biocompatibility make them suitable for applications where a soft and flexible material is required [16]. Furthermore, they can be functionalised with specific biomolecules for detecting analytes or biological markers. These functional hydrogels can be used in biosensors to monitor the presence or concentration of substances in biological samples [17].

In this analysis, diverse studies on controlled drug release systems are constantly growing because they aim to improve and prolong the control of drug administration and avoid uncontrolled use. These hydrogels offer a versatile and effective platform for controlled drug delivery, providing sustained and controlled release over time [18]. This is particularly useful for medications that require constant or prolonged dosing to obtain the desired therapeutic effect [19].

In addition, vitamin D, or cholecalciferol, is a steroid hormone whose main function is to regulate calcium homeostasis, bone formation and reabsorption through its interaction with the parathyroid glands, kidneys and intestines, and it can be incorporated into the hydrogel matrix. However, studies have shown that it is difficult to incorporate this hormone into hydrogels because of its high hydrophobicity [15]. Moreover, evidence obtained in the laboratory indicates that vitamin D3 induces a series of extraskeletal biological responses, including the regulation of skin cell proliferation, effects on the cardiovascular system, and protection against several autoimmune diseases, multiple sclerosis, cancer, obesity and chronic intestinal inflammation [20,21].

Vitamin D deficiency can cause a range of health problems, including rickets in children, osteoporosis in adults and autoimmune diseases. Furthermore, a lack of vitamin D may also be associated with a greater risk of developing chronic diseases such as diabetes, cancer and heart disease. Based on this, many people resort to vitamin D supplementation to ensure adequate levels of vitamin D in the body [22]. Additionally, vitamin D plays a crucial role in calcium adsorption and bone health [22]. Some studies have also suggested that vitamin D deficiency may be associated with irregularities in the menstrual cycle, such as longer or irregular cycles. Vitamin D plays a role in hormonal regulation, and a lack of vitamin D can negatively affect female hormonal balance [23]. Moreover, vitamin D plays an important role in female fertility, and vitamin D deficiency may be associated with ovulation problems and pregnancy difficulties. Furthermore, studies have shown a possible link between vitamin D deficiency and reduced success rates in assisted fertility treatments; for example, research has indicated that vitamin D deficiency is linked to the development and worsening of polycystic ovarian syndrome [24]. Furthermore, a deficiency of this steroid hormone has been linked to an increased risk of ovarian cancer, with reduced vitamin D levels being observed in ovarian cancer patients [25].

Thus, considering how critical it is to produce innovative materials for sustainable development—mainly in the area of new materials for controlled drug release with the potential for tissue regeneration—and for promoting poor Brazilian regions, the present research synthesised new nanocrystals and nanofibres from underutilised fibres from a semiarid region of Brazil and incorporated them into a hydrogel matrix modified with vitamin D3 for use in controlled drug delivery systems. These hydrogels were intensively characterised using different techniques. Biological analyses, including in vitro accumulative kinetic drug release, were subsequently performed to determine the biocompatibility and potential of the novel controlled drug delivery system. Similar systems have been developed in the literature [6,10,12,15,16], but to the best of the authors' knowledge, no systems using nanocellulose with different morphologies of this fibre from the Brazilian region that successfully incorporated vitamin D3, a common hormone with a hydrophobic character, have been evaluated. Thus, this is the first time that these systems have been proposed and evaluated for their high potential as drug delivery systems in diverse treatments where a lack of vitamin D is a major problem.

2. Materials and Methods

2.1. Materials

Sodium salt of carboxymethyl cellulose (CMC) (≥99.5%, Synth, Brazil, DS = 0.7, Mw = 250,000 g·mol^{-1}), citric acid ($C_6H_8O_7$, ≥99.5%, Synth, Brazil), sodium hydroxide (Neon, Brazil, ≥99%, NaOH), ($(CH_3)_2CHOH$), acetic acid (Synth, Brazil, 99.8%, MM = 60.05 g·mol^{-1}, CH_3CO_2H), hydrochloric acid (Neon, Brazil, 36.5–38.0%, HCl), sulfuric acid (Synth, Brazil, 99.8%, H_2SO_4) and sodium chlorate (Synth, Brazil, 99.9%, NaCl) were used. Deionised water (DI water) (Millipore Simplicity™, Merck, Darmstadt, Germany) with a resistivity of 18 MΩ·cm was used. Fibres from *Ceiba speciosa* (a typical plant of the semiarid/Brazilian Cerrado) were used as feedstocks to produce nanocrystals and nanofibres (CNCs). The fibre collected from *Ceiba speciosa* was the 'seed hair' that spreads throughout the seeds in the spring in Brazil. The fibres were collected in the northern region of Minas Gerais (a semiarid region), Brazil, in the municipality of Janaúba at 15°48′10″ S 43°18′32″ O during the spring (a unique time at which the fibres appeared). Cholecalciferol 50,000 UI (50,000 ng·mL^{-1}) from Myralis pharmaceutics (Brazil) was used.

2.2. Carboxymethyl Cellulose Hydrogel Conjugation with Cellulose Nanocrystals and Vitamin D3

The hydrogel was produced using sodium salt carboxymethyl cellulose (CMC) and citric acid (AC); this was done by incorporating cellulose nanocrystals and nanofibres (CNCs) extracted from *Ceiba speciosa* via two different chemical routes, as described in our previous research [2], and using cholecalciferol (vitamin D3) for controlled drug delivery. First, 2% CMC was dissolved in deionised water for 24 h under moderate stirring (CMC_CNC solution). Then, 1 mL of CNCs was added to the CMC solution and left under moderate stirring for 24 h to homogenise the system. Subsequently, 100,000 UI of cholecalciferol (vitamin D3) was added to the CMC_CNC solution and left for 48 h under moderate stirring for full homogenisation. Then, 10% (m/m of polymer) of citric acid was added to the solution and left under moderate stirring for 20 min. Finally, the solution was put into a polystyrene Petri dish and placed in an oven at 40 °C for 24 h to remove the water. The temperature was increased to 80 °C for cross-linking among the citric acid and CMC chains, producing the hydrogel. The procedure was performed on both types of CNCs extracted by diverse chemical routes in the same way.

2.3. Carboxymethyl Cellulose Hydrogel Conjugated with CNCs and Vitamin D3 Characterisation

The hydrogels were characterised by Fourier transform infrared spectroscopy (FTIR) on a Nicolet 6700 device (Thermo Fisher, Waltham, MA, USA), with attenuated total reflectance (ATR) in the wavelength range of 675 to 4000 cm^{-1} and 48 scans. Additionally, SEM analysis was used to characterise the hydrogels. Images were recorded on a Zeiss Ultra Plus system with accelerating voltages of 2–6 kV at a working distance of 4–5 mm and an in-lens detector. EDX spectra were acquired at 15 kV on an Oxford Inca EDX detector. Transmission electron microscopy (TEM) analysis was performed using an FEI Titan G2 80-300FEG S/TEM instrument with a Schottky-type electron gun operated at 300 kV and a Bruker XFlash 6T-30 detector (resolution: 129 eV). In addition, X-ray diffraction (XRD) analysis was carried out using a PANalytical Empyrean (UK) diffractometer (Cu-Kα radiation with λ = 1.5418° Å). Measurements were performed in the 2θ range from 15.0021 to 69.9941° with steps of 0.04° at 2 °C/min. Moreover, the absorbance wavelength was determined by ultraviolet visible spectroscopy with a BEL UV-Visible spectrometer (BEL UV-MX, Italy) in the range of 190 to 800 nm.

To determine the physiological adsorption in tissue and the kinetic dissolution of the hydrogel, the swelling (SD) and gel fraction (GF) procedures were used. Then, all hydrogels were then put into a phosphate-buffered saline (PBS) solution (pH = 7.0). The kinetic intervals were 1, 2, 3, 4, 6, 8, 24 and 48 h to reach equilibrium for both procedures.

Altogether, 14 samples were used for each system (n = 14, with 7 samples in two different syntheses). The results were statistically equivalent (ANOVA, one-way included Tukey's test, $p < 0.05$, software Origin v.8.1, OriginLab Corporation, Northampton, MA, USA).

2.4. Biological Analyses of Carboxymethyl Cellulose Hydrogels Conjugated with CNCs and Vitamin D3

2.4.1. Cell Culture

Human dermofibroblast (HDF) cells were used as a cell line to evaluate drug delivery in tissue regeneration. The cells were cultured in Dulbecco's modified Eagle medium (DMEM) supplemented with 10% (v/v) foetal bovine serum (FBS), streptomycin sulphate (10 mg/mL), sodium penicillin G (10 units/mL) and amphotericin-b (0.025 mg/mL), all of which were supplied by Gibco BRL (New York, NY, USA), in an oven with 5% CO_2 and at a temperature of 37 °C.

2.4.2. Cell Viability Assay with Resazurin

HDFs were trypsinised and plated (3×10^4 cells/well) in 24-well plates. As a reference control, cells plus culture medium (DMEM + 10% SFB) were used, and as a positive control, Triton X-100 (2% v/v in phosphate-buffered saline) and PBS (Gibco BRL, New York, NY, USA) were used. A total of 500 μL of hydrogel was added to each well, and the culture medium was added. After 24 h of treatment, the medium was aspirated and replaced with culture medium (DMEM + 10% SFB) supplemented with 10% resazurin solution in PBS at a proportion of 5%, then incubated for 3 h in an oven with 5% CO_2 at 37 °C. Then, 100 μL of this solution was transferred to a flat 96-well plate and subjected to spectrophotometric analysis at wavelengths of 530 nm and 590 nm using Varioskan equipment (Thermo Fisher). The cell viability was measured using Equation (1).

$$Cell\ viability\ (\%) = (Abs\ of\ sample\ and\ cells / Abs\ of\ control) \times 100 \tag{1}$$

Additionally, LIVE/DEAD® Viability/Cytotoxicity was used. The HDF cells at passages 13 and 8 were synchronised in serum-free medium for 24 h. Sequentially, the cells were trypsinised, seeded (3×10^5 cells/well) on hydrogels (1 mg of hydrogel and 200 μL of medium, w/v) and placed in a 96-well plate. The reference controls were cultured in DMEM supplemented with 10% FBS. Subsequently, all the media were aspirated, and the cells were washed twice with 10 mL of PBS (Gibco BRL, NY, USA). The cells were treated with the LIVE/DEAD® kit (Life Technologies of Brazil Ltd., São Paulo, Brazil) for 30 min according to the manufacturer's specifications. Images of fluorescent emissions were acquired with an inverted optical microscope (Leica DMIL LED, Wetzlar, Germany) and revealed calcein at 530 ± 12 nm and EthD-1 (ethidium homodimer-1) at 645 ± 20 nm.

2.5. In Vitro Accumulative Kinetic Drug Delivery in Carboxymethyl Cellulose Hydrogels Conjugated with CNCs and Vitamin D3

Kinetic experiments were performed to determine the time to equilibrium for vitamin D3 delivery in tissue. The accumulative kinetics of drug delivery were determined in triplicate at 37.0 ± 0.1 °C in ethanol (pH 7.4). Hydrogels with an area of 1 cm^2 were placed inside a plastic basket immersed in 15 mL of ethanol under magnetic stirring, and drug release was monitored for 96 h (n = 3). At each time interval, 1 mL of ethanol medium was collected and analysed by UV-VIS to determine the vitamin D3 concentration based on the Beer–Lambert correlation curve (λ = 246 nm).

3. Results and Discussion

3.1. Visual Characterisation of Hydrogels

A visual qualitative evaluation was performed on the hydrogels with incorporated nanocrystals and nanofibres. In this analysis, complete solubilisation of carboxymethylcellulose occurred, as characterised by the clear and transparent appearance of the hydrogels (Figure 1A). After the hydrogels were dried, a smooth, homogeneous and transparent hydrogel was observed (Figure 1). In the case of hydrogels with vitamin D, a more flexible film was observed, which was also translucent and homogeneous (Figure 1B). This can be justified by the fact that vitamin D also cross-linked the hydrogel, enabling greater elasticity of the hydrogel, although future mechanical analysis is necessary to evaluate this elasticity.

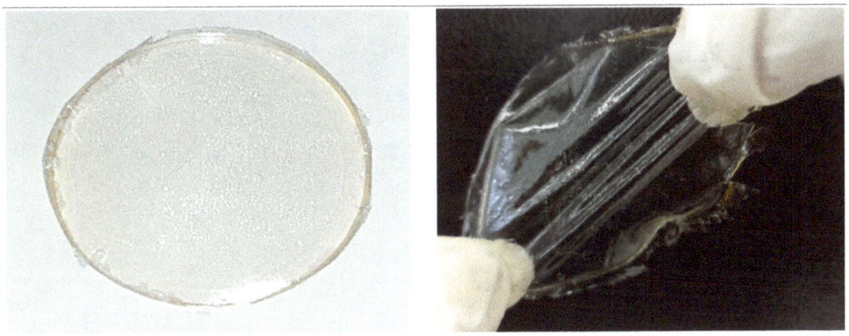

(A) (B)

Figure 1. (**A**) Carboxymethyl cellulose hydrogel with CNCs without vitamin D3 and (**B**) carboxymethyl cellulose hydrogel with CNCs and vitamin D.

3.2. Spectroscopy Analyses

In our previous research [2], the FTIR spectra of pure *Ceiba speciosa* showed bands at 3100 cm^{-1}, 2982 cm^{-1}, 1282 cm^{-1} and 1109 cm^{-1} related to OH, -C-H, -CH_2 stretching, -C-O-C and -C-H vibrations, respectively, associated with cellulose chemical groups [2,26,27]. In the case of the CNCs obtained by two different chemical routes, slight differences were detected at 3010 cm^{-1}, 2975 cm^{-1}, 2987 cm^{-1}, 1382 cm^{-1}, 1119 or 1109 cm^{-1} and 898 cm^{-1}, associated with -OH, -C-H, -CH_2 stretching, -C-O-C and -C-H vibrations, respectively [2]; for example, the disappearance of hydroxyl groups may be caused by the formation of acid sulphate (O-SO_3H) groups in the hydrolysis step [2,28]. This formation should help with the incorporation of vitamin D because it may provide a more hydrophobic character to the hydrogel [29].

Additionally, in the CMC, bands related to OH groups at 3100–3300 cm^{-1} were observed. Moreover, for all hydrogels (CMC_AC, CMC_CNC_H_2SO_4, CMC_CNC_H_2SO_4/HCl, CMC_CNC_H_2SO_4_D, and CMC_CNC_H_2SO_4/HCl_D), this band increased in intensity compared with that of pure CMC because of the interaction between citric acid and CMC (Figure 2). Moreover, in the case of hydrogels with vitamin D (CMC_CNC_H_2SO_4_D, CMC_CNC_H_2SO_4/HCl_D), a slight increase was observed, probably because of the incorporation of vitamin D (Figure 2 inset) [2,30]. Furthermore, the intensity of the band at 2916 cm^{-1} decreased for hydrogels with vitamin D and almost disappeared, probably because of the incorporation of vitamin D [30,31]. In addition, bands at 1651 and 1719 cm^{-1} related to C=O groups were more intense in hydrogels with vitamin D associated with fatty acids [31]. The band related to the (-N+ $(CH_2)_3$) group of vitamin D at 1060 cm^{-1} was blueshifted in the hydrogels (CMC_CNC_H_2SO_4_D, CMC_CNC_H_2SO_4/HCl_D) [32,33].

UV-VIS spectroscopic analysis was performed for both CNCs in our previous research [2], which showed that bands at 230 and 280 nm were linked with cellulose I and n–π^* (surface functional groups), which are associated with the C=C, carbonyl and hydroxyl groups in CNC structures [2,34]. For the hydrogels (CMC-AC, CMC-CNC-H_2SO_4, CMC-CNC-H_2SO_4/HCl, CMC-CNC-H_2SO_4/D, CMC-CNC-H_2SO_4/D and CMC-CNC-H_2SO_4/HCl/D), the same bands were observed but with slight differences in intensity (Figure 3), mainly in the band at 280 nm, which was more intense for the hydrogels with vitamin D; this is probably associated with chromophores in the chemical structure of vitamin D. Additionally, the opacity of all the hydrogels was calculated based on Equation (2) to evaluate the passage of light through the hydrogels, which can affect the chemical structure of vitamin D in the presence of sunlight [34,35]. The values of transparency found based on Equation (2) for all hydrogels were very close: 80 ± 1%, 81 ± 1%, 78 ± 1%, 79 ± 1% and 74 ± 1% for CMC-AC, CMC-CNC-H_2SO_4, CMC-CNC-H_2SO_4/HCl, CMC-CNC-H_2SO_4/D and CMC-CNC-H_2SO_4/HCl-D, respectively. These results indicate good transparency for the passage of sunlight through hydrogels for skin applications.

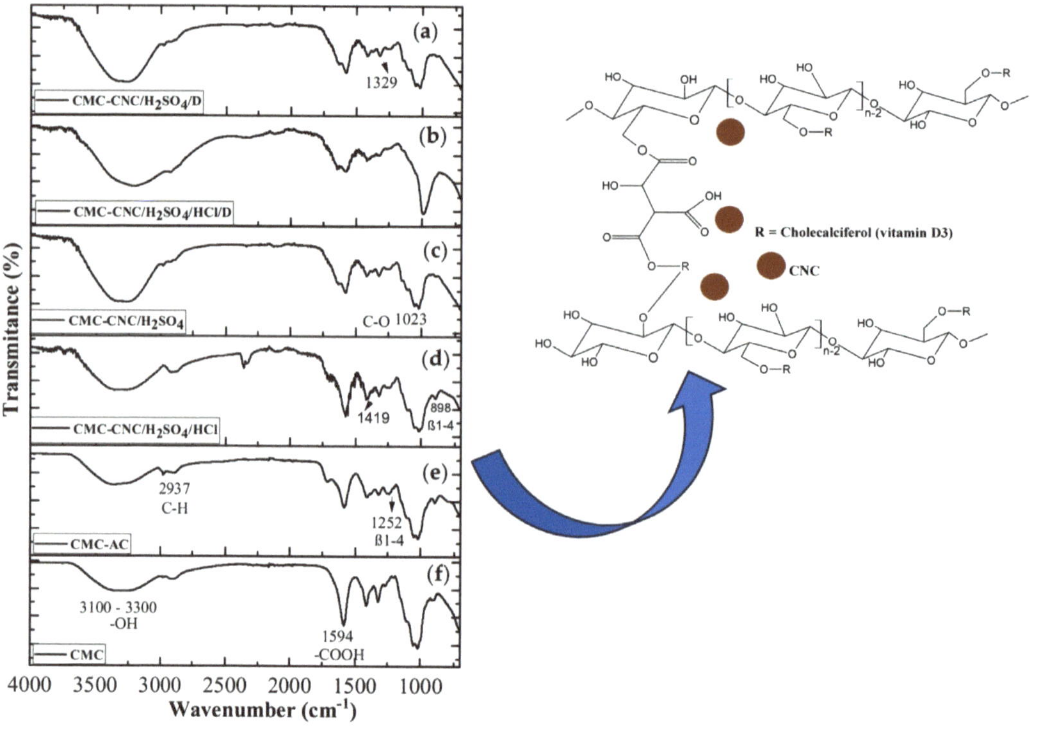

Figure 2. FTIR spectra of the (**a**) CMC, (**b**) CMC-AC, (**c**) CMC-CNC-H_2SO_4, (**d**) CMC-CNC_H_2SO_4/HCl, (**e**) CMC-CNC-H_2SO_4/D and (**f**) CMC-CNC-H_2SO_4/HCl/D hydrogels (inset, schematic of the chemical linkage in the hydrogel with vitamin D).

$$Transparency\ (\%) = log\ (\%T_{600})/x \quad (2)$$

here, $\%T_{600}$ is the percentage of transmittance of light at 600 nm and x is the thickness of the hydrogel [36].

3.3. Microscopy Analysis

SEM analyses were performed on the CNCs and all the hydrogels (CMC, CMC-AC, CMC-CNC-H_2SO_4, CMC-CNC-H_2SO_4/HCl, CMC-CNC-H_2SO_4/D and CMC-CNC-H_2SO_4/HCl/D). In the case of the CNCs, as previously shown in our research [2], SEM images did not reveal the real morphology of the CNCs, and TEM analysis was subsequently performed, which revealed that the presence of nanocrystals and nanofibres depended on the chemical route used (Figure 4G,H). The SEM images of the hydrogels revealed a homogenous hydrogel (Figure 4), even though slight differences were detected between the hydrogels with citric acid (Figure 4B) and those with CNCs (Figure 4C,D). In the hydrogel with citric acid (chemical cross-link) (Figure 4B), some roughness was not detected in the hydrogel with pure CMC (Figure 4A); this morphology is likely because of the cross-linking process performed with citric acid, which slowly evaporates (48 h). In the hydrogels with nanocrystals (Figure 4C) and nanofibres (Figure 4D), slight differences in morphological agglomeration with different colours were observed between the two hydrogels, which may be caused by the clustering of nanocrystals and nanofibres in the hydrogel during cross-linking. Additionally, TEM images of the nanocrystals and nanofibres showed that, in the case of the CNCs produced by the chemical route using H_2SO_4, the nanocrystals were similar to those reported in the literature (Figure 4H), but in the case of the chemical route with the introduction of HCl, nanofibres were predominant (Figure 4I). Moreover, in the hydrogels with vitamin D3 (Figure 4E,F), no difference was observed, indicating that vitamin incorporation promoted a more homogeneous spread of CNCs, independent of the type of morphology.

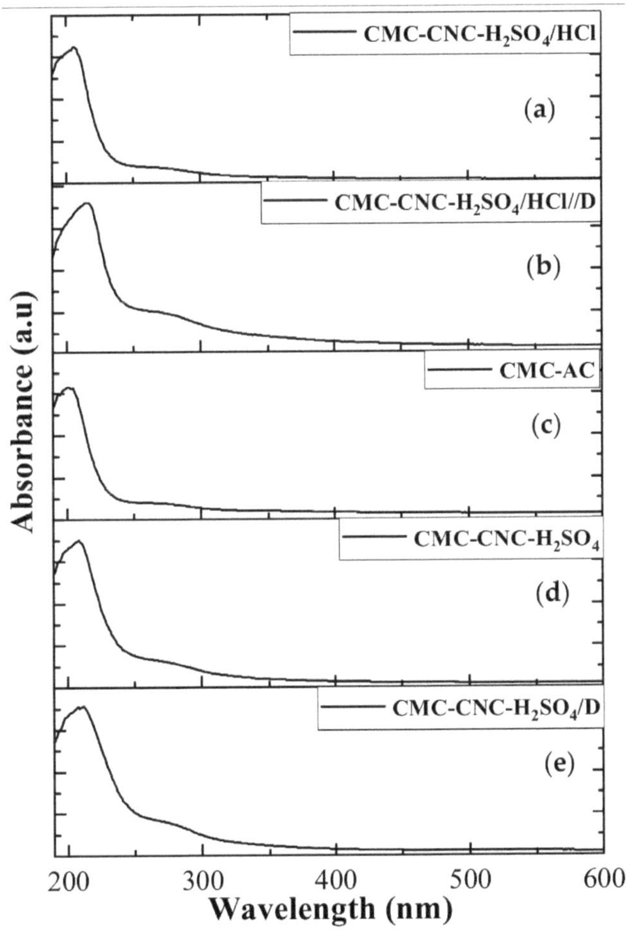

Figure 3. UV-VIS spectra of the (**a**) CMC-CNC-H_2SO_4/HCl, (**b**) CMC-CNC-H_2SO_4/HCl/D, (**c**) CMC-AC, (**d**) CMC-CNC-H_2SO_4 and (**e**) CMC-CNC-H_2SO_4/D hydrogels.

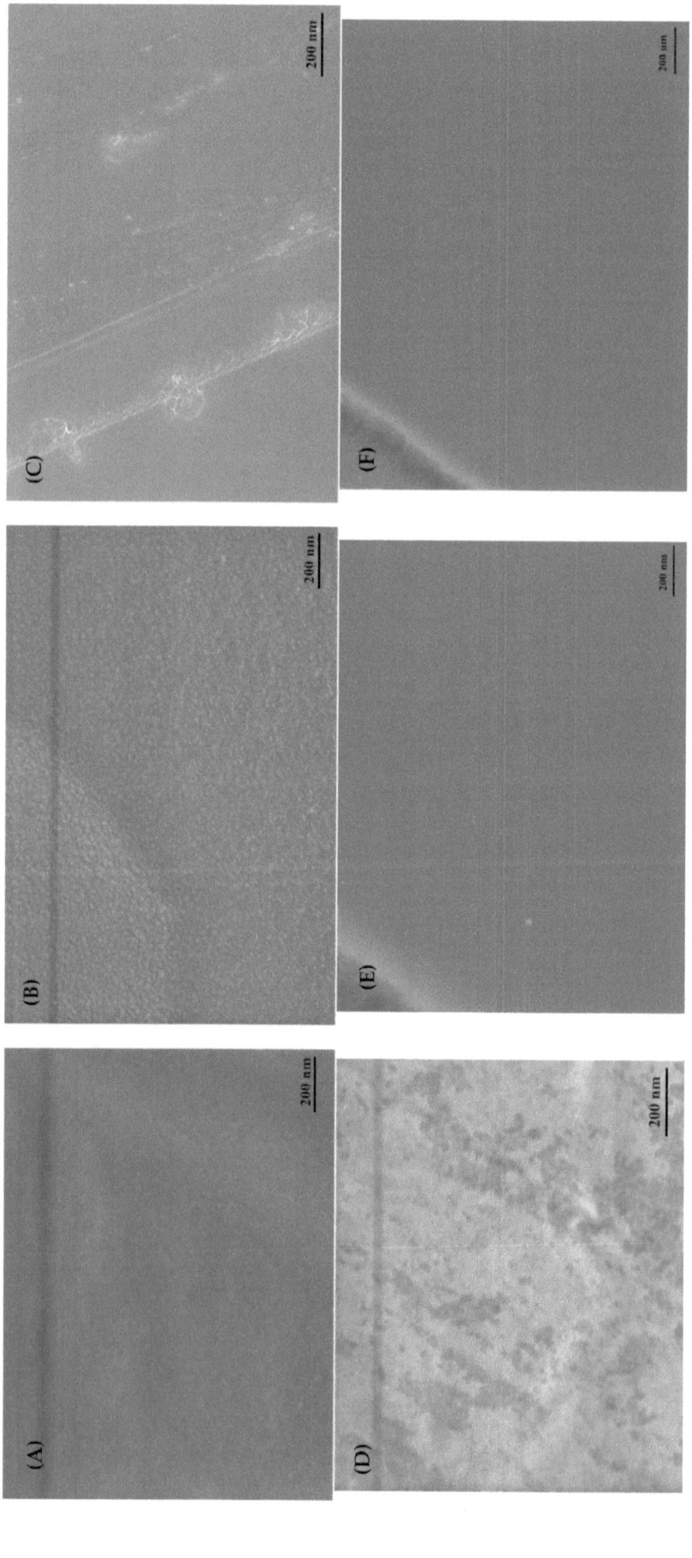

Figure 4. *Cont.*

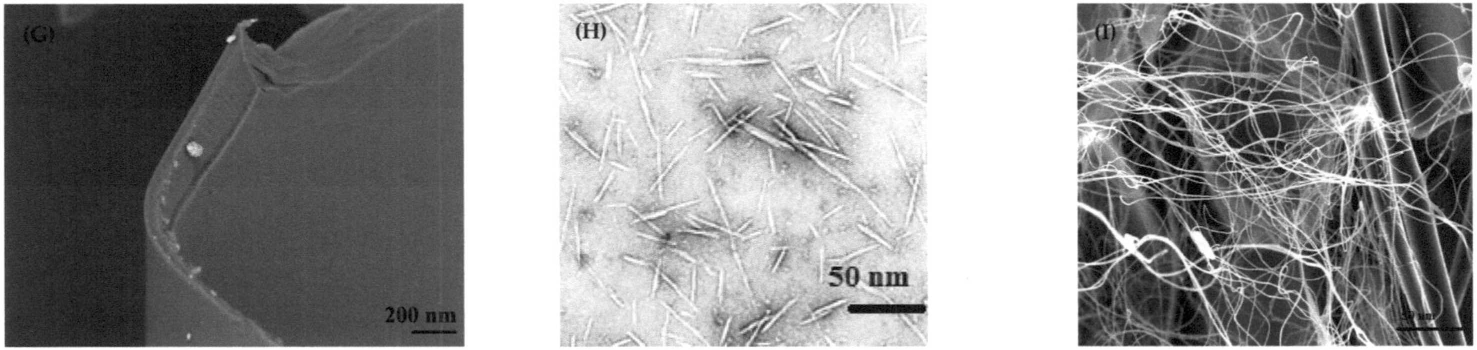

Figure 4. SEM images of hydrogels: (**A**) CMC, (**B**) CMC-AC, (**C**) CMC-CNC-H_2SO_4, (**D**) CMC-CNC-H_2SO_4/HCl, (**E**) CMC-CNC-H_2SO_4/D and (**F**) CMC-CNC-H_2SO_4/HCl/D. (**G**) Thickness of hydrogels and TEM images of hydrogels with CNCs (copyright 5801880110187 by Springer [2]), (**H**) CMC-CNC_H_2SO_4 and (**I**) CMC-CNC-H_2SO_4/HCl. In biomaterial applications, promoting the cell adhesion and biocompatibility of materials is critical. In controlled drug delivery, adequate adhesion of the biomaterials to membrane cells is the first step in targeted drug release [37,38]. Moreover, homogeneous spreading of nanoparticles in the hydrogel matrix should improve the properties of the hydrogel [39]. Thus, when vitamin D3 was incorporated into hydrogels with nanocrystals and nanofibres, a more homogeneous spread was perceived, indicating that the incorporation of the drug improved the hydrogel. As shown in Figure 4, SEM images revealed a homogeneous spread of nanocrystals and nanofibres because no clusters were observed in hydrogels with vitamin D3. Furthermore, roughness was visualised only in the hydrogels with citric acid without CNCs and vitamin D3, showing that the incorporation of these two compounds improved the hydrogels. Additionally, the thickness of the hydrogels was homogeneous, as shown in Figure 3.

3.4. XRD Analysis of Hydrogels Loaded with Nanocrystals and Nanofibres Modified with Vitamin D

An XRD analysis of the CNCs revealed peaks at 15°, 17° and 34.3° (Figure 5), which were associated with cellulose I [2,39,40], and a prominent peak at 22.6° linked to the pattern (0 0 2) [2,40–42]. Additionally, the sizes found according to Equation (3) were 10 nm and 20 nm for CNC_H_2SO_4 and CNC_H_2SO_4/HCl, respectively, which were very close to the sizes determined via TEM analysis. Moreover, the crystallinity based on the total area and the amorphous area was determined, and the values were 81% and 77% for CNC_H_2SO_4 and CNC_H_2SO_4/HCl, respectively.

$$Crystallinity\ (\%) = (crystalline\ area/total\ area) \times 100 \quad (3)$$

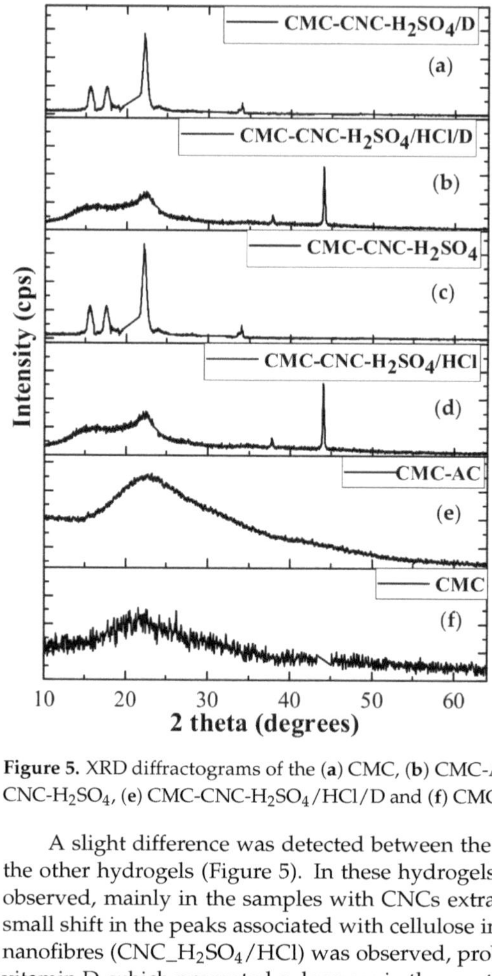

Figure 5. XRD diffractograms of the (**a**) CMC, (**b**) CMC-AC, (**c**) CMC-CNC-H_2SO_4/HCl, (**d**) CMC-CNC-H_2SO_4, (**e**) CMC-CNC-H_2SO_4/HCl/D and (**f**) CMC-CNC-H_2SO_4/D hydrogels.

A slight difference was detected between the hydrogels containing vitamin D and the other hydrogels (Figure 5). In these hydrogels, a slight amorphisation process was observed, mainly in the samples with CNCs extracted with H_2SO_4 and HCl because a small shift in the peaks associated with cellulose in the nanocrystals (CNC_H_2SO_4) and nanofibres (CNC_H_2SO_4/HCl) was observed, probably because of the incorporation of vitamin D, which promoted a decrease in the crystallinity of the hydrogels. Moreover, a peak at 44° was observed, which was attributed to the NaOH that remained in the CNCs after extraction, even after dialysis. This accumulation in hydrogels with vitamin D was expected considering the visual analysis described in Section 3.1, where greater elasticity was visualised in these hydrogels than in the other hydrogels. However, as mentioned before, future mechanical analyses should be performed.

3.5. Swelling and Gel Fraction of Hydrogels Loaded with Nanocrystals and Nanofibres Modified with Vitamin D

An important parameter for biomaterials is physiological fluid adsorption and degradation/dissolution in the physiological environment. Thus, swelling (SD) and GF (solvent dissolution) tests were performed on the hydrogels. The results are presented in Figure 6. In the case of GF, the behaviours of the hydrogels without vitamin D were very similar, showing good chemical stability when compared with the hydrogels in the literature [43–45]. In the case of hydrogels with vitamin D, greater swelling was observed, mainly in the hydrogels with nanofibres (CMC-CNC-H_2SO_4/HCl/D), probably because of the presence of CNCs with a nanofibre morphology, which may enhance the entanglement of vitamin D3 and promote faster fluid release compared to the other hydrogels (CMC-CNC-H_2SO_4/D). However, the dissolution of fluid in the medium indicated by GF was greater after the first few hours, which may be a good indicator of the controlled release of the drug.

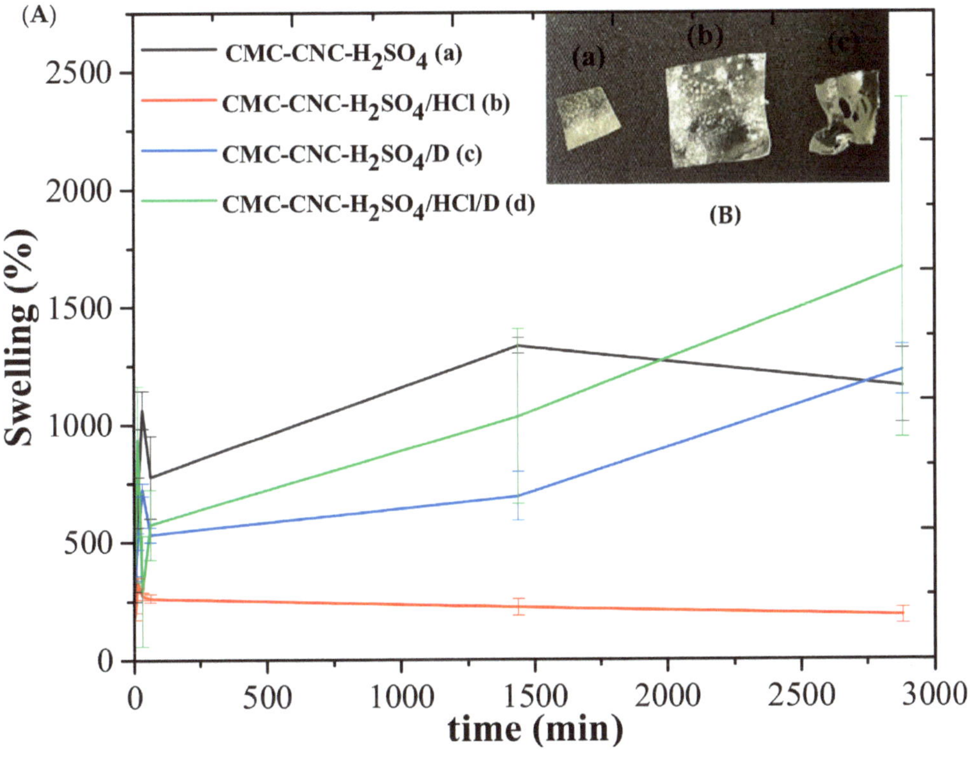

Figure 6. *Cont.*

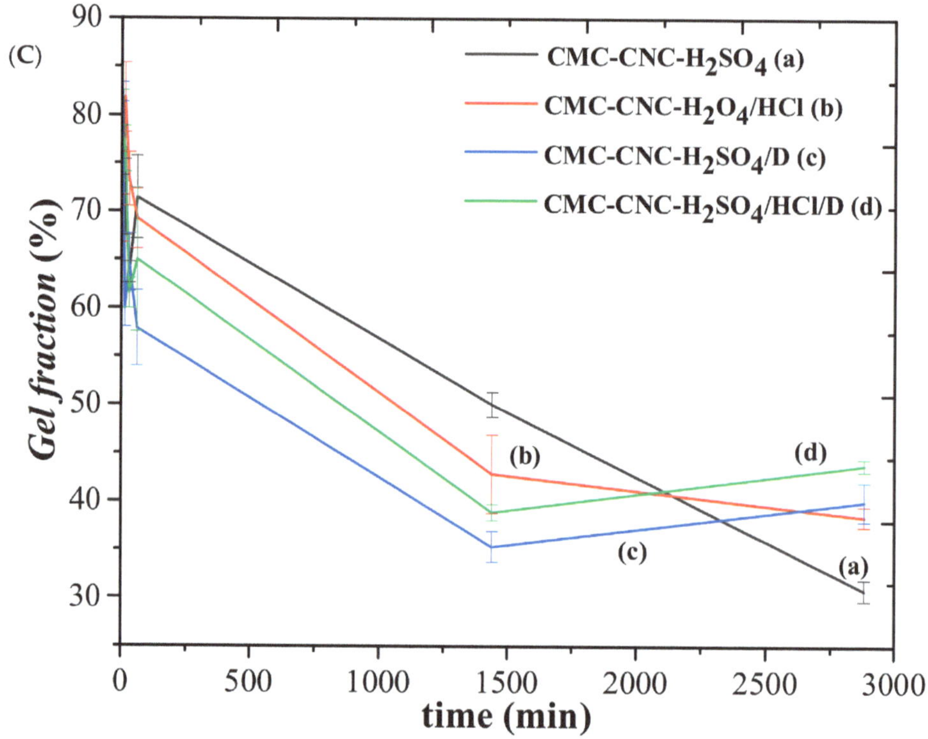

Figure 6. (**A**) Swelling of hydrogels (a) CMC-CNC-H_2SO_4, (b) CMC-CNC-H_2SO_4/HCl, (c) CMC-CNC-H_2SO_4/D and (d) CMC-CNC-H_2SO_4/HCl/D. (**B**) Photography of hydrogels with vitamin D (a) before swelling, (b) after swelling and (c) after gel fraction. (**C**) Gel fractions of hydrogels (a) CMC-CNC-H_2SO_4, (b) CMC-CNC-H_2SO_4/HCl, (c) CMC-CNC-H_2SO_4/D and (d) CMC-CNC-H_2SO_4/HCl/D.

3.6. In Vitro Kinetic Studies of Vitamin D Release

Vitamin D3—or cholecalciferol—is a precursor of the steroid calcitriol, a regulator of calcium and phosphate metabolism in the body. This prohormone is soluble in lipids and plays a fundamental role in the development of bone tissue, skeletal muscles and several other areas of the body [46]. A lack of vitamin D in the body causes several problems, including low bone density, bone fractures, osteopenia, osteoporosis and muscle injuries, and some studies have indicated infertility problems in women [47]. Furthermore, studies have indicated that vitamin D has anti-inflammatory effects on immune system cells and can affect several processes [46–48]. These anti-inflammatory effects have been linked to the regulation of proinflammatory cytokines. Additionally, the skin has been shown to play an important role in the modification of 7-dehydrocholesterol in cholecalciferol when exposed to ultraviolet radiation [49]. Furthermore, studies have proposed a strong role for vitamin D3 in protecting against colon and breast cancer, diabetes and other health problems [15].

In this context, promoting biomaterials to control vitamin D3 release is a great idea for diverse biomedical applications. Thus, one of the main proposals of the present research was to evaluate the potential of incorporating CNCs and vitamin D3 into hydrogels produced from carboxymethyl cellulose and citric acid via a friendlier chemical route using a raw material from a semiarid region in Brazil. For this purpose, hydrogels with innovative cellulose nanocrystals and nanofibres conjugated with vitamin D3 were evaluated for their in vitro kinetics of vitamin D3 release, determining the cumulative amount of vitamin D3 released over time, here using 96 h of analysis (Figure 7).

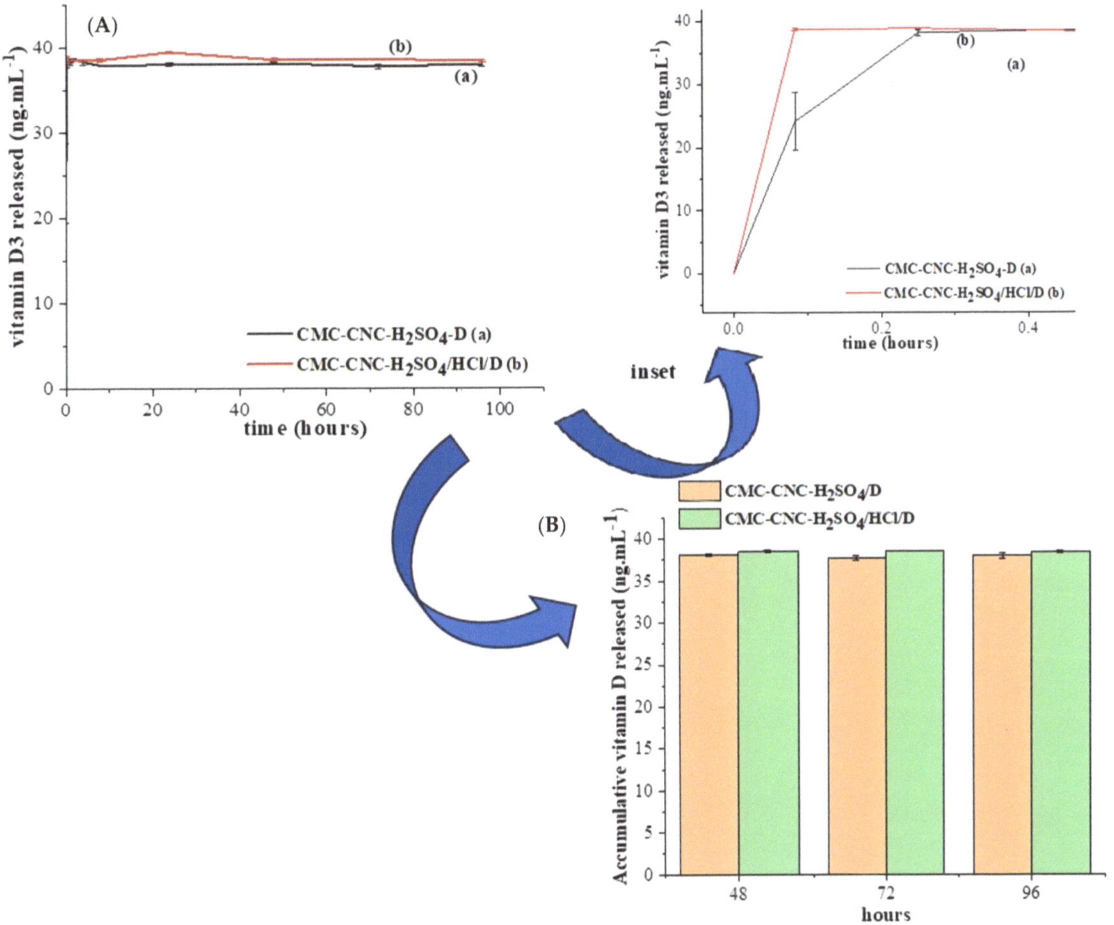

Figure 7. (**A**) Kinetics of vitamin D3 release in vitro over time in (a) CMC-CNC-H_2SO_4/D and (b) CMC-CNC-H_2SO_4/HCl/D hydrogels; (**B**) cumulative vitamin D released at 48, 72 and 96 h in both hydrogels with vitamin D (CMC-CNC-H_2SO_4/D and CMC-CNC-H_2SO_4/HCl/D) (inset shows the vitamin D3 release from the two hydrogels in the first 30 min).

The results of vitamin D3 release showed a great discharge in the first 30 min, but there was a slight difference between the two hydrogels (inset Figure 7). Additionally, these releases reached nearly 40 ng·mL^{-1} of vitamin D3 accumulated in the medium, a value slightly higher than that recommended by the National Institutes of Health (NIH) ($\geq$20 ng·mL^{-1}) but lower than an excess daily intake of the vitamin (>50 ng·mL^{-1}) [50]. Moreover, the release of vitamin D3 from the CMC_CNC_H_2SO_4/HCl_D hydrogel was greater than that from the CMC_CNC_H_2SO_4_D hydrogel and occurred faster in the first 15 min (Figure 7 inset). Most research [45–50] has shown slower vitamin D release, over a period close to 2 days, at which point the release equilibrium of the vitamin is reached. Although hydrogels are not suitable for conjugation with hydrophobic drugs, as is the case for vitamin D [48,51], the present research showed strong results because the release of vitamin D reached its equilibrium in hours, not days, and the daily value of equilibrium approached that recommended by the NIH. Additionally, there was a slight difference between the hydrogels (38.0 ± 0.5 ng·mL^{-1} and 39.0 ± 0.5 ng·mL^{-1} for CMC-CNC-H_2SO_4/D and CMC-CNC-H_2SO_4/HCl/D, respectively), even though the equilibrium results were very close. This slight difference in vitamin release may be related to the CNC morphology in the hydrogel with nanofibres (CMC-CNC-H_2SO_4/HCl/D), which promoted better vitamin D release.

3.7. Cytotoxicity Evaluation of Hydrogels Loaded with Nanocrystals and Nanofibres Modified with Vitamin D

In our previous research [2], it was shown that different concentrations of CNCs were nontoxic when the nanocrystals and nanofibres were formed via two chemical routes. However, because the present research focused on hydrogels loaded with nanocrystals and nanofibres modified with vitamin D, it was important to evaluate the cytotoxicity of the hydrogels. Therefore, for this evaluation, HDF cells were used. As described above, vitamin D is crucial to the health of many organs in the body, including effects on fertility and some cancers that affect reproductive organs in women [21,24,25,46–48,52].

Various studies have indicated that hydrogels have promising features, such as mechanical performance, biodegradability, chemical stability and tissue compatibility [53], most of which are associated with their polymeric matrices, which can also improve drug delivery linked with physiochemical characteristics, allowing for better controlled release and pharmacokinetic behaviour in biomedical areas [53,54]. In the current study, the cellular viability of hydrogels with and without vitamin D3 was assessed using an assay with resazurin and LIVE/DEAD in HDFs to evaluate the cytotoxicity of the nanomaterial (Figure 8). Figure 8G shows that the positive control performed with Triton X-100 (a nonionic surfactant for cell lysis) was the only one that showed a significant difference when compared with the reference control. All hydrogels, when in contact with HDFs, did not show toxicity when compared with the control group. Furthermore, the incorporation of CNCs and vitamin D3 into the hydrogel matrix did not cause toxicity, as indicated by the LIVE/DEAD ratio and cell viability, indicating that these hydrogels can be used for controlled drug release applications.

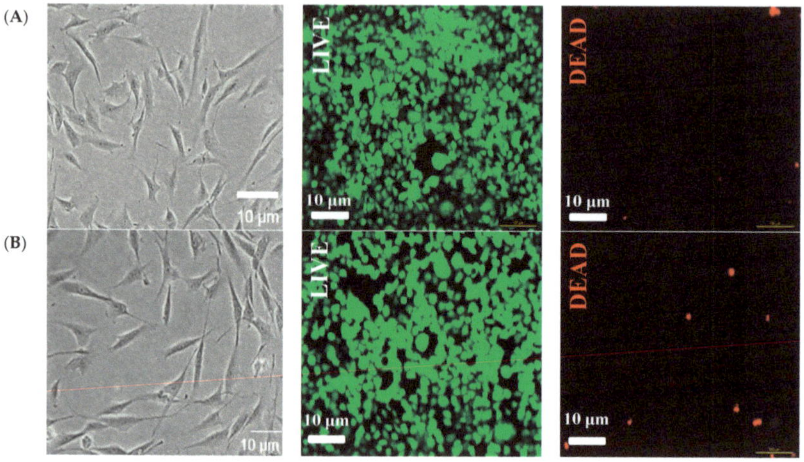

Figure 8. *Cont.*

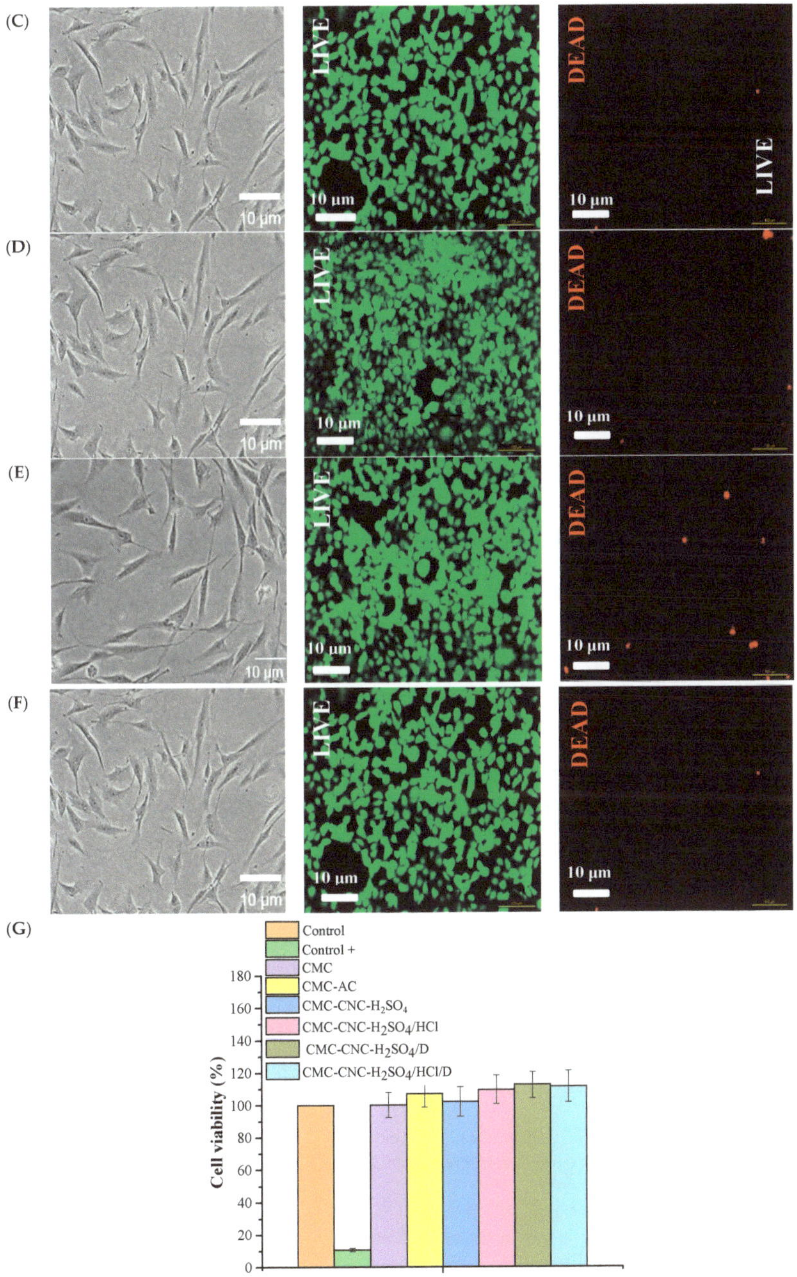

Figure 8. (**A**) Micrographs and LIVE/DEAD® assays of HDFs after 168 h of incubation (live cells (green) and dead cells (red) in the control group at 200× magnification) with (**A**) CMC, (**B**) CMC-AC, (**C**) CMC-CNC-H_2SO_4, (**D**) CMC-CNC-H_2SO_4/HCl, (**E**) CMC-CNC-H_2SO_4/D and (**F**) CMC-CNC-H_2SO_4/HCl/D. (**G**) Histogram of the viability of HDFs in the presence of all hydrogels and in the control groups.

4. Conclusions

The present research described the production of new CMC hydrogels with the incorporation of different CNCs (nanocrystals and nanofibres) obtained from a new Brazilian semiarid biomass source and loaded with vitamin D3 for controlled drug release. These innovative hydrogels produced with the incorporation of CNCs from *Ceiba speciosa* fibres are a potential sustainable alternative for improving the semiarid Brazilian region both socially and economically; thus, they can promote poverty eradication, zero hunger and sustainable agriculture in this area. Additionally, the results for these hydrogels showed that the different physiochemical characteristics and morphologies of the CNCs changed the vitamin D3 release performance. Moreover, the swelling and GF behaviour of the hydrogels treated with vitamin D3 differed, probably because of the incorporation of vitamin D3, as shown by spectroscopy analysis. Furthermore, no differences in cytotoxicity were detected, indicating the high potential of applying these hydrogels for controlled vitamin D3 release in diverse biomedical applications.

Supplementary Materials: The following supporting information can be downloaded at https://www.mdpi.com/article/10.3390/pr12071437/s1. Figure S1: Calibration curve of vitamin D3.

Author Contributions: N.d.C.S.: conceptualisation, methodology, investigation. C.J.T.S.: conceptualisation, methodology, investigation. M.P.G.: writing—original draft, validation, writing—review and editing. F.G.L.M.B.: conceptualisation, supervision, project administration, writing—original draft, validation, writing—review and editing, funding acquisition. All authors have read and agreed to the published version of the manuscript.

Funding: This research was funded by BIOSEM-LESMA from the Universidade Federal dos Vales do Jequitinhonha e Mucuri (UFVJM) for all chemical experiments and by the FAPEMIG (APQ-02565-21), FINEP/MCTI (0 1 22 0528 00), CAPES, and CNPq.

Data Availability Statement: The authors confirm that the data supporting the findings of this study are available within the article and can be shared if necessary. Calibration curve of vitamin D3 is shown in Figure S1.

Conflicts of Interest: The authors declare no conflicts of interest.

References

1. UN. The ODS in Brazil. 2024. Available online: https://brasil.un.org/pt-br/sdgs (accessed on 29 April 2024).
2. Silva, N.C.; Rodrigues, J.S.; Bernardes, M.S.; Gonçalves, M.P.; Medeiros Borsagli, F.G.L. Cellulose nanocrystals and nano-fiber from sub-wear out Brazilian semiarid source for biological applications. *J. Cluster Sci.* **2024**, *1*, 1–10.
3. Junges, A.; Denti, A.F.; Bernardi, J.L.; Polina, C.C.; Meregalli, M.M.; Vanz, J.B.; Mignoni, M.L. Application of nanotechnology in food engineering: A review. *Res. Soc. Dev.* **2022**, *11*, 1–10.
4. Borsagli, F.G.L.M.; Borsagli, A. Luminescent carbon dots for environmental photocatalytic. In *Inorganic-Organic Composites for Water and Wastewater Treatment*; Lichtfouse, E., Muthu, S.S., Khadir, A., Eds.; Springer: Singapore, 2021; pp. 201–228.
5. Paiva, A.E.; Guin, J.P.; Vasquez, J.F.B.; Thampi, K.R.; Sullivan, J.A.; Borsagli, F.G.L.M.; Morris, M.A. Degradation of an antibiotic via a photo-Fenton process on block copolymernanopatterned iron oxide coated substrates. *Chem. Eng. J.* **2023**, *466*, 142925. [CrossRef]
6. Rodrigues, J.S.; Silva, C.J.T.; Silva, N.C.; Aguiar, R.A.; Bernardes, M.S.; Soares, C.M.; Madureira, J.A.; Borsagli, F.G.L.M. Carbohydrate polymer hydrogels: An environmental and eco-friendly choice for biomedical applications? *J. Polym. Res.* **2023**, *30*, 382. [CrossRef]
7. Thomas, P.; Duolikun, T.; Rumjit, N.P.; Moosavi, S.; Lai, C.W.; Johan, M.R.B.; Fen, L.B. Comprehensive review on nanocellulose: Recent developments, challenges and future prospects. *J. Mech. Behav. Biomed. Mater.* **2020**, *110*, 103884. [CrossRef]
8. Newman, L.; Jasim, D.A.; Prestat, E.; Lozano, N.; Lazaro, I.; Nam, Y.; Assas, B.M.; Pennock, J.; Haigh, S.J.; Bussy, C.; et al. Splenic capture and in vivo intracellular biodegradation of biological-grade graphene oxide sheets. *ACS Nano* **2020**, *14*, 10168–10186. [CrossRef] [PubMed]
9. Park, K.; Choi, H.; Kang, K.; Shin, M.; Son, D. Soft stretchable conductive carboxymethylcellulose hydrogels for wearable sensors. *Gels* **2022**, *8*, 92. [CrossRef] [PubMed]
10. Barhoum, A.; Sadak, O.; Ramirez, I.A.; Iverson, N. Stimuli-bioresponsive hydrogels as new generation materials for implantable, wearable, and disposable biosensors for medical diagnostics: Principles, opportunities, and challenges. *Adv. Colloid Interface Sci.* **2023**, *317*, 102920. [CrossRef] [PubMed]

11. Fadl, F.I.A.E.; Hegazy, D.E.; Maziad, N.A.; Ghobashy, M.M. Effect of nano-metal oxides (TiO_2, MgO, CaO, and ZnO) on antibacterial property of (PEO/PEC-co-AAm) hydrogel synthesized by gamma irradiation. *Int. J. Biol. Macromol.* **2023**, *250*, 126248. [CrossRef] [PubMed]
12. Borsagli, F.G.L.M.; Souza, A.J.M.; Paiva, A.E. Ecofriendly multifunctional thiolated carboxymethyl chitosan-based 3D scaffolds with luminescent properties for skin repair and theragnostic of tissue regeneration. *Int. J. Biol. Macromol.* **2020**, *165*, 3051–3064. [CrossRef] [PubMed]
13. Parfenyuk, E.V.; Dolinina, E.S. Silica hydrogel composites as a platform for soft drug formulations and cosmetic compositions. *Mater. Chem. Phys.* **2022**, *287*, 126160. [CrossRef]
14. Ijaz, F.; Tahir, H.M.; Ali, S.; Ali, A.; Khan, H.A.; Muzamil, A.; Manzoor, H.H.; Qayyum, K.A. Biomolecules based hydrogels and their potential biomedical applications: A comprehensive review. *Int. J. Biol. Macromol.* **2023**, *253*, 127362. [CrossRef]
15. Ghorbanzadeh, M.; Golmohammadzadeh, S.; Karimi, M.; Farhadian, N. Evaluation of vitamin D3 serum level of microemulsion based hydrogel containing Calcipotriol drug. *Mater. Todays Commun.* **2022**, *33*, 104409. [CrossRef]
16. Franco, P.; Marco, I. Contact Lenses as Ophthalmic Drug Delivery Systems: A review. *Polymers* **2021**, *17*, 1102. [CrossRef] [PubMed]
17. Zheng, C.; Lu, K.; Lu, Y.; Zhu, S.; Yue, Y.; Xu, X.; Mei, C.; Xiao, H.; Wu, Q.; Han, J. A stretchable, self-healing conductive hydrogels based on nanocellulose supported graphene towards wearable monitoring of human motion. *Carbohyd. Polym.* **2020**, *250*, 116905. [CrossRef] [PubMed]
18. Cai, Y.; Xin, L.; Li, H.; Sun, P.; Liu, C.; Fang, L. Mussel-inspired controllable drug release hydrogel for transdermal drug delivery: Hydrogen bond and ion-dipole interactions. *J. Control. Release* **2024**, *365*, 161–175. [CrossRef] [PubMed]
19. Narayanaswamy, R.; Torchilin, V.P. Hydrogels and Their Applications in Targeted Drug Delivery. *Molecules* **2019**, *24*, 603. [CrossRef] [PubMed]
20. Lee, W.K.; Chng, A.L.B.; Tan, K.H.; Shub, A.; Tan, T.; Hian, T.K.; Lee, R.W.K.; Ling, L.S.; Kuma, K.; Siew, C.Y. Clinical practice of vitamin D screening and supplementation in pregnancy in Asia-pacific countries: A cross-sectional study. *Heliyon* **2023**, *9*, 21186. [CrossRef] [PubMed]
21. Qi, D.; Nie, X.-L.; Wu, S.; Cai, J. Vitamin D and hypertension: Prospective study and meta-analysis. *PLoS ONE* **2017**, *12*, e0174298. [CrossRef]
22. Haitchi, S.; Moliterno, P.; Widhalm, K. Prevalence of vitamin D deficiency in seniors—A retrospective study. *Clin. Nutr. ESPEN* **2023**, *57*, 691–696. [CrossRef] [PubMed]
23. Holick, M. The vitamin D deficiency pandemic: Approaches for diagnosis, treatment and prevention. *Rev. Endocr. Metab. Disord.* **2017**, *18*, 153–165. [CrossRef] [PubMed]
24. Berry, S.; Seidler, K.; Neil, J. Vitamin D deficiency and female infertility: A mechanism review examining the role of vitamin D in ovulatory dysfunction as a symptom of polycystic ovary syndrome. *J. Reprod. Immunol.* **2022**, *151*, 103633. [CrossRef] [PubMed]
25. Seraphin, G.; Reiger, S.; Hewison, M.; Capobianco, E.; Lisse, T.S. The impact of vitamin D on cancer: A mini review. *J. Steroid Biochem. Mol. Biol.* **2023**, *231*, 106308. [CrossRef] [PubMed]
26. Mousa, H.M.; Fahmy, H.S.; Abouzeid, R.; Abdel-Jaber, G.T.; Ali, W.Y. Polyvinylidene fluoride-cellulose nanocrystals hybrid nanofiber membrane for energy harvesting and oil-water separation applications. *Mater. Lett.* **2022**, *306*, 130965. [CrossRef]
27. Sobrinho, M.S.; Tabatinga, G.M.; Machado, I.C.; Lopes, A.V. Pollination biology and mating system of Angelonia hirta (Scrophulariaceae), a neotropical shrub with uncommon floral traits. *Acta Bot. Bras.* **2013**, *27*, 456. [CrossRef]
28. Costa, L.A.S.; Fonsêca, A.F.; Pereira, F.V.; Druzian, J.I. Biodegradability of nanocomposites based on cassava starch and clay. *Cellul. Chem. Technol.* **2015**, *49*, 127–134.
29. Medeiros Borsagli, F.G.L.; Rodrigues, J.S.; Aguiar, R.A.; Paiva, A.E.; Vasquez, J.F.B.; Ramos, W.T.S.; Allibrandini, P.; Rocha, E.P.A.; Gonçalves, M.P.; Souza, F.E. Low-cost luminescent scaffolds-based on thiol chitosans by microwave radiation for vertebral disc repair/theragnostic. *Int. J. Biol. Macromol.* **2022**, *209*, 2109–2118. [CrossRef] [PubMed]
30. Bampidis, V.; Azimonti, G.; Bastos, M.L.; Christensen, H.; Dusemund, B.; Durjava, M.K.; Kouba, M.; Lopez-Afonso, M.; Puente, S.L.; Marcon, F.; et al. Safety and efficacy of sodium carboxymethyl cellulose for all animal species. *EFSA J.* **2020**, *18*, 6211.
31. Paiva, A.E.; Medeiros Borsagli, F.G.L. Ecofriendly multiphase aqueous colloidal based on carboxymethylcellulose nanoconjugates with luminescence properties for potential bioimaging cancer cells. *J. Polym. Environ.* **2020**, *23*, 3076–3096. [CrossRef]
32. Molaveisi, M.; Shahidi-Noghabi, M.; Naji-Tabasi, S. Vitamin D3-loaded nanophytosomes for enrichment purposes: Formulation, structure optimization, and controlled release. *Food Process Eng.* **2020**, *2020*, 13560. [CrossRef]
33. Semalty, A.; Semalty, M.; Rawat, M.S.M.; Franceschi, F. Supramolecular phospholipids–polyphenolics interactions: The PHYTOSOME® strategy to improve the bioavailability of phytochemicals. *Fitoterapia* **2010**, *81*, 306–314. [CrossRef] [PubMed]
34. Zhang, Y.; Gao, Z.; Zhang, W.; Wang, W.; Chang, J.; Kai, J. Fluorescent carbon dots as nanoprobe for determination of lidocaine hydrochloride. *Sens. Actuators B Chem.* **2018**, *262*, 928. [CrossRef]
35. Zhang, W.; Zhang, Y.; Cao, J.; Jiang, W. Improving the performance of edible food packaging films by using nanocellulose as an additive. *Int. J. Biol. Macromol.* **2020**, *166*, 288–296. [CrossRef] [PubMed]
36. Zhao, J.; Wang, Y.; Liu, C. Film transparency and opacity measurements. *Food Anal. Methods* **2022**, *15*, 2840–2846. [CrossRef]
37. Kirtane, A.R.; Verma, M.; Karandikar, P.; Furin, J.; Langer, R.; Traverso, G. Nanotechnology approaches for global infectious diseases. *Nat. Nanotechnol.* **2021**, *16*, 369–384. [CrossRef] [PubMed]

38. Raza, H.Z.; Shah, A.A.; Noreen, Z.; Usman, S.; Zafar, S.; Yasin, N.A.; Sayed, S.R.M.; Al-Mana, R.A.; Elansary, H.O.; Ahmad, A.; et al. Calcium oxide nanoparticles mitigate lead stress in Abelmoschus esculentus through improving the key antioxidative enzymes, nutritional content and modulation of stress markers. *Plant Physiol. Biochem.* **2024**, *206*, 108171. [CrossRef]
39. Fekete, T.; Borsa, J.; Takacs, E.; Wojnarovitis, L. Synthesis and characterization of superabsorbent hydrogels based on hydroxyethylcellulose and acrylic acid. *Carbohyd. Polym.* **2017**, *166*, 300–308. [CrossRef] [PubMed]
40. French, A.D. Correction to: Increment in evolution of cellulose crystallinity analysis. *Cellulose* **2020**, *27*, 5445–5448. [CrossRef]
41. French, A.D.; Santiago Cintrón, M. Cellulose polymorphy, crystallite size, and the Segal Crystallinity Index. *Cellulose* **2020**, *20*, 583–588. [CrossRef]
42. Liu, Y.Y.; Yu, N.Y.; Fang, W.D.; Tan, Q.-G.; Ji, R.; Yang, L.-Y.; Wei, S.; Zhang, X.-W.; Miao, A.-J. Photodegradation of carbon dots cause cytotoxicity. *Nat. Commun.* **2021**, *12*, 812. [CrossRef] [PubMed]
43. Schmitz, T.; Grabovac, V.; Palmberger, T.F.; Hoffer, M.H.; Bernkop-Schnurch, A. Synthesis and characterization of a chitosan-N-acetyl cysteine conjugate. *Int. J. Pharm.* **2008**, *347*, 79–85. [CrossRef] [PubMed]
44. Chen, Y.; Zhang, Y.; Mensaha, A.; Li, D.; Wang, Q.; Wei, Q. A plant-inspired long-lasting adhesive bilayer nanocomposite hydrogel based on redox-active Ag/Tannic acid-Cellulose nanofibers. *Carbohydr. Polym.* **2021**, *255*, 117508. [CrossRef] [PubMed]
45. Alizadeh, R.; Zarrintaj, P.; Kamrava, S.K.; Bagher, Z.; Farhadi, M.; Heidari, F.; Komeili, A.; Gutierrez, T.J.; Saeb, M.R. Conductive hydrogels based on agarose/alginate/chitosan for neural disorder therapy. *Carbohydr. Polym.* **2019**, *224*, 115161. [CrossRef] [PubMed]
46. Kim, D.-S.; Kim, J.H.; Baek, S.-W.; Lee, J.-K.; Park, S.-Y.; Choi, B.; Kim, T.-H.; Min, K.; Han, D.K. Controlled vitamin D delivery with injectable hyaluronic acid-based hydrogel for restoration of tendinopathy. *J. Tissue Eng.* **2022**, *13*, 1–15. [CrossRef] [PubMed]
47. Chu, C.; Tsuprykov, O.; Chen, X.; Elitok, S.; Krämer, B.K.; Hocher, B. Relationship between vitamin d and hormones important for human fertility in reproductive-aged women. *Front. Endocrinol.* **2021**, *12*, 666667. [CrossRef] [PubMed]
48. Simioni, Y.R.; Ricatti, F.; Salvay, A.G.; Jerez, H.E.; Schilrreff, P.; Romero, E.L.; Morilla, M.J. Activity of hydrogel-vitamin D3/bacterioruberin nanoparticles on imiquimod-induced fibroblasts-keratinocytes spheroids. *J. Drug Deliv. Sci. Technol.* **2024**, *97*, 105738. [CrossRef]
49. Eslami, M.; Shahedi, M.; Fath, M. Development of hydrogels for entrapment of vitamin D3: Physicochemical characterization and release study. *Food Biophys.* **2018**, *13*, 284–291. [CrossRef]
50. National Institutes of Health (NIH). Vitamin D. Available online: https://ods.od.nih.gov/factsheets/VitaminD-HealthProfessional/ (accessed on 30 April 2024).
51. Desfrançois, C.; Auzély, R.; Texier, I. Lipid nanoparticles and their hydrogel composites for drug delivery: A review. *Pharmaceuticals* **2018**, *11*, 118. [CrossRef] [PubMed]
52. Goodwin, E.C.; Yang, E.; Lee, C.-J.; Lee, H.-W.; DiMaio, D.; Hwang, E.-S. Rapid induction of senescence in human cervical carcinoma cells. *Proc. Natl. Acad. Sci. USA* **2000**, *97*, 10978–10983. [CrossRef] [PubMed]
53. Yuan, M.; Ruibo, Z.; Gao, H.; Li, W.; Yun, X.; Liu, J.; Zhao, X.; Zhao, G.; Zhang, F. One-step, green, and economic synthesis of water-soluble photoluminescent carbon dots by hydrothermal treatment of wheat straw and their bio-applications in labeling, imaging, and sensing. *Appl. Surf. Sci.* **2015**, *355*, 1136–1144. [CrossRef]
54. Thankam, F.G.; Muthu, J.; Sankar, V.; Gopal, R.K. Growth and survival of cells in biosynthetic poly vinyl alcohol–alginate IPN hydrogels for cardiac applications. *Colloids Surf. B Biointerfaces* **2013**, *107*, 137–145. [CrossRef]

processes

MDPI

Article

The Improved Cytotoxic Capacity of Functionalized Nanodiamonds with Metformin in Breast and Ovarian Cancer Cell Lines

Lucero Evelia Acuña-Aguilar [1], Alain Salvador Conejo-Dávila [1,2], Mario Miki-Yoshida [1], Olga N. Hernández-de la Cruz [3], Gricelda Sánchez-Sánchez [3], César López-Camarillo [3], Joan Sebastian Salas-Leiva [1,4], Erasto Armando Zaragoza-Contreras [1], Reyna Reyes-Martínez [5] and Erasmo Orrantia-Borunda [1,*]

1 Departamento de Medio Ambiente y Energía, Centro de Investigación en Materiales Avanzados, S.C. Av. Miguel de Cervantes Saavedra 120, Chihuahua 31136, Mexico; lucero.acuna@cimav.edu.mx (L.E.A.-A.); alain.conejo@cimav.edu.mx (A.S.C.-D.); mario.miki@cimav.edu.mx (M.M.-Y.); joan.salas@cimav.edu.mx (J.S.S.-L.); armando.zaragoza@cimav.edu.mx (E.A.Z.-C.)
2 Centro de Innovación Aplicada en Tecnologías Competitivas, A.C. Calle Omega No. 201, Delta, León 37545, Mexico
3 Posgrado en Ciencias Genómicas, Universidad Autónoma de la Ciudad de México, Ciudad de México 06720, Mexico; ediacara79@yahoo.com.mx (O.N.H.-d.l.C.); gricelda.sanchez.sanchez@estudiante.uacm.edu.mx (G.S.-S.); cesar.lopez@uacm.edu.mx (C.L.-C.)
4 CONAHCyT-Departamento de Medio Ambiente y Energía, Centro de Investigación en Materiales Avanzados, S.C. Av. Miguel de Cervantes 120, Chihuahua 31136, Mexico
5 Departamento de Química, Facultad de Ciencias Químicas, Universidad Autónoma de Chihuahua, Ciudad Universitaria, Campus 2, Chihuahua 31110, Mexico; rreyesm@uach.mx
* Correspondence: erasmo.orrantia@cimav.edu.mx; Tel.: +614-439-1111 (ext. 2412)

Citation: Acuña-Aguilar, L.E.; Conejo-Dávila, A.S.; Miki-Yoshida, M.; Hernández-de la Cruz, O.N.; Sánchez-Sánchez, G.; López-Camarillo, C.; Salas-Leiva, J.S.; Zaragoza-Contreras, E.A.; Reyes-Martínez, R.; Orrantia-Borunda, E. The Improved Cytotoxic Capacity of Functionalized Nanodiamonds with Metformin in Breast and Ovarian Cancer Cell Lines. *Processes* **2023**, *11*, 2616. https://doi.org/10.3390/pr11092616

Academic Editor: Paolo Trucillo

Received: 29 June 2023
Revised: 8 August 2023
Accepted: 14 August 2023
Published: 2 September 2023

Abstract: Nanodiamonds (ND-COOH) are used as drug delivery systems because of their attractive properties, as they allow for optimized transport of therapeutic agents in cellular models. Metformin (MET) is a drug used in diabetes mellitus therapy and exhibits anti-cancer properties. In this study, dispersed nanodiamonds were functionalized with metformin by directly binding them to 1,6-hexanediol (ND-MET), and their effects on the cytotoxicity of breast and ovarian cancer cells were evaluated in vitro. A simple synthesis of ND-MET was performed and characterized using FT-IR, XPS, Boehm titration, RAMAN, XDR, TEM, and dynamic light scattering (DLS). Data showed an increased intensity of the C-N bond band, indicating the presence of metformin in ND-MET. We detected signals at 1427 cm^{-1} and 1288 cm^{-1} corresponding to the C-N and C-H bonds, and adsorptions at 1061 cm^{-1} and 3208 cm^{-1} corresponding to the N-O and N-H bonds, respectively. The deconvolution of the C1s binding energy was also found at 286.24 eV. The biological effects of ND-MET were tested in both SKOV3 ovarian cancer and Hs-578T and MDA-MB-231 triple-negative breast cancer cell lines. In SKOV3 cells, the IC_{50} for the ND-MET complex was 35 ± 14 μg/mL, while for Hs-578T and MDA-MB-231 breast cancer cells, the IC_{50} for ND-MET was 759 ± 44 μg/mL and 454 ± 49 μg/mL, respectively. Our data showed that ND-MET could be used as an intracellular delivery system for metformin in cancer cells. Cell viability assays evidenced a reduced viability of all cell lines in a time and dose-dependent manner, with a significant sensitivity observed in SKOV3 ovarian cancer cells treated with ND-MET.

Keywords: nanodiamonds; biguanide; functionalization; ovarian cancer; breast cancer

1. Introduction

Nanodiamonds (ND-COOH) are extensively used as drug delivery systems. Due to their chemical and physical surface properties, nanodiamonds allow for optimized delivery of various therapeutic agents [1–4], and can diffuse within cells due to their lipophilicity [5–7]. In addition, longer drug retention has been observed when associated with nanodiamonds compared to the drug alone. Since metformin (MET) can be conjugated via ionic bonds, covalent bonds, or physical adsorption to the nanodiamond, it is possible to have a prolonged and controlled pH-mediated release. This conjugation minimizes anti-cancer drug side effects and maximizes delivery efficiency [8].

MET is a drug used in treating type II diabetes since it inhibits gluconeogenesis [9], and also seems to exert a cytotoxic effect on cancer cells. Its potential use in anti-cancer processes is currently being investigated [10–16]. The effect of MET on primary breast cancer cells (PBCC) has been studied, and assays show a dose-dependent reduction of viability and proliferation in ER+, PR+, and HER2- type cancers. MET, at an effective dose (25 mM, 24 h), can inhibit tumor cell proliferation and metastasis by modulating the expression of metalloproteinases MMP-2 and MMP-9, and interfering with NF-kB protein signaling in PBCC [17]. Also, MET has been delivered to graphene oxide nanoparticles, loaded with hyaluronic acid (HA-GO-MET), and targeted to both CD44 receptors and overexpressing cancer cells, with MET as the therapeutic agent and hyaluronic acid (HA) as the active targeting system. Results indicated an inhibition of cell migration of MDA-MB-231, a triple-negative breast cancer (TNBC) line, when treated with significantly lower doses of HA-GO-MET (20 μg/mL) compared to metformin alone (300 μg/mL); it attained 50% viability. The GO-MET system evaluated at a 50 μg/mL dose, and showed no significant inhibition in cell viability [18].

MET requires high doses to inhibit cancer cell growth depending on the type and subtype of cell line [19], the metformin derivative [12], and the duration of treatment being evaluated. For example, the IC_{50} at 48 h of MET treatment for SKOV3 was 14.92 mM [16], and for Hs578t and MDA-MB-231, it was 16.3 mM and 51.4 mM, respectively [19]. This high dose is a disadvantage; however, it can be overcome using efficient drug delivery systems that selectively transport MET to the target cells [20]. Therefore, functionalizing nanodiamonds with MET is an interesting way to evaluate their effect on cancer cell lines [21,22].

Since we are interested in the search for a compound that can serve as a treatment against tumor cells, we decided to work with two types of cancer that are difficult to treat because they present resistance to conventional therapies. We looked at triple-negative breast cancer and serous ovarian cancer subtypes. Triple-negative breast cancer is resistant to hormonal therapies and serious breast cancer to chemical therapies. Considering this, we chose representative and widely used cell lines to study these tumor subtypes, MDA and HS (triple negative), and SKOV (serous).

The resistance of tumor cells to specific therapies represents a significant challenge in cancer treatment. In breast cancer, the subtype that typically exhibits a lower response to therapies and has a poorer prognosis is triple-negative breast cancer (TNBC) [23]. TNBC is characterized by the absence of specific molecular markers on the cell surface: estrogen, progesterone, and HER2/neu receptors. These markers are crucial in targeted therapy against other breast cancer subtypes. The lack of these receptors in TNBC renders this type of cancer unresponsive to hormone-targeted therapies, such as tamoxifen, which are effective in tumors expressing estrogen or progesterone receptors. Moreover, due to the lack of HER2/neu receptor expression, triple-negative breast cancer does not respond to HER2-targeted therapies like trastuzumab [24–26].

Consequently, the inefficacy of these specific therapeutic options makes TNBC more challenging to treat when compared to other subtypes of breast cancer. Hence, we decided to evaluate nano-compounds' effects on the viability of TNBC tumor cells (MDA-MB-231 and Hs578t). Additionally, TNBC is highly aggressive, and cell lines of this subtype serve as excellent models for studying invasion and metastasis processes, making them relevant models for advanced-stage cancer.

In ovarian cancer (OC), the serous subtype is highly prevalent. It can exhibit high aggressiveness and be associated with a poor prognosis and reduced responsiveness to therapies compared to other OC subtypes. SKOV3 is a cell line derived from serous ovarian carcinoma. It has been extensively employed as a study model due to its resistance to certain chemical compounds traditionally used for OC treatment, such as Paclitaxel and Adriamycin [27,28].

This study evaluated the cytotoxic capacity of nanodiamonds (ND-COOH) modified with MET in different cancer cell lines. The objective was to verify whether the action of ND-MET can reduce the amount of MET necessary to induce appreciable biological effects. Moreover, the functionalization of nanodiamonds with metformin was achieved using a straightforward method.

2. Materials and Methods

2.1. Regents

Carboxylated nanodiamonds (ND-COOH) monodispersed in water with a concentration of 10 mg mL^{-1} and a particle size of 5–6 nm were obtained from Sigma-Aldrich (Milwaukee, WI, USA). 1,6-hexanediol ($C_6H_{14}O_2$), N, N-dimethylformamide (DMF), water (H_2O), and sodium hydroxide (NaOH) were obtained from MACRON Fine Chemicals (Xalostoc, Estado de México, Mexico). Acetone (CH_3COCH_3) was obtained from Química Tech SA de CV (MX). Ethanol (EtOH) was obtained from CTR Scientific (MX). Sulfuric acid (H_2SO_4) and potassium permanganate ($KMnO_4$) were obtained from Fermont (MX). Hydrochloric acid (HCl) was obtained from JT Baker (Phillipsburg, NJ, USA), and the Metformin 850 mg tablet was obtained from the Pisa pharmaceutical company (San Diego, CA, USA).

2.2. ND-MET Synthesis

2.2.1. Metformin Hydrochloride Extraction (MET*HCl)

Eight metformin tablets, each containing 850 mg, were triturated and added to an Erlenmeyer flask with 400 mL of ethanol, forming a suspension, and left in constant agitation at 25 °C for 2 h. It was filtered to separate the solids, then the solution was recovered and concentrated at a constant pressure until a white crystalline solid was obtained. Metformin was extracted by the Universidad Autónoma de Chihuahua Department of Chemistry.

2.2.2. Obtention of ND-MET Complex

The functionalization of the nanodiamonds started with a Fisher esterification reaction using 1,6-hexanediol and ND-COOH in a 1:10 mass ratio. In total, 2 mL of N, N-dimethylformamide (DMF) and 20 µL of H_2SO_4 were added. The mixture was placed in an oil bath at 100 °C and stirred for 6 h. It was then cooled to 25 °C, centrifuged at 900 rpm, washed twice with deionized water, and once with acetone. The dispersion was sonicated between each wash. Then, the solvent was evaporated at 25 °C to obtain ND_{hexa} (Figure 1a). Liu et al. functionalized nanodiamonds with 1,6-hexanediol using $SOCl_2$ to bind them to biopolymers [29] covalently.

The functionalization of the nanodiamonds was continued with the oxidation of the primary alcohols from 1,6-hexanediol to obtain the carboxylic functional groups (-COOH) [30]. A solution of NaOH 1 N was mixed with 100 mg of ND_{HEXA}. Afterward, a solution of $KMnO_4$ 0.17 equivalents relative to 1,6-hexanediol was prepared and added dropwise to the previous solution in an ice bath, then stirred for 30 min. The ice bath was removed to stir for 18 h. The pH was adjusted to 3–4 with HCl 0.1 N, centrifuged at 900 rpm, washed with deionized water with 5 min sonication between each wash, and dried at 25 °C to obtain the ND_{HEXA}-CO_2H, as shown in Figure 1b.

MET was linked to ND_{HEXA}-CO_2H through a neutralization reaction between the carboxylic acids group (which has a pKa of 4.75) and the MET's amino group (pKa of 10.73). The reaction produces electrostatic interactions between the drug and the nanodiamond, which allows its transport and easy delivery. For drug binding, 50 mg of ND_{HEXA}-CO_2H was mixed with 50 mg of metformin on agitation for 12 h at 25 °C, washed with deionized water, then dried at room temperature to obtain ND-MET.

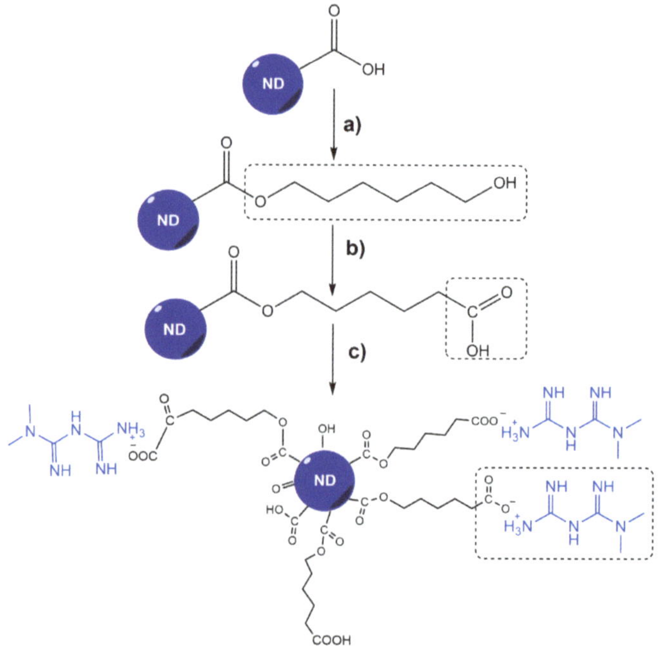

Figure 1. Reaction sequence for the synthesis of the ND-MET complex, (**a**) 1:10 ND-COOH:1,6-hexanediol, DMF, and H_2SO_4, (**b**) 1N NaOH, $KMnO_4$ and HCl, (**c**) Metformin.

2.3. *Characterization*

The characterizations of ND-MET complex synthesis were performed by FT-IR, XDR, XPS, and RAMAN. The particle size and shape help determine nanoparticle chemical and physical stability.

2.3.1. Fourier Transform Infrared Spectroscopy

Functional groups were identified using an FT-IR spectrometer (IR Affinity 1S, Shimadzu, Kyoto, Japan). Spectra were obtained by reflectance using a Smiths ATR (Total Attenuated Reflectance) accessory, model Quest, with a one-step diamond window.

2.3.2. X-ray Photoelectron Spectroscopy

X-ray photoelectron spectroscopy (XPS) analysis was performed on the Escalab 250Xi (ThermoFisher, Waltham, MA, USA) to confirm the presence of carboxyl groups. An Al K alpha source, with a 650 µm spot size, was used in CAE analysis mode with a pass energy of 200 eV and an energy step size of 1.000 eV.

2.3.3. Bohem Titration

The Bohem titration was performed with an automatic titrator Metrohm model 848 Titrino Plus. An HCl 0.1 M solution was prepared to titrate the $NaHCO_3$ 0.05 M base containing the sample and quantify the carboxyl groups of ND-COOH.

2.3.4. RAMAN Spectroscopy

A RAMAN spectrometer (LabRamHR Vis-633, Horiba, Kyoto, Japan) complemented functional group characterization. The sample was excited to analyze the molecular structure with a laser $\lambda = 325$ nm.

2.3.5. X-ray Diffraction Analysis

X-ray diffraction (XRD) was run on the X'PertPRO (PANalytical, Boston, MA, USA) diffractometer using CuKα radiation with λ = 1.54056 Å in the scanning range 2θ = 10°–100°, with a step of 0.01 and a time of 100 s/step, with Bragg Brentano. The analysis was performed to identify the nanodiamond crystalline structure and crystallite size.

2.3.6. Transmission Electron Microscopy

Transmission electron microscopy (TEM) with an energy dispersive system (EDS) was performed on JEM 2200FS+CS (JEOL, Ciudad de México, MX) equipment, with a spherical aberration corrector of the condenser lens for the scanning transmission mode (STEM), operated at 200 kV. With this technique, the nanoparticles' morphology and particle sizes were obtained.

2.3.7. Dynamic Light Scattering

The hydrodynamic radius and Z-potential measurements were carried out with the Zetasizer nano series (Malvern Panalytical's, Worcestershire, UK) to analyze the particle size and charge. Nine measurements were performed using distilled water as the solution medium.

2.4. Biological Assays

2.4.1. Cell Culture

SKOV3 ovarian cancer cells, MDA-MB-231, and Hs578t breast cancer cells were purchased from ATCC (Manassas, VA, USA) and maintained in a DMEM/F-12 (Dulbecco's Modified Eagle Medium/Nutrient Mixture F-12; Gibco, BRL, Grand Island, NY, USA) medium supplemented with 10% fetal bovine serum (FBS; Gibco BRL, Grand Island, NY, USA) and 1% penicillin/streptomycin antibiotics (Gibco BRL, Grand Island, NY, USA). The cells were cultured in a humidified atmosphere with 5% CO_2 at 37 °C and subcultured every 2–3 days.

2.4.2. Cytotoxicity Assay

The cytotoxic capacity of the ND-MET bioconjugates on the breast and ovarian cancer cell lines was evaluated using a 3-(4,5-dimethylthiazol-2-yl)-2,5-diphenyltetrazolium bromide (MTT) colorimetric assay. The different cell lines were cultured in vitro in 96-well plates (10,000 cells per well) until they reached 70–80% confluency. Subsequently, the cells were treated with different concentrations (15, 30, 60, 125, 250, 500, and 750 μg/mL) of the ND-MET nanoparticles, which were then dissolved in a DMEM-F12 complete medium. Afterward, they were incubated at different times (24, 48, and 72 h) at 37 °C and 5% CO_2. Cells without treatment were used as an experimental control. When post-treatment times were complete, cell viability was analyzed using MTT assays. Thus, the medium with the treatment was removed and replaced with 100 μL of an MTT solution (0.5 μg/mL in DMEM-F12 medium), which was then incubated for 4 h at 37 °C and 5% CO_2. After the medium with MTT was aspirated, 100 μL of extraction buffer (0.1 N HCl dissolved in isopropanol) was added to each well to dissolve the resulting purple formazan crystals. The absorbance of the resulting solution was measured at 570 nm in a microplate reader (Elx800-BioTek Instruments). A plot of absorbance versus concentration was generated to determine the IC_{50} value using Quest Graph TM IC_{50} 2023 software (ATT Bioquest) [31]. The values obtained from the control cells, without treatment, were considered 100% viable and were used to calculate the viability of the tumor cells treated with ND-MET nanoparticles.

For IC50 data, a one-way analysis of variance (ANOVA) and a Turkey post-hoc test were conducted to investigate differences in the IC50 across three treatment groups [32].

3. Results and Discussion

3.1. ND-MET Characterization

Figure 1 illustrates the reaction sequence to obtain the ND-MET complex, where an electrostatic interaction occurs between the carboxylic group of the ND_{HEXA}-CO_2H reaction intermediate and the amino groups of the MET's molecule, since an attractive force is occurring between two opposite ionic species.

The FT-IR spectrum was used to identify the functional groups of the nanodiamond particles (ND-COOH), the products of the different functionalization steps, and the ND-MET complex. Figure 2 shows the transmission bands of ND-COOH attributed to the stretching bond of the hydroxyl group at 3387 cm^{-1}, the stretching of the carbonyl group at 1763 cm^{-1}, the bending of the hydroxyl bond at 1630 cm^{-1}, and the stretching bond of C-OH at 1245 cm^{-1}, which all confirm the presence of carboxylic acids in the structure of the nanodiamond. It is worth mentioning that nanodiamonds present various oxygen-functional groups on the surface, such as aldehydes, ketones, hydroxyls, and carboxylic acids, making it ideal for surface functionalization [33].

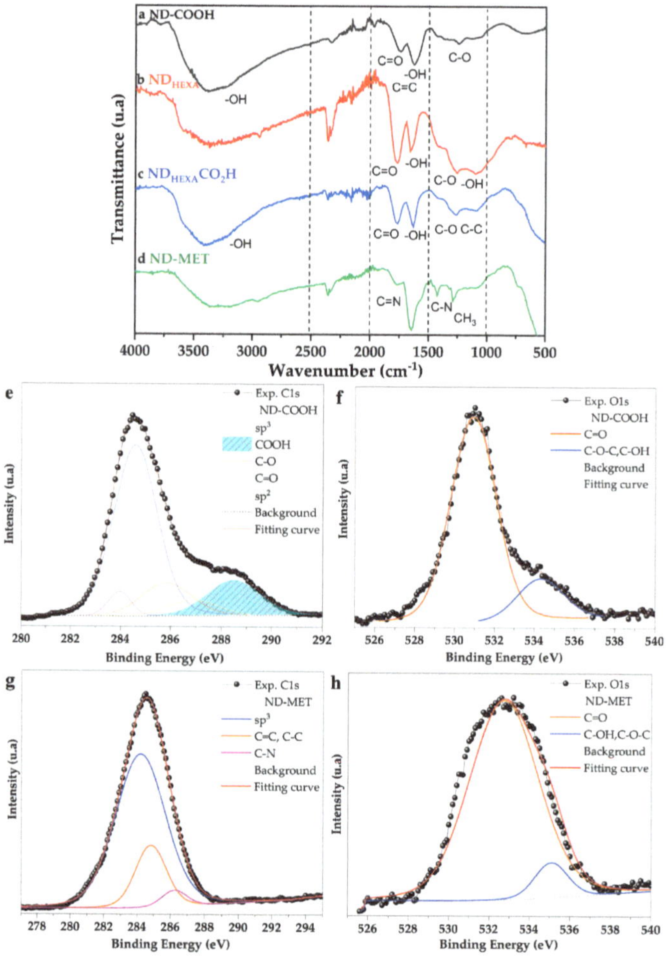

Figure 2. (**a–d**) show infrared spectra of nanodiamonds (ND-COOH) functionalized with MET, (**e**) deconvolution of C1s of nanodiamonds, (**f**) deconvolution of O1s of nanodiamonds, (**g**) deconvolution of C1s of ND-MET and (**h**) deconvolution of O1s of ND-MET.

When the nanodiamonds were esterified with 1,6-hexanediol, the stretching of the C-O bond of the ester formed can be observed at 1259 cm^{-1}, and the signal from the primary -OH of 1,6-hexanediol can be observed at 1075 cm^{-1}, as shown in Figure 2b. The surface oxidation reaction of the primary alcohols present in the nanodiamond to generate carboxylic acid groups (Figure 2c) was carried out using potassium permanganate ($KMnO_4$). The formation of the acid groups is confirmed by the presence of the signals at 1778 and 1632 cm^{-1} corresponding to the stretching of the carbonyl group and the bending of the hydroxyl, respectively.

MET presents several primary amines ($-NH_2$) in its structure (Figure 1c). These functional groups show signals between 1640–1560 cm^{-1} and 3500–3400 cm^{-1}. When metformin reacts with $ND_{HEXA}CO_2H$, the intensity of the signals coming from the primary amines decreases. In addition, a new signal appears at 2850 cm^{-1} due to the quaternization of the nitrogen of the amine that reacted with the acid groups of the $ND_{HEXA}CO_2H$. Secondary amines appear at 1427 cm^{-1}. Moreover, the ND-MET spectrum (Figure 2d) presents the signals from the bond stretching of the carbonyl group that decreases in intensity at 1778 cm^{-1}. The two signals at 1427 and 1288 cm^{-1} corresponding to the C-N and CH_3 bonds, respectively, confirm the formation of the ND-MET.

XPS analysis was performed to confirm the abundance of carboxylic groups and to determine the presence of bonds in ND-COOH and ND-MET. The C1s and O1s of both samples were deconvoluted using the Lorentzian fit. Figure 2e shows the deconvolution of the ND-COOH C1s peak into five binding energy bands at 288.48 eV (COOH), 286.89 eV (C=O), 285.86 eV (C-O), 284.60 eV (sp^3), and 283.60 eV (sp^2). In Figure 2f, the O1s peak was deconvoluted into two binding energies at 530.94 eV (C=O, OH-) and 534.32 eV (C-OH, C-O-C, H_2O).

It is possible to observe that the large area of the carbonyl deconvolution curve for the C1s and O1s of ND-COOH shown at 286.89 eV and 530.94 eV (C=O) binding energies, respectively, indicate a high density of the carbonyl functional group on the ND-COOH surface [33].

Figure 2g indicates the deconvolution of the C1s of the ND-MET complex at three binding energies: 284.19 eV (sp^3), 284.83 eV (C=C, C-C), and 286.24 eV (C-N). Figure 2h shows the deconvolution of the O1s of ND-MET at two binding energies: 532.09 eV (C=O) and 534.28 eV (C-OH, C-O-C, H_2O). Comparing Figure 2e,g, it is observed that the area of COOH at 288.49 eV no longer appears in the ND-MET sample, which could be attributed to the binding of metformin to this functional group. In addition, an atomic percentage of 3.96% belongs to the C-N bond, which presents a binding energy at 286.24 eV in Figure 2g.

Boehm titration was used to quantify the carboxylic groups on the ND-COOH surface. Table 1 shows the COOH sites for both the nanodiamond (ND-COOH) and the nanodiamond oxidized with $KMnO_4$, increasing from 1.10 ± 0.04 COOH per nm^{-2} to 9.36 ± 1.42 COOH per nm^{-2}, respectively. The acid-oxidized nanodiamonds reported average carboxylic group values of 0.85 COOH per nm^{-2} after treatment [34,35]. Using $KMnO_4$ to oxidize the primary alcohol to carboxylic acid proved more efficient for enhancing this functional group.

Table 1. COOH sites for ND-COOH and ND_{HEXA}-CO_2H.

Sample	COOH Sites (COOH/nm^2)
ND-COOH	1.10 ± 0.04
ND_{HEXA}-CO_2H	9.36 ± 1.42

RAMAN spectroscopy was used to complement molecular structure characterization. Figure 3a–e show the Raman spectra for ND-COOH, MET, and each of the derivatives obtained during the functionalization, indicating the changes in the molecule. ND-COOH exhibits stretching modes for C-Hx and O-H near 3000 and 3500 cm^{-1}, respectively. As seen in Figure 3a, around 1330 cm^{-1}, the nanodiamond shows an asymmetric peak, denoted as the D-band, representing the sp^3 hybridization of the diamond. Amorphous carbonaceous compounds show a band denoted G in the range 1400–1800 cm^{-1}; this band does not appear when the sp^3 carbon form is pure. Some interpretations for this band are the existence of sp^2 carbon, an sp^2 cluster, an sp^2/sp^3 mixture, surface hydroxyl groups, and C=C as defects within the diamond structure [36,37]. Mochalin et al. assigned it to surface O-H bending vibrations or absorbed water. For the ND-MET complex in Figure 3d, two bands would indicate the binding of metformin to the nanodiamond; the vibrational mode at 1061 cm^{-1} as ascribed to the N-O bond, which is the one that would be formed, and the presence of N-H at 3208 cm^{-1}, evidences the metformin molecule [38]. The RAMAN spectrum of MET in Figure 3e includes the C-N-C deformation at 480 cm^{-1} and 724 cm^{-1}, the C=N stretching vibration at 1592 cm^{-1}, the methyl group -CH_3 stretching at 2951 cm^{-1}, the presence of nitrogen near the methyl group which reduces the CH_3 stretching symmetry, and at 3183 cm^{-1}, the N-H bond of the C=N-H stretching. This vibration frequency usually decreases in the presence of the hydrogen bond [39]. Other bands at 3300–3800 cm^{-1} represent C-H and N-H [40,41]; the bands at 3355 cm^{-1} and 3183 cm^{-1} are specific for the asymmetric and symmetric N-H stretching vibrations, respectively [39].

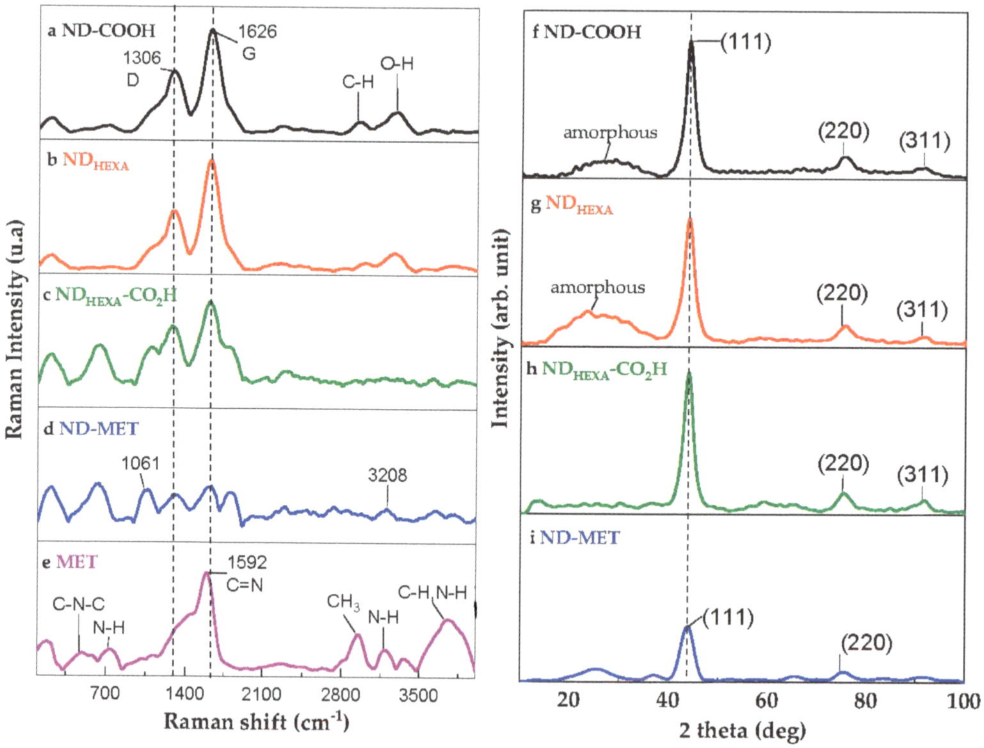

Figure 3. (**a–e**) Raman spectrum of ND-COOH, MET, and its derivatives and (**f–i**) XDR of ND-COOH and its derivatives from functionalization of ND-MET.

The crystallinity of ND-COOH and ND-MET was investigated using X-ray diffraction. The diffraction patterns are shown in Figure 3f–i. The XRD spectrum of ND-COOH (Figure 3f) indicates three dominant peaks at $2\theta = 43.81°$, $75.54°$, and $91.81°$, corresponding to the (111), (220), and (311) planes, respectively. These results evidence the crystallinity of ND-COOH, corresponding to a cubic structure (PDF ICDD 03-065-6329). In addition, the interplanar distance (*d*) measured on the high-resolution micrograph (Figure 4a), $d = 2.10 \pm 0.02$, corresponds to the most intense line (111) of the XDR.

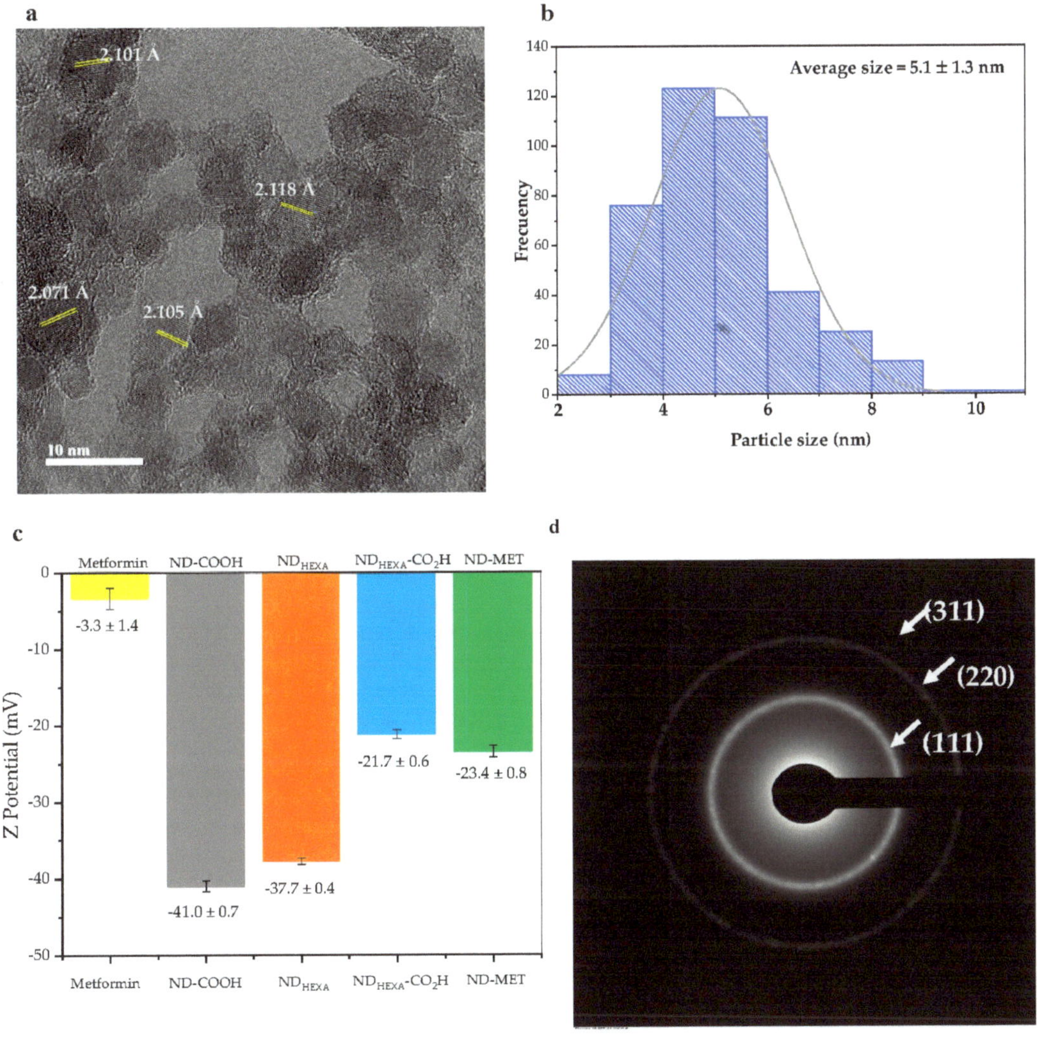

Figure 4. (**a**) High resolution micrograph of ND-MET, (**b**) particle size distribution of ND-MET, (**c**) Z-potential of ND-COOH, ND-MET, and its derivatives, and (**d**) diffraction pattern of ND-MET.

A band in the 20°–35° range is observed; this band is associated with amorphous carbon [42,43] because the crystalline nanodiamond's surface can undergo a phase transition during synthesis. This phase transition is also verified using XPS analysis in Figure 2e, where the area under the curve, corresponding to the sp^2 hybridization of the nanodiamond, is shown. In Figure 3g, the intensity of the amorphous band is increased by adding 1,6-hexanediol; this occurs because it is an amorphous organic compound that contributes to the increase of the intensity of the 20°–35° band. In ND$_{HEXA}$CO$_2$H, oxidizing the alcohol with $KMnO_4$ also oxidizes the surface of graphene, forming oxidized graphene (GO), which shows a peak around 10° [43], as seen in Figure 3h.

When the drug is added, the intensity of the (111) and (311) planes of the nanodiamond decreases, which indicates that the percentage of crystallinity decreases. For all derivatives, the peaks remain constant in the same position during the process, so it is possible to determine that the cubic phase of the nanodiamond is maintained.

High-resolution TEM, Figure 4a, presents the morphology of ND-MET. As noted, the material exhibits a globular morphology and interplanar distance of about 2.059 ± 0.020 Å, corresponding to the (111) plane of the nanodiamond. The sample shows a group of particles of uniform size, Figure 4b. The individual size of the ND-MET averages 5.1 ± 1.3 nm in diameter, similar to that obtained for ND-COOH, which was 5.3 ± 1.3 nm.

The particle size determines nanoparticles' biocompatibility, toxicity, and surface area. The charge and size of the nanoparticle also depends on their particle size distribution. Positively charged particles are retained in the liver, spleen, and lungs, while the neutral or slightly negatively charged ones show prolonged circulation in the blood [44].

Figure 4c shows the Z potential, obtained by DLS, of ND-COOH, ND-MET, and their functionalization derivatives obtained by DLS. The ND-COOH presented a hydrodynamic radius of 13.80 ± 0.04 nm (±SD) on average, and a Z potential of −41.0 ± 0.7 mV (±SD); this value indicates a stable system, free of agglomeration [45]. The degree of ionization of the carboxylic groups determines the strength of the interparticle interaction. When the carboxylic groups are ionized, the ND-COOH particles become negatively charged, and the repulsion between them becomes significant, resulting in a size decrease [33].

Figure 4c shows the decrease of the Z potential value for the functionalization derivatives (ND_{HEXA} and $ND_{HEXA}COOH$). This change in the Z potential is attributed to the conversion of anionic carboxylate groups on the surface of the nanodiamond to charge the neutral 1,6-hexanediol, giving a network of decreasing negative charge on the particle's surface. The Z potential varies if the pH changes. The Z potential for pH 7 corresponds to −21 mV for the carboxylated nanodiamond. This change indicates the deprotonation of the -COOH group, causing the nanodiamond (ND-COOH) surface to become negatively charged [33], making it possible to obtain better functionalization.

On the other hand, Figure 4d displays the diffraction pattern of the ND-MET, showing the rings associated with interplanar distances 2.11 Å, 1.30 Å, and 1.10 Å, corresponding to the (111), (220), and (311) planes, respectively.

3.2. Cytotoxicity Evaluation of ND-MET in Breast and Ovarian Cancer Cell Lines

Antitumor properties of nanodiamonds (ND-COOH), drug MET alone, and ND-MET complex on ovarian cancer (SKOV3) and triple-negative breast cancer (Hs578t and MDA-MB-231) cell lines were evaluated using a MTT assay. The rationale for the use of these cancer cell lines have been documented in many publications. Hs578t and MDA-MB-231 breast cancer cells exhibit a lower response to therapies and have a poorer prognosis; they are an example of triple-negative breast cancer (TNBC). The lack of estrogen, progesterone, and HER2/neu receptors in TNBC renders this type of cancer unresponsive to hormone-targeted therapies, such as tamoxifen, which are effective in tumors expressing estrogen or progesterone receptors. Moreover, due to the lack of HER2/neu receptor expression, TNBC does not respond to HER2-targeted therapies, like trastuzumab. Consequently, the inefficacy of these specific therapeutic options makes TNBC more challenging to treat when compared to other subtypes of breast cancer. Hence, we decided to evaluate the effect of nano-compounds as a novel therapeutic tool. For ovarian cancer (OC), the serous subtype is highly prevalent and can exhibit high aggressiveness, and is also associated with a poor prognosis and reduced responsiveness to therapies. SKOV3 is a cell line derived from serous ovarian carcinoma and has been extensively employed as a study model due to its resistance to certain chemical compounds.

We found that SKOV3 cells treated for 24 and 48 h, with different concentrations (15, 30, 60, 125, 250, 500, and 750 μg/mL) of the compounds, showed a significant reduction in cell viability in a time- and concentration-dependent manner. However, the most notable effect was observed when the cells were treated with the ND-MET complex, as shown in Figure 5a,b. Some studies indicate that the inhibition of cell viability also depends on the size and shape of the nanoparticles [1,4,6]. It has been reported that ND does not exhibit toxicity for SKOV3 with sizes of 95 nm at a concentration of 10 μM/mL [46].

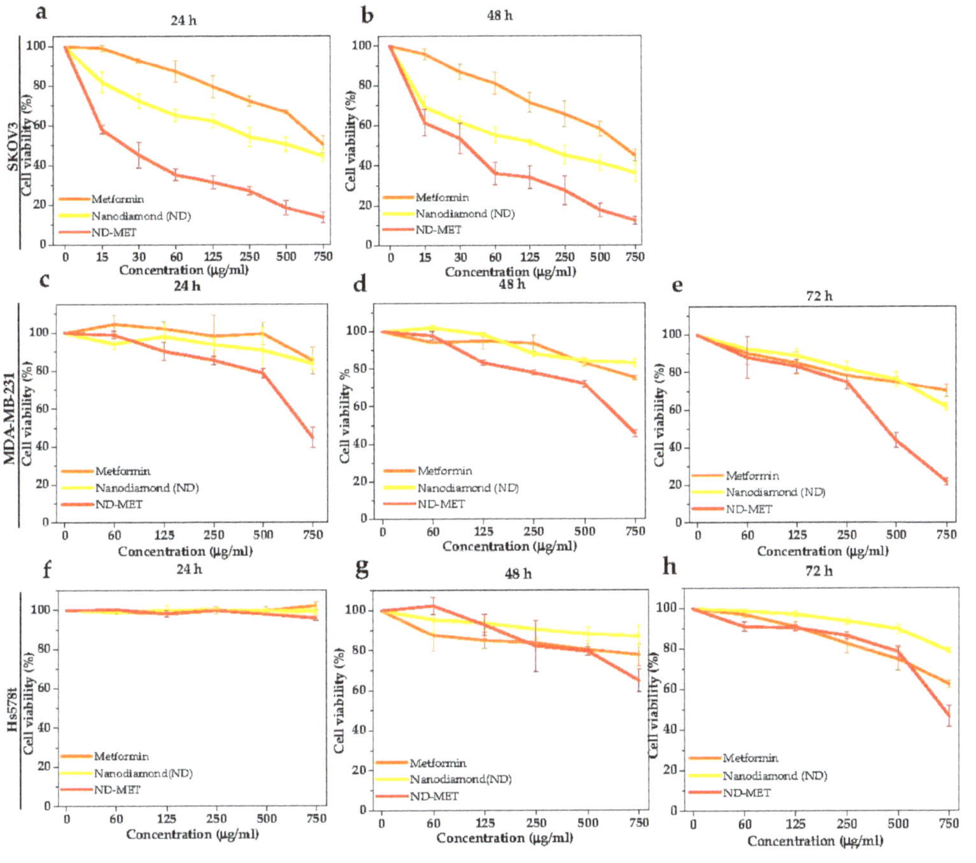

Figure 5. (**a**–**h**) Dose-response curves were generated for three cancer cell lines (SKOV3, MDA-MB-231, and Hs578t) and treated with different concentrations of metformin, ND-COOH, and ND-MET for 24, 48, and 72 h. All the values are expressed as average and standard deviation (n = 3).

The half maximal inhibitory concentration (IC_{50}) of MET alone at 24 and 48 h was 914 ± 127 μg/mL and 632 ± 116 μg/mL, respectively. The IC_{50} of ND-COOH at 24 and 48 h was 458 ±139 μg/mL and 157 ± 64 μg/mL, respectively. The ND-MET complex was the one that most effectively decreased the cell viability of ovarian tumor cells since IC_{50} was 24 ± 7 μg/mL and 35 ± 14 μg/mL at 24 and 48 h, respectively (Figure 5a,b; Table 2).

Table 2. Half-maximum inhibitory concentration (IC_{50}) values for metformin, nanodiamonds (ND-COOH), and nanodiamond-coupled metformin (ND-MET) in three cancer cell lines at 24, 48, and 72 h.

Cancer Cell Lines	Time								
	24 h			48 h			72 h		
SKOV3	MET	ND-COOH	ND-MET	MET	ND-COOH	ND-MET	MET	ND-COOH	ND-MET
	914 ± 127	458 ± 139	24 ± 7 **	632 ± 116	157 ± 64	35 ± 14 **	UD	UD	UD
MDA-MB-231	UD	UD	842 ± 81	UD	UD	801 ± 37	UD	UD	454 ± 49
Hs578T	UD	UD	UD	UD	UD	UD	UD	UD	759 ± 44

UD = Undetermined. Statistical analysis was performed using a one-way ANOVA test (** $p < 0.001$).

Preliminary tests that we carried out with the two breast cancer lines (MDA-MB-231 and Hs578t) showed that the effect of the different compounds on these lines is less compared to the ovarian line (SKOV3). Therefore, we decided to carry out the cytotoxicity assays at an additional time point (72 h), and the range of concentrations analyzed was increased from 60 to 750 µg/mL (Figure 5c–h).

As observed in Figure 5c–e and Table 2, only ND-MET resulted in an inhibition percentage greater than 50% for the MDA-MB-231 cell line under the study conditions.

The IC_{50} values calculated for ND-MET in MDA-MB-231 at 24, 48, and 72 h were 842 ± 81, 801 ± 37, and 454 ± 49 µg/mL, respectively. Likewise, in the Hs578t cell line, only the ND-MET complex achieved an inhibition percentage of 50% up to 72 h, with an IC_{50} value of 759 ± 44 µg/mL (Figure 5h; Table 2). For the other compounds alone (MET and ND-COOH), 50% cell inhibition was not reached even after incubating with the maximum concentration of the compounds and the maximum exposure time (Figure 5e,h). These analyses show that the cytotoxicity of the different nano-compounds depends on the concentration, treatment time, and cell type.

Finally, the IC_{50} values for ND-MET were analyzed (Table 2) and compared with the IC_{50} of a PLGA-PEG nanoparticle complex loaded with metformin [16], both evaluated in the ovarian cell line SKOV3. We observed that ND-MET presented a 50% inhibition of 35 ± 14 µg/mL at 48 h of treatment, while with the PLGA-PEG complex loaded with MET, it was 1295.51 (no deviation standard reported) µg/mL for the same treatment duration. This result shows a high percentage of inhibition of the synthesized complex.

4. Conclusions

A simple method to functionalize dispersed nanodiamonds with metformin by direct binding to 1,6-hexanediol was proposed. The cytotoxic effect of nanodiamonds coupled with MET on cancer cell lines is higher than their components. This cytotoxicity is attributed to the effect of metformin and the characteristics of the ND-MET complex. The size of the complex also makes it a biocompatible material, which is an essential factor for various biomedical applications. Furthermore, it is a compound with a homogeneous and monodisperse solubility. This property is because the functionalization makes the 1,6-hexanediol act as an anionic surfactant, which improves its biodistribution and makes it more stable. Additionally, functional groups like carboxyl and amino increase the solubility of MET, which have low bioavailability, thus allowing the molecule to interact electrostatically with the nanodiamond system. Some drugs are efficiently incorporated with this type of electrostatic interaction because the ND-COOH, given its liposolubility, can carry it to the interior of the cell, and the inhibition obtained is attributed to these characteristics. All these factors allow an effective and directed biomolecular interaction, so it is estimated that once evaluated in vivo, it will reduce the side effects.

We found that SKOV3 cancer cells treated for 24 and 48 h with different concentrations of the compounds showed a significant reduction in cell viability in a time- and concentration-dependent manner. However, the most notable effect was when the cells were treated with the ND-MET complex. Some studies indicate that the inhibition of cell viability also depends on the size and shape of the nanoparticles. Unexpectedly, our findings suggest that the carrier exhibits a pronounced effect on SKOV3 cells, surpassing that observed in breast cancer cell lines, which has been reported in different studies; thus we expected to find some toxicity with nanoparticles alone. Therefore, the use in preclinical models such as mouse models will require additional standardization, as the carrier's impact in a clinical setting will depend on a variety of factors including new doses, inoculation method, weight, and immune system status.

Author Contributions: Conceptualization, L.E.A.-A. and E.O.-B.; methodology, L.E.A.-A. and A.S.C.-D.; validation, O.N.H.-d.l.C. and G.S.-S.; formal analysis, L.E.A.-A., M.M.-Y. and J.S.S.-L.; investigation, L.E.A.-A., A.S.C.-D., O.N.H.-d.l.C. and G.S.-S.; resources, C.L.-C., E.A.Z.-C. and R.R.-M.; writing—original draft preparation, L.E.A.-A.; writing—review and editing, L.E.A.-A., E.O.-B., M.M.-Y., A.S.C.-D., O.N.H.-d.l.C., J.S.S.-L., C.L.-C. and E.A.Z.-C.; visualization, L.E.A.-A.; supervision, E.O.-B.; project administration, L.E.A.-A.; funding acquisition, E.O.-B. All authors have read and agreed to the published version of the manuscript.

Funding: This research was funded by the Consejo Nacional de Humanidades, Ciencias y Tecnologias (CONAHCYT), grant number 296401.

Data Availability Statement: Not applicable.

Acknowledgments: This project was carried out at the Centro de Investigación en Materiales Avanzados, S.C., and the Universidad Autónoma de la Ciudad de México. LEAA was a recipient of the 754406 CONAHCYT scholarship.

Conflicts of Interest: The authors declare no conflict of interest.

References

1. Zhu, Y.; Li, J.; Li, W.; Zhang, Y.; Yang, X.; Chen, N.; Sun, Y.; Zhao, Y.; Fan, C.; Huang, Q. The Biocompatibility of Nanodiamonds and Their Application in Drug Delivery Systems. *Theranostics* **2012**, *2*, 302–312. [CrossRef]
2. Gvozdev, D.A.; Ramonova, A.A.; Slonimskiy, Y.B.; Gudkova, V.R.; Nikelshparg, E.I.; Moisenovich, A.M.; Moisenovich, M.M.; Zaitsev, A.V.; Olshevskaya, V.A.; Paschenko, V.Z.; et al. Nanodiamonds as a Platform for Targeted Delivery of Chlorin-Based Photosensitizers to Cancer Cells. *Diam. Relat. Mater.* **2021**, *120*, 108676. [CrossRef]
3. Chow, E.K.; Zhang, X.-Q.; Chen, M.; Lam, R.; Robinson, E.; Huang, H.; Schaffer, D.; Osawa, E.; Goga, A.; Ho, D. Nanodiamond Therapeutic Delivery Agents Mediate Enhanced Chemoresistant Tumor Treatment. *Sci. Transl. Med.* **2011**, *3*, 73. [CrossRef] [PubMed]
4. Xu, J.; Chow, E.K.-H. Biomedical Applications of Nanodiamonds: From Drug-Delivery to Diagnostics. *SLAS Technol.* **2023**, *4*, S2472. [CrossRef] [PubMed]
5. Perevedentseva, E.; Hong, S.-F.; Huang, K.-J.; Chiang, I.-T.; Lee, C.-Y.; Tseng, Y.-T.; Cheng, C.-L. Nanodiamond Internalization in Cells and the Cell Uptake Mechanism. *J. Nanoparticle Res.* **2013**, *15*, 1834. [CrossRef]
6. Chipaux, M.; van der Laan, K.J.; Hemelaar, S.R.; Hasani, M.; Zheng, T.; Schirhagl, R. Nanodiamonds and Their Applications in Cells. *Small* **2018**, *14*, 1704263. [CrossRef]
7. Dai, L.; Liu, M.; Long, W.; Hu, X.; Ouyang, H.; Feng, Y.; Deng, F.; Wen, Y.; Zhang, X.; Wei, Y. Synthesis of Water Dispersible and Biocompatible Nanodiamond Composite via Photocatalytic Surface Grafting of Zwitterionic Polymers for Intracellular Delivery of DOX. *Mater. Today Commun.* **2022**, *30*, 103010. [CrossRef]
8. Perevedentseva, E.; Lin, Y.-C.; Cheng, C.-L. A Review of Recent Advances in Nanodiamond-Mediated Drug Delivery in Cancer. *Expert Opin. Drug Deliv.* **2021**, *18*, 369–382. [CrossRef]
9. Szymczak-Pajor, I.; Wenclewska, S.; Śliwińska, A. Metabolic Action of Metformin. *Pharmaceuticals* **2022**, *15*, 810. [CrossRef]
10. Tądel, K.; Wiatrak, B.; Bodetko, D.; Barg, E. Metformin and Proliferation of Cancer Cell Lines. *Pediatr. Endocrinol. Diabetes Metab.* **2020**, *26*, 159–166. [CrossRef]
11. El-khayat, S.M.; Abouegylah, M.; Abdallah, D.; Geweil, A.G.; Elenbaby, A.M.; Zahra, O.S. The Effect of Metformin When Combined with Neoadjuvant Chemotherapy in Breast Cancer Patients. *Med. Oncol.* **2022**, *39*, 1. [CrossRef]
12. Koh, M.; Lee, J.-C.; Min, C.; Moon, A. A Novel Metformin Derivative, HL010183, Inhibits Proliferation and Invasion of Triple-Negative Breast Cancer Cells. *Bioorg. Med. Chem.* **2013**, *21*, 2305–2313. [CrossRef] [PubMed]
13. Amable, G.; Martínez-León, E.; Picco, M.E.; Rey, O. Metformin and Colorectal Cancer. *Biocell* **2021**, *46*, 51–59. [CrossRef]
14. Joentausta, R.M.; Rannikko, A.; Murtola, T.J. Prostate Cancer–Specific Survival after Radical Prostatectomy Is Improved among Metformin Users but Not among Other Antidiabetic Drug Users. *Eur. Urol. Open Sci.* **2021**, *34*, 86–93. [CrossRef] [PubMed]
15. Mogheri, F.; Jokar, E.; Afshin, R.; Akbari, A.A.; Dadashpour, M.; Firouzi-amandi, A.; Serati-Nouri, H.; Zarghami, N. Co-Delivery of Metformin and Silibinin in Dual-Drug Loaded Nanoparticles Synergistically Improves Chemotherapy in Human Non-Small Cell Lung Cancer A549 Cells. *J. Drug Deliv. Sci. Technol.* **2021**, *66*, 102752. [CrossRef]
16. Faramarzi, L.; Dadashpour, M.; Sadeghzadeh, H.; Mahdavi, M.; Zarghami, N. Enhanced Anti-Proliferative and pro-Apoptotic Effects of Metformin Encapsulated PLGA-PEG Nanoparticles on SKOV3 Human Ovarian Carcinoma Cells. *Artif. Cells Nanomed. Biotechnol.* **2019**, *47*, 737–746. [CrossRef] [PubMed]
17. Yenmis, G.; Yaprak Sarac, E.; Besli, N.; Soydas, T.; Tastan, C.; Dilek Kancagi, D.; Yilanci, M.; Senol, K.; Karagulle, O.O.; Ekmekci, C.G.; et al. Anti-Cancer Effect of Metformin on the Metastasis and Invasion of Primary Breast Cancer Cells through Mediating NF-KB Activity. *Acta Histochem.* **2021**, *123*, 151709. [CrossRef]
18. Basu, A.; Upadhyay, P.; Ghosh, A.; Bose, A.; Gupta, P.; Chattopadhyay, S.; Chattopadhyay, D.; Adhikary, A. Hyaluronic Acid Engrafted Metformin Loaded Graphene Oxide Nanoparticle as CD44 Targeted Anti-Cancer Therapy for Triple Negative Breast Cancer. *Biochim. Biophys. Acta (BBA) Gen. Subj.* **2021**, *1865*, 129841. [CrossRef]
19. Lee, K.-M.; Lee, M.; Lee, J.; Kim, S.W.; Moon, H.-G.; Noh, D.-Y.; Han, W. Enhanced Antitumor Activity and Cytotoxic Effect on Cancer Stem Cell Population of Metformin-Butyrate Compared with Metformin HCl in Breast Cancer. *Oncotarget* **2016**, *7*, 38500–38512. [CrossRef]
20. Chen, Y.; Shan, X.; Luo, C.; He, Z. Emerging Nanoparticulate Drug Delivery Systems of Metformin. *J. Pharm. Investig.* **2020**, *50*, 219–230. [CrossRef]

21. Jariwala, D.H.; Patel, D.; Wairkar, S. Surface Functionalization of Nanodiamonds for Biomedical Applications. *Mater. Sci. Eng. C* **2020**, *113*, 110996. [CrossRef] [PubMed]
22. Reina, G.; Zhao, L.; Bianco, A.; Komatsu, N. Chemical Functionalization of Nanodiamonds: Opportunities and Challenges Ahead. *Angew. Chem. Int. Ed.* **2019**, *58*, 17918–17929. [CrossRef] [PubMed]
23. Bai, X.; Ni, J.; Beretov, J.; Graham, P.; Li, Y. Triple-negative breast cancer therapeutic resistance: Where is the Achilles' heel? *Cancer Lett.* **2021**, *497*, 100–111. [CrossRef] [PubMed]
24. Dawson, S.J.; Provenzano, E.; Caldas, C. Triple negative breast cancers: Clinical and prognostic implications. *Eur. J. Cancer* **2009**, *45*, 27–40. [CrossRef]
25. Waks, A.G.; Winer, E.P. Breast Cancer Treatment: A review. *JAMA* **2019**, *321*, 288. [CrossRef]
26. Bianchini, G.; Balko, J.M.; Mayer, I.A.; Sanders, M.E.; Gianni, L. Triple-negative breast cancer: Challenges and opportunities of a heterogeneous disease. *Nat. Rev. Clin. Oncol.* **2016**, *13*, 674–690. [CrossRef]
27. Hirst, J.; Crow, J.; Godwin, A. Ovarian Cancer Genetics: Subtypes and Risk Factors. In *Ovarian Cancer—From Pathogenesis to Treatment*; InTech Open: London, UK, 2018.
28. Elias, K.M.; Emori, M.M.; Papp, E.; MacDuffie, E.; Konecny, G.E.; Velculescu, V.E.; Drapkin, R. Beyond genomics: Critical evaluation of cell line utility for ovarian cancer research. *Gynecol. Oncol.* **2015**, *139*, 97–103. [CrossRef]
29. Liu, R.; Zhao, F.; Yu, X.; Naito, K.; Ding, H.; Qu, X.; Zhang, Q. Synthesis of Biopolymer-Grafted Nanodiamond by Ring-Opening Polymerization. *Diam. Relat. Mater.* **2014**, *50*, 26–32. [CrossRef]
30. Ciufolini, M.A.; Swaminathan, S. Synthesis of a Model Depsipeptide Segment of Luzopeptins (BBM 928), Potent Antitumor and Antiretroviral Antibiotics. *Tetrahedron. Lett.* **1989**, *30*, 3027–3028. [CrossRef]
31. AAT Bioquest, Inc. Quest Graph IC50 Calculator. AAT Bioquest. Available online: https://www.aatbio.com/tools/ic50-calculator (accessed on 15 August 2023).
32. Wawruszak, A.; Luszczki, J.J.; Grabarska, A.; Gumbarewicz, E.; Dmoszynska-Graniczka, M.; Polberg, K.; Stepulak, A. Assessment of Interactions between Cisplatin and Two Histone Deacetylase Inhibitors in MCF7, T47D and MDA-MB-231 Human Breast Cancer Cell Lines—An Isobolographic Analysis. *PLoS ONE* **2015**, *10*, e0143013. [CrossRef]
33. Jeong, S.H.; Lim, D.; Kim, K.; Kang, E.; Lim, S.; Ricci, J.; Sung, S.; Kwon, M. Comprehensive Evaluation of Carboxylated Nanodiamond as a Topical Drug Delivery System. *Int. J. Nanomed.* **2016**, *11*, 2381. [CrossRef] [PubMed]
34. Arnault, J.C. Surface Modifications of Nanodiamonds and Current Issues for Their Biomedical Applications. In *Novel Aspects of Diamond. Topics in Applied Physics*; Yang, N., Ed.; Springer: Cham, Switzerland, 2015; Volume 121, pp. 85–122. [CrossRef]
35. Ullah, M.; Kausar, A.; Siddiq, M.; Subhan, M.; Abid Zia, M. Reinforcing Effects of Modified Nanodiamonds on the Physical Properties of Polymer-Based Nanocomposites: A Review. *Polym. Plast. Technol. Eng.* **2015**, *54*, 861–879. [CrossRef]
36. Mermoux, M.; Chang, S.; Girard, H.A.; Arnault, J.-C. Raman Spectroscopy Study of Detonation Nanodiamond. *Diam. Relat. Mater.* **2018**, *87*, 248–260. [CrossRef]
37. Mochalin, V.; Osswald, S.; Gogotsi, Y. Contribution of Functional Groups to the Raman Spectrum of Nanodiamond Powders. *Chem. Mater.* **2009**, *21*, 273–279. [CrossRef]
38. Espectroscopía Vibracional. Available online: https://independent.academia.edu/MauricioSandy (accessed on 28 May 2023).
39. Srinivasan, S.; Renganayaki, V. Computational Studies of Vibration Spectra and Thermodynamic Properties of Metformin Using HF, DFT Methods. *Mater. Sci. Res. India* **2011**, *8*, 165–172. [CrossRef]
40. Bayramov, F.B.; Toporov, V.V.; Chakchir, O.B.; Anisimov, V.N.; Rud, V.Y.; Glinushkin, A.P.; Bairamov, B.H. Raman Spectroscopic Characterization of Anti-Diabetic Drug Metformin Hydrochloride. *J. Phys. Conf. Ser.* **2019**, *1400*, 033010. [CrossRef]
41. Arafat, M.; Sakkal, M.; Yuvaraju, P.; Esmaeil, A.; Poulose, V.; Aburuz, S. Effect of Excipients on the Quality of Drug Formulation and Immediate Release of Generic Metformin HCl Tablets. *Pharmaceuticals* **2023**, *16*, 539. [CrossRef]
42. Fan, J.; Li, T.; Gao, Y.; Wang, J.; Ding, H.; Heng, H. Comprehensive Study of Graphene Grown by Chemical Vapor Deposition. *J. Mater. Sci. Mater. Electron.* **2014**, *25*, 4333–4338. [CrossRef]
43. Johra, F.T.; Lee, J.-W.; Jung, W.-G. Facile and Safe Graphene Preparation on Solution Based Platform. *J. Ind. Eng. Chem.* **2014**, *20*, 2883–2887. [CrossRef]
44. Cabral, H.; Miyata, K.; Osada, K.; Kataoka, K. Block Copolymer Micelles in Nanomedicine Applications. *Chem. Rev.* **2018**, *118*, 6844–6892. [CrossRef]
45. Riddick, T.M. *Control of Colloid Stability Through Zeta Potential*; With a Closing Chapter on Its Relationship to Cardiovascular, Disease; Zeta-Meter, Incorporated; Livingston Publishing Company: Wynnewood, PA, USA, 1968; Volume 1, p. 172.
46. Saed, M.E. The Use of Nanotechnology for Treatment of Multidrug Resistant Ovarian Cancer Cells. Ph.D. Thesis, Trinity College Dublin, Dublin, Germany, 2019.

processes

Article

Development of Tea Tree Oil Based Nanoemulgel Loaded with Azithromycin for Enhancing the Antibacterial Activity

Nasrin E. Khalifa [1,2], Marwa H. Abdallah [1,3,*], Hanaa A. Elghamry [3], Weam M. A. Khojali [4,5], El-Sayed Khafagy [6,7], Hemat El-Sayed El-Horany [8,9] and Seham Shawky [10]

1 Department of Pharmaceutics, College of Pharmacy, University of Ha'il, Ha'il 81442, Saudi Arabia; n.aldirdiri@uoh.edu.sa
2 Department of Pharmaceutics, Faculty of Pharmacy, University of Khartoum, Khartoum 11115, Sudan
3 Department of Pharmaceutics and Industrial Pharmacy, Faculty of Pharmacy, Zagazig University, Zagazig 44519, Egypt; hanaaelghamry@yahoo.com
4 Department of Pharmaceutical Chemistry, College of Pharmacy, University of Ha'il, Ha'il 81442, Saudi Arabia; we.ali@uoh.edu.sa
5 Department of Pharmaceutical Chemistry, Faculty of Pharmacy, Omdurman Islamic University, Omdurman 14415, Sudan
6 Department of Pharmaceutics, College of Pharmacy, Prince Sattam bin Abdulaziz University, Al-Kharj 11942, Saudi Arabia; e.khafagy@psau.edu.sa
7 Department of Pharmaceutics and Industrial Pharmacy, Faculty of Pharmacy, Suez Canal University, Ismailia 41552, Egypt
8 Department of Biochemistry, College of Medicine, University of Ha'il, Ha'il 81442, Saudi Arabia
9 Department of Medical Biochemistry, Faculty of Medicine, Tanta University, Tanta 31511, Egypt
10 Department of Pharmaceutics and Pharmaceutical Technology, Faculty of Pharmacy, Al-Azhar University, Cairo 11651, Egypt
* Correspondence: mh.abdallah@uoh.edu.sa

Citation: Khalifa, N.E.; Abdallah, M.H.; Elghamry, H.A.; Khojali, W.M.A.; Khafagy, E.-S.; El-Sayed El-Horany, H.; Shawky, S. Development of Tea Tree Oil Based Nanoemulgel Loaded with Azithromycin for Enhancing the Antibacterial Activity. *Processes* **2023**, *11*, 1836. https://doi.org/10.3390/pr11061836

Academic Editor: Paolo Trucillo

Received: 28 April 2023
Revised: 10 June 2023
Accepted: 14 June 2023
Published: 17 June 2023

Abstract: Azithromycin (AZ) is an azalide macrolide antibiotic that is frequently employed for treating bacterial skin infections. It suffers from limited oral bioavailability, which results from incomplete absorption or extensive first-pass metabolism. Therefore, preparing azithromycin formulations for topical administration is highly recommended to avoid first-pass metabolism and to boost the concentration of the drug on the skin. The objective of our investigation was to formulate and evaluate the efficacy of AZ-loaded nanoemulgel as an antimicrobial drug. The physical appearance, spreadability, viscosity, particle size, in vitro drug release, ex vivo permeation investigations, and antimicrobial efficiency of the prepared formulations were evaluated. The prepared formulation loaded with AZ exhibited good physical quality. AZ-loaded nanoemulgel had a greater ex vivo drug permeation across rabbit skin than other formulations (AZ-loaded gel and AZ-loaded emulgel), revealing improved drug permeation and greater transdermal flux in addition to enhanced antibacterial efficacy ($p < 0.05$). Overall, our findings imply that tea-tree-oil-based nanoemulgel would be a promising delivery system for enhancing the antimicrobial efficiency of azithromycin.

Keywords: Azithromycin; anti-bacterial activity; nanoemulgel; emulgel; tea tree oil

1. Introduction

E. coli is a Gram-negative bacterium that is typically found in feces and the lower intestine of warm-blooded animals. The most common cause of food- and water-borne diarrhea in people is *E. coli*. Moreover, *E. coli* is a frequent reason for infection in surgical wounds, especially those following abdominal operations, where it is frequently found combined with other gut bacteria. Compared to a number of multi-drug resistant bacteria (MDR), *E.coli* (*Escerichia coli*) strains have developed progressively stronger resistance to antibiotics [1]. Rai et al. argue that infections caused by MDR bacteria are more challenging to treat and need broad-spectrum antibiotics, which are harmful and expensive [2].

Azithromycin is an azalide, which is a subclass of macrolide antibiotics with a broad spectrum that has bacteriostatic activity against a variety of Gram-negative and Gram-positive bacteria [3]. Azithromycin inhibits the synthesis of bacterial proteins by blocking both the assembly of the 50S ribosomal subunit and the transpeptidation/ translocation process of protein synthesis [4]. Unlike erythromycin, azithromycin has a methyl-substituted nitrogen atom in the macrolide ring; this property makes it more stable, tolerated, and effective than erythromycin [5]. Azithromycin has limited oral bioavailability, which results from incomplete absorption or extensive first-pass metabolism [6]. Therefore, preparing azithromycin formulation for topical administration is highly recommended to avoid systemic side effects, including nausea, diarrhea, and abdominal pain. It also avoids first-pass metabolism, enhances patient acceptance, and reduces dose via direct contact with the pathologic site [7].

Essential oils are natural compounds derived from natural sources, and because of their efficiency and safety, they have been widely used in many formulations [8]. Tea tree oil is an essential oil that is derived from the leaves of *Melaleuca alternifolia*, a small tree that is indigenous to New South Wales and Queensland, Australia [9]. It is commonly known as melaleuca oil, which refers to its place of origin. Tea tree oil demonstrated broad-spectrum action, including effects against bacterial and fungal skin conditions, preventing infection, and promoting healing [10]. It demonstrates anticancer and antioxidant activity, in addition to its known antimicrobial and disinfecting activity [11], which can be attributed to terpoen-4-ol, which is the major constituent of tea tree oil. The formulation of this essential oil in a suitable drug delivery system could be beneficial because of the several advantages of tea tree oil.

Many pharmacological substances, especially those that are poorly water-soluble and have a narrow therapeutic index, have been transported via the skin using the transdermal drug delivery system (TDDS) [12,13]. TDDS enables the pharmacological agent to cross the epidermis reach the circulatory system, then be transported to the body via the bloodstream [14]. The transdermal administration route may offer benefits, such as lengthening the duration of action, avoiding first-pass hepatic metabolism, improving pharmacological efficacy, reducing undesirable side effects, and enhancing patient compliance. [15]. Therefore, the transdermal system development for the treatment of numerous clinical disorders, such as skin infection, has a feasible potential.

TDDSs include transdermal patches, gel, emulgel, and nanoemulgel [16]. Conventional dosage forms, including ointments, creams, gels, emulsions, and emulgels, have limited therapeutic applications due to the low drug permeability through the skin because of their large particle size. Therefore, the nanoemulgel concept was developed as an approach to solving the permeability issue [17]. Nanoemulgels effectively help the delivery of both lipophilic and hydrophilic medications, compared to hydrogels, which have a limitation in the transportation of lipophilic pharmaceuticals [18]. Additionally, nanoemulgels have a larger surface area than emulgels, which permits the rapid permeation of the drug through the pores of the skin and more effective transport of the drug into the circulatory system.

In the current study, we investigated the effectiveness of a nanoemulgel in strengthening the antimicrobial activity of azithromycin transdermal administration. The inclusion of azithromycin into nanoemulgel effectively increases its percutaneous penetration, which, in turn, enhances its antibacterial activity against skin infection. As far as we are aware, there has not been much information regarding the combination of tea tree oil and azithromycin in a nanoemulgel formulation meant to treat skin infections. The combination of azithromycin and tea tree oil into a nanoemulgel formulation intended for skin infection.

2. Materials and Methods

2.1. Materials

Tea tree oil was obtained from NOW® Foods, USA. Azithromycin (AZ), hydroxypropylmethyl cellulose (HPMC) ethyl alcohol, PEG 400, and Tween 80 were purchased from Sigma-Aldrich (St Louis, MO, USA).

2.2. Formulation of Topical Preparations Loaded with Azithromycin

2.2.1. Formulation of Gel

A suitable amount of HPMC (4% *w/w*), a synthetic gelling agent [19], was added to water, and the mixture was agitated until the formation of a hydrogel base using a magnetic stirrer. A known quantity of the medication was mixed with ethanol and vortexed for five minutes. To create a uniform gel, the initial mixture was subsequently added to the prepared hydrogel base [20].

2.2.2. Emulgel Preparation

Emulgel is a type of gelled emulsion made by adding gelling agents to an emulsion. For the oily phase preparation of emulsion, a known quantity of Azithromycin was dispersed in tea tree oil. The aqueous phase was created by adding the weighted amount of Tween 80 (surfactant), PEG 400 (co-surfactant), and ethanol (solvent) to water. This mixture was vortexed for five minutes. The primary emulsion was then created by continuously stirring the aqueous phase with the oily phase. On the other hand, as was previously noted, a gelling agent was used to prepare the gel base. A homogeneous emulgel was developed by adding the previously prepared emulsion that was loaded with AZ to the gel base and thoroughly mixing it for five minutes [21].

2.2.3. Nanoemulgel Preparation

For the nanoemulgel, the nanoemulsion is mixed with a gelling agent. The same method that was used to create the emulsion loaded with AZ was also applied here, followed by the homogenization with a homogenizer mixer for 10 min at 6000 rpm (20 high-speed digital, Ika-Eurostar, Staufen, Germany) to obtain nanoemulsion. A homogenous AZ-loaded nanoemulgel was created by adding the AZ-loaded nanoemulsion to the hydrogel base (4% HPMC) and blending for 10 min. The different topical AZ-loaded formulations' compositions are outlined in Table 1.

Table 1. Constituents of 2% *w/w* AZ-loaded formulations.

Formulation	Azithromycin (%w/w)	Gelling Agent (%w/w)	Tea Tree Oil (mL)	Ethyl Alcohol (mL)	Tween 80 (mL)	PEG400 (mL)	Water Up to (mL)
Gel	2	4	–	1	–	–	50
Emulgel	2	4	5	1	1	1	50
Nanoemulgel	2	4	5	1	1	1	50

2.3. Characterization of the Formulated AZ-Loaded Topical Formulations

2.3.1. Organoleptic Assessment

The color, homogeneity, and appearance of the topical preparation loaded with azithromycin were examined visually. Moreover, pH values were checked at RT, utilizing a pH meter (PCT-407, Taipei City, Taiwan) [22].

2.3.2. Viscosity

The Brookfield viscometer (Model DV-II, USA, Spindle number 06 at 10 rpm, at 25 °C) was used to assess the formulations' viscosity. The viscosity was measured in triplicate [23].

2.3.3. Spreadability Measurement

The spreadability regulates the area on the skin where the formulation can spread freely after use. One gram of the developed preparations was sandwiched between two glass slides under a standard load. The spreadability value was determined by calculating the formulations' spreading areas diameter [21].

2.3.4. Extrudability Determination

The extrudability is the weight in grams required to extrude a preparation from a collapsible tube of at least a half-centimeter ribbon in 10 s. Ten grams of the AZ-loaded formulations was packaged in a collapsible tube, and the preparation was forced out when the cap was removed by pressing the tube's crimped end [24]. The following equation was used to determine the extrudability (g/cm^2):

$$\text{Extrudability} = \frac{\text{Applied weight to extrude 0.5 cm ribone like gel from collipsible tube in 10 seconds (Gram)}}{\text{Area (cm}^2\text{)}}$$

2.3.5. Size and Size Distribution

The polydispersity (PDI) and particle size of emulgel and nanoemulgel loaded with AZ were measured using Malvern Zetasizer Apparatus (Worcestershire, UK). The assessments were carried out at a scattering angle of 90° and 25 °C [25].

2.3.6. Morphological Assessment

One milliliter of distilled water was employed to dilute ten-milligram samples, and then, a sample of one drop was coated with gold after being dried at room temperature and checked using a scanning electron microscopy [21].

2.3.7. Determination of Drug Content

Half a gram of the generated formulations was dissolved in 100 mL of PBS, pH 7.4, and stirred continuously for a half-hour. The final product was filtered through Whatman filter paper. The ultraviolet-visible spectrophotometer was used to detect the solution's absorbance at 232 nm, and the following equation was used to determine the amount of AZ included [22].

$$\text{Drug content percentage} = \frac{\text{Drug actual amount in the formuation}}{\text{Drug theoretical amount in the formulation}} \times 100$$

2.4. In Vitro Release of Azithromycin-Loaded Preparations

Using the previously described process by Abdallah et al. [22], with certain modifications, USP dissolution apparatus II was used for assessing the azithromycin in in vitro release profiles. Briefly, a half gram of the generated preparation containing ten milligrams of AZ was placed within glass tubes with a cellophane membrane (MWCO 14,000) attached to one end. The tubes were attached to the dissolution apparatus, and they were permitted to revolve in 100 mL of phosphate buffer at 50 rpm while maintaining a temperature of 37 ± 0.5 °C (pH 7.4). Samples were taken out at predetermined intervals and replaced with fresh media to maintain the sink condition throughout the investigation. Spectrophotometry was used to analyze samples at a maximum wavelength of 232 nm [26].

2.5. Stability Studies of the Developed AZ-Loaded Formulations

The samples were kept in tightly closed containers for three months at 4 ± 0.5 °C in the refrigerator and at 25 ± 0.5 °C. The samples' physicochemical properties and in vitro drug release at the predefined duration were assessed.

2.6. Ex Vivo Skin Permeation Experiment

2.6.1. Rabbit Skin Preparation

Skin permeation experiments were conducted using white albino male rabbit skin because of its comparatively low cost and availability. An electric clipper was used to shave the white albino male rabbits' abdominal skin. The abdominal skin was detached, and the adipose tissue was then removed after scarifying the animals. The skin samples were stored at 4 °C overnight in PBS (pH 7.4).

2.6.2. Azithromycin Permeation from Various Formulations

Using locally fabricated diffusion cells, a skin permeation investigation of AZ was conducted through male albino rabbits' skin [22]. The donor compartment faced the stratum corneum side of the mounted rabbit skin, which was placed in between the donor and receptor compartments. Half a gram of the formulation corresponding to 2.5 mg of AZ was applied to the skin in the donor compartment. The receptor medium, which contained PBS pH 7.4 (100 mL) and sodium azide (0.02% *w/v*), was constantly swirled at 100 rpm while being maintained at a temperature of 37 ± 0.5 °C using the dissolution apparatus II [27]. At regular intervals, samples (5 mL) were collected and analyzed spectrophotometrically at λ_{max} 232 nm, using the appropriate receptor media as a blank to prevent any interference. To keep a constant volume of receptor media, an equal volume of fresh buffer was added. Ex vivo permeation characteristics of AZ through rabbit skin were estimated, including

$$J_{ss} = \frac{Amount\ of\ permeated\ drug}{area\ of\ permeation\ X\ time}$$

$$ER = \frac{J_{ss\ (test)}}{J_{ss\ (control)}}$$

where J_{ss} is steady state transdermal flux (μg/cm^2.h), and *ER* is the enhancement ratio.

2.7. Antibacterial Investigation

The antimicrobial efficiency of the generated nanoemulgel loaded with AZ against *Escherichia coli* (*E. coli*), Gram-negative bacteria was evaluated utilizing the agar diffusion technique according to Zafar et al. with some adjustment [28]. Initially, the Petri dishes were sterilized at 160 °C for 60 min in a hot-air oven [29]. Afterward, ten milliliters of sterile nutrient agar media, sterilized in an autoclave at 121 °C for 15 min, were added to each plate. The microorganism was introduced into the plates after solidification under aseptic conditions [28]. A cork borer was used to make three millimeter-diameter cups in the plate, and the samples (placebo nanoemulgel, AZ-loaded gel, and AZ-loaded nanoemulgel) were then inserted to test their effectiveness [30]. The Petri dishes were left for thirty minutes at room temperature for efficient drug diffusion; then, they were incubated for a twenty-four-hour period at 37 °C [31]. The antibacterial efficacy was assessed by measuring the diameter of the inhibitory zone using a graded scale.

2.8. Statistics

Mean ± standard deviation (n = 3) was used to express the collected data. One-way ANOVA (GraphPad Prism version 5) was used for the statistical analysis. The difference was considered statistically significant if p was less than 0.05.

3. Results

3.1. Determination of pH and Organoleptic Assessment

The physical characteristics of the various topical AZ preparations (gel, emulgel, and nanoemulgel) that were effectively generated are provided in Table 2. Homogenous, smooth, and white formulations were produced, free from coarse particles. All formulations had pH levels that ranged from 5.8 to 6.4, which were deemed to be appropriate and would not irritate the skin when applied.

Table 2. Organoleptic evaluation of AZ-loaded formulations.

Properties	AZ-Loaded Gel	AZ-Loaded Emulgel	AZ-Loaded Nanoemulgel
Color and homogeneity	White homogenous	White homogenous	White homogenous
pH	5.8 ± 0.1	6.4 ± 0.2	6.3 ± 0.3
Viscosity (cP)	70,700 ± 1868	99,100 ± 1015 *	84,500 ± 1400 *,#
Spreadability (mm)	60.33 ± 2.5	51.33 ± 2.08 *	42.67 ± 1.53 *,#
Extrudability (g/cm^2)	154.67 ± 4.62	135.33 ± 5.03 *	106.67 ± 6.11 *,#
Drug content (%)	97.76 ± 2.92	98.92 ± 1.75	96.74 ± 3.23

compared to AZ-loaded emulgel, * compared to AZ-loaded gel ($p < 0.05$).

3.2. Viscosity Determination

The data regarding viscosity determination are demonstrated in Table 2. It was demonstrated that the emulgel's viscosity (99,100 ± 1015 cP) was much higher than that of gel and nanoemulgel (70,700 ± 1868 and 84,500 ± 1400 cP). These results were consistent with those of Mohamed et al., who concluded the highest viscosity of the topical emulgel loaded with atorvastatin compared to gel and nanoemulgel [32].

3.3. Spreadability and Extrudability Determination

The spreadability and extrudability are important criteria for topical preparations. They are considered essential aspects for the uniform application of topical preparations and patient compliance. The AZ-loaded gel, emulgel, and nanoemulgel formulations had spreadability values of 60.33 ± 2.5, 51.33 ± 2.08, and 42.67 ± 1.53 mm, respectively, and extrudability values of 154.67 ± 4.62, 135.33 ± 5.03, and 106.67 ± 6.11 g/cm^2, respectively (Table 2). These results indicate that the generated formulations have great skin spreadability and can be easily extruded by pressing down with the thumb. These findings are equivalent to those attained by Ali et al., who estimated extrudability and spreadability importance as the crucial factors for uniform distribution and patient compliance in the gel preparations [24,33].

3.4. Assessment of Drug Content Percent

The AZ-loaded gel has a 97.76 ± 2.92% drug content percentage; the AZ-loaded emulgel has a 98.92 ± 1.75 drug content percentage, while AZ-loaded nanoemulgel has a 96.74 ± 3.23 drug content percentage. This outcome demonstrated that the drug content was within the specified range (100 ± 10%). This revealed that the drug was dispersed uniformly throughout the generated formulations.

3.5. Polydispersity Index and Particle Size

The average particle size of emulgel loaded with azithromycin was 707.3 ± 17 nm, while the average particle size of AZ-loaded nanoemulgel was 227.8 ± 12 nm (Figure 1). Additionally, the polydispersity index (PDI) of the nanoemulgel and the emulgel loaded with AZ were estimated. AZ-loaded emulgel had a PDI of 0.631, while AZ-loaded nanoemulgel had a PDI of 0.395. Therefore, our findings supported the uniform size distribution of the nanoemulgel formulations. Additionally, the nanoemulgel's smaller particle size and lower PDI value are considered positive indicators of formulation stability and contribute to the enhanced drug penetration in the skin [34].

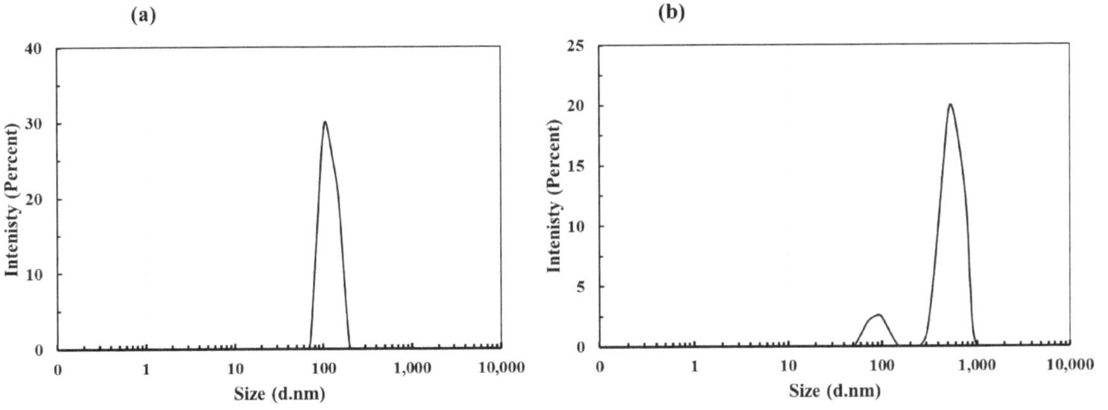

Figure 1. Size distribution of formulated (**a**) AZ-loaded nanoemulgel; (**b**) AZ-loaded emulgel.

3.6. Morphological Assessment

The morphology of the nanoemulgel loaded with AZ was estimated using SEM. Small spherical vesicles of nanoemulsion are dispersed throughout the macromolecular polymer network, as shown in Figure 2. Additionally, no visible AZ crystals were found, indicating adequate drug solubility.

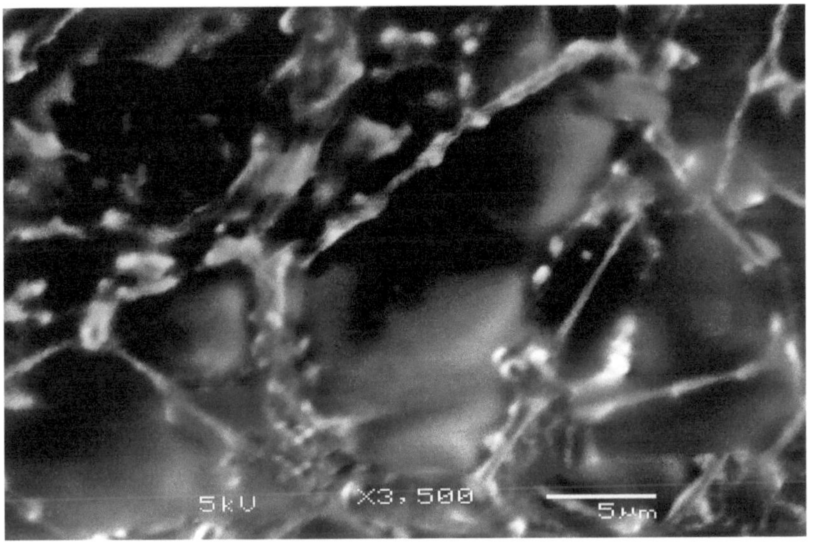

Figure 2. Observation of AZ-loaded nanoemulgel using a scanning electron microscope (SEM).

3.7. Azithromycin In Vitro Release Investigation

The percentage of AZ released after 6 h was 70.93 ± 2.21, 60.33 ± 2.82%, and 51.78 ± 2.70 for the gel, nanoemulgel, and emulgel, respectively, while AZ suspension showed a 97.94 ± 2.04% release after 4 h (Figure 3). In comparison to the free drug, all of the formulations' percentages of AZ released were significantly ($p < 0.05$) reduced. In addition, a significantly greater AZ amount was released from gel than from other formulations ($p < 0.05$) (nanoemulgel or emulgel). This is a result of the increased water proportion of the gel formula, which facilitates the drug's diffusion into the releasing medium. However, the significantly decreased release of AZ from nanoemulgel and emulgel, in contrast to the gel formulation, was explained by the higher viscosity, which could hinder the drug diffusion in addition to the decreased aqueous composition and tea tree oil incorporation [35]. It is interesting to note that the nanoemulgel's significantly larger drug release ($p < 0.05$) may be due to its smaller average particle size than the emulgel, as illustrated in Figure 1.

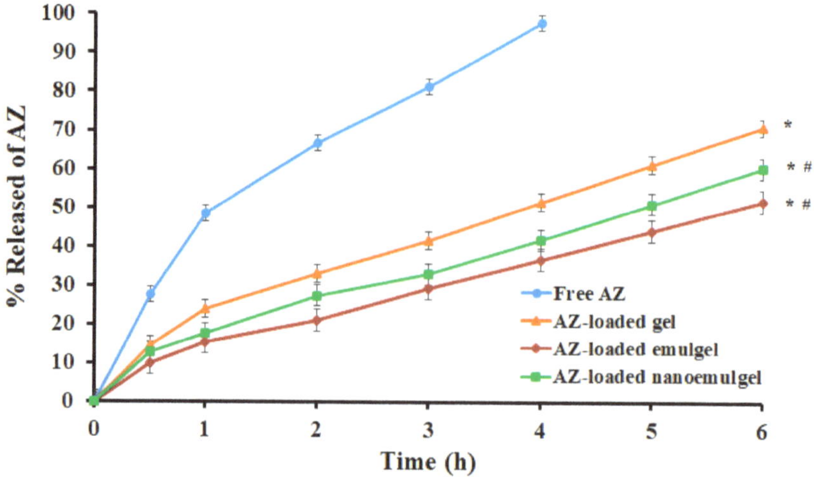

Figure 3. In vitro release study of AZ from different AZ-loaded formulations. # compared to AZ-loaded gel, * compared to free AZ ($p < 0.05$).

3.8. Study of Stability

The physical characteristics and the percent of AZ released from all preparations under investigation were calculated during a 6-month period at 25 °C and 4 °C; the findings are demonstrated in Table 3 and Figure 4. Our findings demonstrated insignificant variation in the homogeneity, color, viscosity, extrudability, or spreadability after six months of storage at either temperature. Furthermore, compared to the corresponding fresh preparations, there were no noticeable variations ($p < 0.05$) in the released amount of AZ from the preserved formulations at either 25 °C or 4 °C (Figure 4). In fact, HPMC used for gel preparation may be responsible for the stability of the AZ-loaded gel. Elmataeeshy et al. reported similar results and claimed that thickening agents are essential for the stability of the preparation [36].

Table 3. Physical properties of AZ-loaded formulations after storage for 6 months at different storage conditions.

Condition	At 4 °C			At 25 °C		
Formulation	Color and Homogeneity	Spreadability (mm)	Viscosity (cp)	Color and Homogeneity	Spreadability (mm)	Viscosity (cp)
AZ-loaded gel	White homogenous	60.67 ± 1.53	72,100 ± 1418	White homogenous	62.67 ± 1.15	71,000 ± 1127
AZ-loaded emulgel	White homogenous	53.33 ± 0.58 *	100,100 ± 1386 *	White homogenous	54.33 ± 0.58 *	98,600 ± 964 *
AZ-loaded nanoemulgel	White homogenous	44.00 ± 1.73 *,#	85,100 ± 1039 *,#	White homogenous	43.67 ± 1.15 *,#	83,500 ± 361 *,#

compared to AZ-loaded emulgel; * compared to AZ-loaded gel ($p < 0.05$).

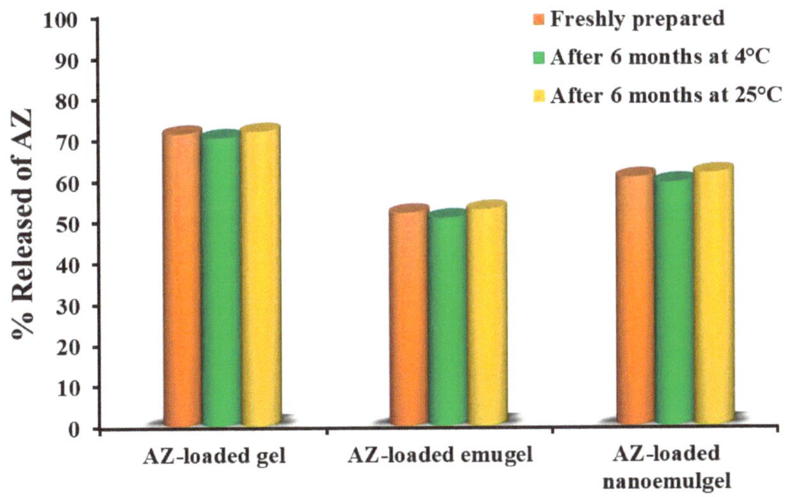

Figure 4. Study of the AZ in vitro release from different formulations upon storage for 6 months at different storage conditions.

3.9. Ex Vivo Study

Figure 5 estimates the skin permeation of AZ from different topical formulations compared to AZ suspension permeation. It was obtained that the cumulative amount of AZ (μg/cm^2) penetrated from nanoemulgel through rabbit skin was significantly larger compared to the AZ-loaded emulgel, AZ-loaded gel, or AZ suspension (Figure 5). The SSTF flux (J_{ss}) of the AZ suspension was noticeably lower than those of the other generated formulations under evaluation, as indicated in Table 4. Contrarily, gel formulation increases AZ permeation by 1.72 folds with a J_{ss} of 148.04 ± 6.09 μg/cm^2.h compared to free drugs. This increase could be related to the gel's colloidal characteristics [37]. Interestingly, nanoemulgel formulation could significantly ($p < 0.05$) enhance the skin permeability properties of AZ compared to AZ-loaded gel and emulgel since it had the largest SSTF value (206.68 ± 7.99 μg/cm^2.h) and the largest ER (2.41 ± 0.17). Actually, the AZ flux from emulgel and nanoemulgel might be improved by the addition of the penetration enhancer (oil) and surfactant. Additionally, the external water phase may hydrate the stratum corneum, cause cell swelling, and facilitate the drug transport [21]. Moreover, the improved permeation of nanoemulgel containing AZ could related to the nano-scaled particles, which provide a greater surface area for AZ penetration and facilitate the release of a large portion of the drug [36]. Our findings were consistent with those of Shehata et al., who found that niosomal emulgel considerably increased the skin penetration of insulin compared to insulin solution and niosomal gel [38].

Table 4. Skin permeation characteristics of various AZ-loaded preparations after ex vivo studies.

Formula	SSTF μg/cm^2.h	ER
Free AZ	85.93 ± 5.29	1
AZ-loaded gel	148.04 ± 6.09 *	1.72 ± 0.05 *
AZ-loaded emulgel	176.95 ± 4.38 *,#	2.06 ± 0.11 *,#
AZ-loaded nanoemulgel	206.68 ± 7.99 *,#,$	2.41 ± 0.17 *,#,$

Compared to AZ-loaded gel. * compared to AZ suspension. $ compared to AZ-loaded emulgel ($p < 0.05$).

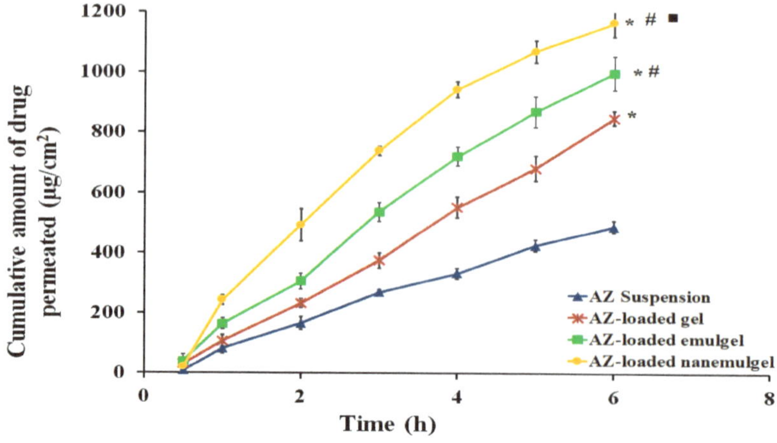

Figure 5. Permeability pattern of AZ from various AZ-loaded formulations. # compared to AZ-loaded gel. * compared to AZ suspension. ■ compared to AZ-loaded emulgel ($p < 0.05$).

3.10. Investigation of Antibacterial Properties

Figure 6 displays the results of the antibacterial activity of the generated formulations against *Escherichia coli* (Gram-negative bacteria) using the agar diffusion technique by measuring the zones of inhibition. The macrolide azithromycin works by preventing the production of proteins in bacterial cells [39]. Azithromycin suppresses protein production by reversibly trussing to 50S ribosomal subunits of affected microorganisms, which effectively prevents the growth of bacteria. Azithromycin has the capacity to penetrate bacterial outer membranes, which is a kind of bacterial self-defense [39].

The results showed that AZ-loaded nanoemulgel was effective against *Escherichia coli* and that it produced a zone of inhibition that was significantly greater than that produced by placebo nanoemulgel and the AZ-loaded gel formulation ($p < 0.05$). The AZ-loaded nanoemulgel displayed zones of inhibition of 16 ± 1.8 mm (after 12 h) and 22 ± 1.7 mm (after 24 h) against *E.coli*. The zones of inhibition against *E.coli* for the AZ-loaded gel were 10 ± 1.3 mm (twelve hours) and 15 ± 1 mm (twenty-four hours). Furthermore, the placebo formulation (nanoemulgel free from AZ) showed that the inhibition zones against *E.coli* were 6 ± 0.5 mm and 8 ± 1 mm after twelve hours and twenty-four hours, respectively. It is interesting that the placebo nanoemulgel formulation with tea tree oil demonstrated a considered bacterial growth inhibition, which was undoubtedly caused by the antibacterial effects of the tea tree oil. The use of tea tree oil, which has antibacterial properties, may have contributed to the fact that the placebo formulation (nanoemulgel free from AZ) showed a certain suppression of bacterial growth. Moreover, the higher antibacterial activity of AZ-loaded nanoemulgel may be due to the antibacterial synergism between AZ and tea tree oil. The action of terpinen-4-ol compound, the main constituent in tea tree oil, may be responsible for the observed antibacterial efficacy of the placebo formulation [40]. It has been demonstrated that tea tree oil prevents respiration in the cells of *E. coli*, and by rupturing the microbial membrane's permeability, the oil inhibits bacterial cell growth and causes the cells to die [41]. According to previous studies, tea tree oil combined with different antibiotics, such as doxycycline [42] and neomycin [43], exhibits additional efficiency against *E. coli*. These conclusions were confirmed by the results of our current investigation, which also demonstrated the antibacterial efficiency of tea tree oil combined with Azithromycin.

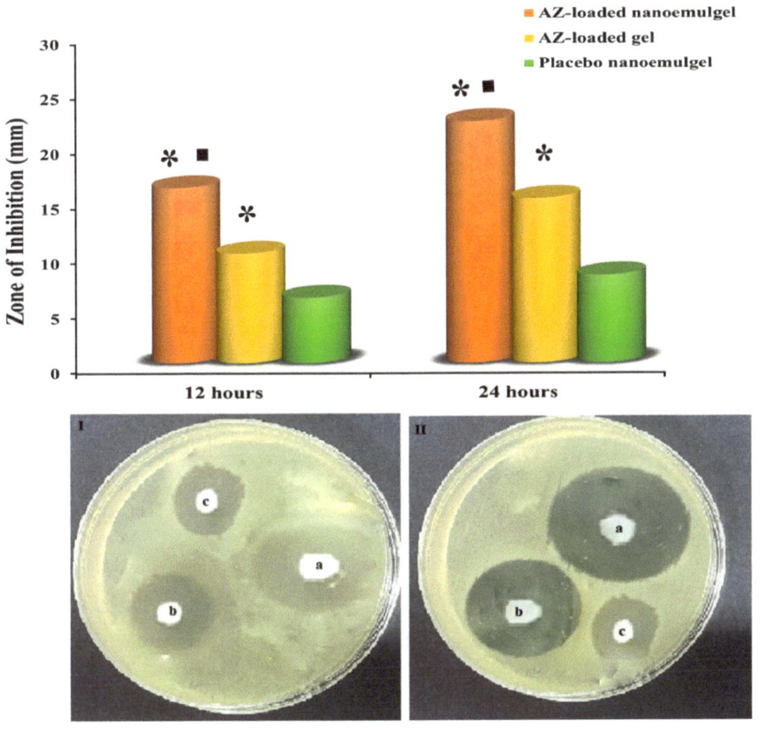

Figure 6. Antimicrobial assessment of (a) AZ-loaded nanoemulgel, (b) AZ-loaded gel, and (c) placebo nanoemulgel against *E.coli* after (**I**) 12 and (**II**) 24 h. ■ versus AZ-loaded gel ($p < 0.05$); and * versus placebo nanoemulgel ($p < 0.05$).

4. Conclusions

In the current study, a nanoemulgel formulated with azithromycin was evaluated for its ability to reduce bacterial growth. The developed azithromycin-loaded nanoemulgel displayed favorable physical properties, viscosity, pH, extrudability, spreadability, and particle size to be applied topically. Additionally, compared to AZ-emulgel or AZ-gels, AZ-nanoemulgel demonstrated better antibacterial efficiency. Therefore, the current study may offer a novel method for the efficient therapy of bacterial infection using azithromycin and a natural product (tea tree oil) formulated into nanoemulgel.

Author Contributions: M.H.A., conceptualization, methodology, writing—review and editing, and supervision; H.A.E., N.E.K. and W.M.A.K., software, data curation, validation, writing—review and editing; E.-S.K., S.S. and H.E.-S.E.-H., methodology, software and writing—original draft preparation, formal analysis, investigation. All authors have read and agreed to the published version of the manuscript.

Funding: This research has been funded by Scientific Research Deanship at the University of Ha'il, Saudi Arabia, through project number RG-22 022.

Institutional Review Board Statement: Our study has been approved by the Research Ethics Committee (REC) at the University of Hail, No. (H-2022-291), dated 13 June 2022.

Informed Consent Statement: Not applicable.

Data Availability Statement: Not applicable.

Acknowledgments: The authors thank Scientific Research Deanship at the University of Ha'il, Saudi Arabia, for funding this research through project number RG-22 022.

Conflicts of Interest: The authors declare no conflict of interest.

References

1. Kibret, M.; Abera, B. Antimicrobial susceptibility patterns of *E. coli* from clinical sources in northeast Ethiopia. *Afr. Health Sci.* **2011**, *11*, 40–45. [CrossRef] [PubMed]
2. Rai, M.K.; Deshmukh, S.D.; Ingle, A.P.; Gade, A.K. Silver nanoparticles: The powerful nanoweapon against multidrug-resistant bacteria. *J. Appl. Microbiol.* **2012**, *112*, 841–852. [CrossRef]
3. McMullan, B.J.; Mostaghim, M. Prescribing azithromycin. *Aust. Prescr.* **2015**, *38*, 87–89. [CrossRef] [PubMed]
4. Imperi, F.; Leoni, L.; Visca, P. Antivirulence activity of azithromycin in *Pseudomonas aeruginosa*. *Front. Microbiol.* **2014**, *5*, 178. [CrossRef] [PubMed]
5. Heidary, M.; Ebrahimi Samangani, A.; Kargari, A.; Kiani Nejad, A.; Yashmi, I.; Motahar, M.; Taki, E.; Khoshnood, S. Mechanism of action, resistance, synergism, and clinical implications of azithromycin. *J. Clin. Lab. Anal.* **2022**, *36*, e24427. [CrossRef] [PubMed]
6. Kong, F.Y.S.; Horner, P.; Unemo, M.; Hocking, J.S. Pharmacokinetic considerations regarding the treatment of bacterial sexually transmitted infections with azithromycin: A review. *J. Antimicrob. Chemother.* **2019**, *74*, 1157–1166. [CrossRef] [PubMed]
7. Al-Saedi, Z.H.F.; Salih, Z.T.; Ahmed, K.K.; Ahmed, R.A.; Jasim, S.A. Formulation and Characterization of Oleogel as a Topical Carrier of Azithromycin. *AAPS PharmSciTech* **2022**, *24*, 17. [CrossRef]
8. Helal, I.M.; El-Bessoumy, A.; Al-Bataineh, E.; Joseph, M.R.P.; Rajagopalan, P.; Chandramoorthy, H.C.; Ben Hadj Ahmed, S. Antimicrobial Efficiency of Essential Oils from Traditional Medicinal Plants of Asir Region, Saudi Arabia, over Drug Resistant Isolates. *BioMed Res. Int.* **2019**, *2019*, 8928306. [CrossRef]
9. Yasin, M.; Younis, A.; Ramzan, F.; Javed, T.; Shabbir, R.; Noushahi, H.A.; Skalicky, M.; Ondrisik, P.; Brestic, M.; Hassan, S. Extraction of essential oil from river tea tree (*Melaleuca bracteata* F. Muell.): Antioxidant and antimicrobial properties. *Sustainability* **2021**, *13*, 4827. [CrossRef]
10. Carson, C.F.; Hammer, K.A.; Riley, T.V. *Melaleuca alternifolia* (tea tree) oil: A review of antimicrobial and other medicinal properties. *Clin. Microbiol. Rev.* **2006**, *19*, 50–62. [CrossRef]
11. Yasin, M.; Younis, A.; Javed, T.; Akram, A.; Ahsan, M.; Shabbir, R.; Ali, M.M.; Tahir, A.; El-Ballat, E.M.; Sheteiwy, M.S. River tea tree oil: Composition, antimicrobial and antioxidant activities, and potential applications in agriculture. *Plants* **2021**, *10*, 2105. [CrossRef] [PubMed]
12. Prausnitz, M.R.; Langer, R. Transdermal drug delivery. *Nat. Biotechnol.* **2008**, *26*, 1261–1268. [CrossRef]
13. Van Hoogevest, P.; Liu, X.; Fahr, A. Drug delivery strategies for poorly water-soluble drugs: The industrial perspective. *Expert Opin. Drug Deliv.* **2011**, *8*, 1481–1500. [CrossRef] [PubMed]
14. Peña-Juárez, M.C.; Guadarrama-Escobar, O.R.; Escobar-Chávez, J.J. Transdermal delivery systems for biomolecules. *J. Pharm. Innov.* **2021**, *17*, 319–332. [CrossRef] [PubMed]
15. Isaac, M.; Holvey, C. Transdermal patches: The emerging mode of drug delivery system in psychiatry. *Ther. Adv. Psychopharmacol.* **2012**, *2*, 255–263. [CrossRef] [PubMed]
16. Mao, Y.; Chen, X.; Xu, B.; Shen, Y.; Ye, Z.; Chaurasiya, B.; Liu, L.; Li, Y.; Xing, X.; Chen, D. Eprinomectin nanoemulgel for transdermal delivery against endoparasites and ectoparasites: Preparation, in vitro and in vivo evaluation. *Drug Deliv.* **2019**, *26*, 1104–1114. [CrossRef]
17. Malavi, S.; Kumbhar, P.; Manjappa, A.; Chopade, S.; Patil, O.; Kataria, U.; Dwivedi, J.; Disouza, J. Topical Emulgel: Basic Considerations in Development and Advanced Research. *Indian J. Pharm. Sci.* **2022**, *84*, 1105–1115. [CrossRef]
18. Choudhury, H.; Gorain, B.; Pandey, M.; Chatterjee, L.A.; Sengupta, P.; Das, A.; Molugulu, N.; Kesharwani, P. Recent Update on Nanoemulgel as Topical Drug Delivery System. *J. Pharm. Sci.* **2017**, *106*, 1736–1751. [CrossRef]
19. Chen, W.-H.; Chen, Q.-W.; Chen, Q.; Cui, C.; Duan, S.; Kang, Y.; Liu, Y.; Liu, Y.; Muhammad, W.; Shao, S. Biomedical polymers: Synthesis, properties, and applications. *Sci. China Chem.* **2022**, *65*, 1010–1075. [CrossRef]
20. Abdallah, M.H.; Lila, A.S.A.; Unissa, R.; Elsewedy, H.S.; Elghamry, H.A.; Soliman, M.S. Brucine-Loaded Ethosomal Gel: Design, Optimization, and Anti-inflammatory Activity. *AAPS PharmSciTech* **2021**, *22*, 269. [CrossRef]
21. Abdallah, M.H.; Elghamry, H.A.; Khalifa, N.E.; Khojali, W.M.; Khafagy, E.-S.; Shawky, S.; El-Horany, H.E.-S.; El-Housiny, S. Development and Optimization of Erythromycin Loaded Transethosomes Cinnamon Oil Based Emulgel for Antimicrobial Efficiency. *Gels* **2023**, *9*, 137. [CrossRef] [PubMed]
22. Abdallah, M.H.; Elghamry, H.A.; Khalifa, N.E.; Khojali, W.M.; Khafagy, E.-S.; Lila, A.S.A.; El-Horany, H.E.-S.; El-Housiny, S. Ginger Extract-Loaded Sesame Oil-Based Niosomal Emulgel: Quality by Design to Ameliorate Anti-Inflammatory Activity. *Gels* **2022**, *8*, 737. [CrossRef]
23. Abdallah, M.H.; Sabry, S.A.; Hasan, A.A. Enhancing transdermal delivery of glimepiride via entrapment in proniosomal gel. *J. Young Pharm.* **2016**, *8*, 335. [CrossRef]
24. Ali, A.; Ali, A.; Rahman, M.A.; Warsi, M.H.; Yusuf, M.; Alam, P. Development of nanogel loaded with lidocaine for wound-healing: Illustration of improved drug deposition and skin safety analysis. *Gels* **2022**, *8*, 466. [CrossRef]
25. Wang, J.-W.; Chen, Q.-W.; Luo, G.-F.; Ji, P.; Han, Z.-Y.; Song, W.-F.; Chen, W.-H.; Zhang, X.-Z. Interference of Glucose Bioavailability of Tumor by Engineered Biohybrids for Potentiating Targeting and Uptake of Antitumor Nanodrugs. *Nano Lett.* **2022**, *22*, 8735–8743. [CrossRef]

26. Thakur, A.; Jain, S.; Pant, A.; Sharma, A.; Kumar, R.; Singla, N.; Suttee, A.; Kumar, S.; Barnwal, R.P.; Katare, O.P. Cyclodextrin derivative enhances the ophthalmic delivery of poorly soluble Azithromycin. *ACS Omega* **2022**, *7*, 23050–23060. [CrossRef] [PubMed]
27. Abdelnabi, D.M.; Abdallah, M.H.; Elghamry, H.A. Buspirone hydrochloride loaded in situ nanovesicular gel as an anxiolytic nasal drug delivery system: In vitro and animal studies. *AAPS PharmSciTech* **2019**, *20*, 134. [CrossRef]
28. Zafar, A.; Imam, S.S.; Yasir, M.; Alruwaili, N.K.; Alsaidan, O.A.; Warsi, M.H.; Mir Najib Ullah, S.N.; Alshehri, S.; Ghoneim, M.M. Preparation of NLCs-Based Topical Erythromycin Gel: In Vitro Characterization and Antibacterial Assessment. *Gels* **2022**, *8*, 116. [CrossRef]
29. Jain, A.; Jain, R.; Jain, S.; Jain, A.; Jain, R.; Jain, S. Sterilization of Glassware; Preparation and Sterilization of Media. In *Basic Techniques in Biochemistry, Microbiology and Molecular Biology: Principles and Techniques*; Springer: Berlin/Heidielberg, Germany, 2020; pp. 93–99.
30. Bonev, B.; Hooper, J.; Parisot, J. Principles of assessing bacterial susceptibility to antibiotics using the agar diffusion method. *J. Antimicrob. Chemother.* **2008**, *61*, 1295–1301. [CrossRef]
31. Hiremath, R. Evaluation of antimicrobial activity of Rasaka Bhasma. *AYU Int. Q. J. Res. Ayurveda* **2010**, *31*, 260–262. [CrossRef]
32. Morsy, M.A.; Abdel-Latif, R.G.; Nair, A.B.; Venugopala, K.N.; Ahmed, A.F.; Elsewedy, H.S.; Shehata, T.M. Preparation and evaluation of atorvastatin-loaded nanoemulgel on wound-healing efficacy. *Pharmaceutics* **2019**, *11*, 609. [CrossRef] [PubMed]
33. Abdallah, M.H.; Lila, A.S.A.; Anwer, M.K.; Khafagy, E.-S.; Mohammad, M.; Soliman, S.M. Formulation, Development and Evaluation of Ibuprofen Loaded Nano-transferosomal Gel for the Treatment of Psoriasis. *J. Pharm. Res.* **2019**, *31*, 1–8. [CrossRef]
34. Tungadi, R.; Wicita, P. Formulation, optimization, and characterization of snakehead fish (*Ophiocephalus striatus*) powder nanoemulgel. *Braz. J. Pharm. Sci.* **2020**, *56*. [CrossRef]
35. Shen, Y.; Ling, X.; Jiang, W.; Du, S.; Lu, Y.; Tu, J. Formulation and evaluation of Cyclosporin A emulgel for ocular delivery. *Drug Deliv.* **2015**, *22*, 911–917. [CrossRef]
36. Elmataeeshy, M.E.; Sokar, M.S.; Bahey-El-Din, M.; Shaker, D.S. Enhanced transdermal permeability of Terbinafine through novel nanoemulgel formulation; Development, in vitro and in vivo characterization. *Future J. Pharm. Sci.* **2018**, *4*, 18–28. [CrossRef]
37. Shah, H.; Nair, A.B.; Shah, J.; Bharadia, P.; Al-Dhubiab, B.E. Proniosomal gel for transdermal delivery of lornoxicam: Optimization using factorial design and in vivo evaluation in rats. *DARU J. Pharm. Sci.* **2019**, *27*, 59–70. [CrossRef]
38. Shehata, T.M.; Nair, A.B.; Al-Dhubiab, B.E.; Shah, J.; Jacob, S.; Alhaider, I.A.; Attimarad, M.; Elsewedy, H.S.; Ibrahim, M.M. Vesicular emulgel based system for transdermal delivery of insulin: Factorial design and in vivo evaluation. *Appl. Sci.* **2020**, *10*, 5341. [CrossRef]
39. Imamura, Y.; Higashiyama, Y.; Tomono, K.; Izumikawa, K.; Yanagihara, K.; Ohno, H.; Miyazaki, Y.; Hirakata, Y.; Mizuta, Y.; Kadota, J.-I. Azithromycin exhibits bactericidal effects on *Pseudomonas aeruginosa* through interaction with the outer membrane. *Antimicrob. Agents Chemother.* **2005**, *49*, 1377–1380. [CrossRef]
40. Wulansari, A.; Jufri, M.; Budianti, A. Studies on the formulation, physical stability, and in vitro antibacterial activity of tea tree oil (*Melaleuca alternifolia*) nanoemulsion gel. *Int. J. Appl. Pharm.* **2017**, *9*, 135–139. [CrossRef]
41. Cox, S.D.; Mann, C.M.; Markham, J.L.; Gustafson, J.E.; Warmington, J.R.; Wyllie, S.G. Determining the antimicrobial actions of tea tree oil. *Molecules* **2001**, *6*, 87–91. [CrossRef]
42. Wei, S.; Tian, Q.; Zhao, X.; Liu, X.; Husien, H.M.; Liu, M.; Bo, R.; Li, J. Tea Tree Oil Nanoemulsion Potentiates Antibiotics against Multidrug-Resistant *Escherichia coli*. *ACS Infect. Dis.* **2022**, *8*, 1618–1626. [CrossRef] [PubMed]
43. Elsewedy, H.S.; Shehata, T.M.; Soliman, W.E. Tea Tree Oil Nanoemulsion-Based Hydrogel Vehicle for Enhancing Topical Delivery of Neomycin. *Life* **2022**, *12*, 1011. [CrossRef] [PubMed]

processes MDPI

Article

Quinine: Redesigned and Rerouted

Chinazom Precious Agbo [1,2,*], Timothy Chukwuebuka Ugwuanyi [1], Osita Christopher Eze [3], Adaeze Linda Onugwu [1], Adaeze Chidiebere Echezona [1], Chinekwu Sherridan Nwagwu [1], Samuel Wisdom Uzondu [1], John Dike Ogbonna [1], Lydia Onyinyechi Ugorji [4], Petra Obioma Nnamani [1], Paul Achile Akpa [1], Joy Nneji Reginald-Opara [1], John Onyebuchi Ogbodo [5], Christopher McConville [2,*], Anthony Amaechi Attama [1,6], Mumuni Audu Momoh [1] and Kenneth Chibuzor Ofokansi [1]

1 Drug Delivery and Nanomedicines Research Laboratory, Department of Pharmaceutics, University of Nigeria, Nsukka 410001, Enugu State, Nigeria; chukwuebuka.ugwuanyi.189655@unn.edu.ng (T.C.U.); adaeze.onugwu@unn.edu.ng (A.L.O.); adaeze.echezona@unn.edu.ng (A.C.E.); chinekwu.nwobi@unn.edu.ng (C.S.N.); samuel.uzondu@unn.edu.ng (S.W.U.); john.ogbonna@unn.edu.ng (J.D.O.); petra.nnamani@unn.edu.ng (P.O.N.); paul.akpa@unn.edu.ng (P.A.A.); joy.achuam@unn.edu.ng (J.N.R.-O.); anthony.attama@unn.edu.ng (A.A.A.); audu.momoh@unn.edu.ng (M.A.M.); kenneth.ofokansi@unn.edu.ng (K.C.O.)

2 School of Pharmacy, College of Medical and Dental Sciences, University of Birmingham, Birmingham B4 6BN, UK

3 Department of Pharmaceutical Microbiology and Biotechnology, Faculty of Pharmaceutical Sciences, University of Nigeria, Nsukka 410001, Enugu State, Nigeria; osy.eze@unn.edu.ng

4 Department of Pharmaceutical Technology and Industrial Pharmacy, Faculty of Pharmaceutical Sciences, University of Nigeria, Nsukka 410001, Enugu State, Nigeria; lydia.ugorji@unn.edu.ng

5 Department of Science Laboratory Technology, University of Nigeria, Nsukka 410001, Enugu State, Nigeria; john.ogbodo@unn.edu.ng

6 Department of Pharmaceutics and Pharmaceutical Technology, Faculty of Pharmaceutical Sciences, Enugu State University of Science and Technology, Agbani, Enugu 400102, Enugu State, Nigeria

* Correspondence: chinazom.agbo@unn.edu.ng (C.P.A.); c.mcconville.2@bham.ac.uk (C.M.); Tel.: +234-8062275858 (C.P.A.); +44-701214143209 (C.M.)

Abstract: Quinine hydrochloride (QHCl) has remained a very relevant antimalarial drug 400 years after its effectiveness was discovered. Unlike other antimalarials, the development of resistance to quinine has been slow. Hence, this drug is to date still used for the treatment of severe and cerebral malaria, for malaria treatment in all trimesters of pregnancy, and in combination with doxycycline against multidrug-resistant malaria parasites. The decline in its administration over the years is mainly associated with poor tolerability due to its gastrointestinal (GIT) side effects such as cinchonism, complex dosing regimen and bitter taste, all of which result in poor compliance. Hence, our research was aimed at redesigning quinine using nanotechnology and investigating an alternative route for its administration for the treatment of malaria. QHCl nanosuspension (QHCl-NS) for intranasal administration was prepared using lipid matrices made up of solidified reverse micellar solutions (SRMS) comprising Phospholipon® 90H and lipids (Softisan® 154 or Compritol®) in a 1:2 ratio, while Poloxamer® 188 (P188) and Tween® 80 (T80) were used as a stabilizer and a surfactant, respectively. The QHCl-NS formulated were in the nanosize range (68.60 ± 0.86 to 300.80 ± 10.11 nm), and highly stable during storage, though zeta potential was low (≤6.95 ± 0.416). QHCl-NS achieved above 80% *in vitro* drug release in 6 h. *Ex vivo* permeation studies revealed that formulating QHCl as NS resulted in a 5-fold and 56-fold increase in the flux and permeation coefficient, respectively, thereby enhancing permeation through pig nasal mucosa better than plain drug solutions. This implies that the rate of absorption as well as ease of drug permeation through porcine nasal mucosa was impressively enhanced by formulating QHCl as NS. Most importantly, reduction in parasitaemia in mice infected with *Plasmodium berghei* ANKA by QHCl-NS administered through the intranasal route (51.16%) was comparable to oral administration (52.12%). Therefore, redesigning QHCl as NS for intranasal administration has great potential to serve as a more tolerable option for the treatment of malaria in endemic areas.

Citation: Agbo, C.P.; Ugwuanyi, T.C.; Eze, O.C.; Onugwu, A.L.; Echezona, A.C.; Nwagwu, C.S.; Uzondu, S.W.; Ogbonna, J.D.; Ugorji, L.O.; Nnamani, P.O.; et al. Quinine: Redesigned and Rerouted. *Processes* **2023**, *11*, 1811. https://doi.org/10.3390/pr11061811

Academic Editors: Paolo Trucillo and Roberto Pisano

Received: 30 April 2023
Revised: 4 June 2023
Accepted: 7 June 2023
Published: 14 June 2023

Keywords: solidified reverse micellar solutions; nanostructured lipid carriers; quinine hydrochloride; intranasal route; severe malaria; cerebral malaria; parasitemia reduction

1. Introduction

The discovery of quinine in the 17th century and its use for malaria treatment was the first successful use of a chemical compound to treat an infectious disease. Unlike other antimalarials, the development of resistance to quinine (QHCl) has been slow, with the first reported case in 1910 [1], hence its use has continued to date. The decline in its administration over the years is mainly associated with poor tolerability due to its gastrointestinal side effects such as cinchonism (nausea, vomiting, dysphoria, tinnitus, and high-tone deafness). Adverse events such as hypoglycemia are also associated with quinine use, especially in pregnancy. Other drawbacks of quinine include its complex dosing regimen and bitter taste, all of which result in poor compliance [1].

Despite these shortcomings of quinine, its use has continued to date because of its defiance to resistance by malaria parasites. Intravenous (IV) quinine still finds use as a second-line treatment for severe malaria, after IV artesunate [2]. Quinine is considered safe for malaria treatment in all trimesters of pregnancy. Quinine in combination with doxycycline is effective against multidrug-resistant parasites [3]. Another important advantage of quinine compared to other antimalarials is its affordability; this is a major deciding factor in most endemic regions which also come under the low- or middle-income countries. For instance, in Nigeria, a complete dose of quinine for a seven-day course costs about $1.2, while a three-day treatment using artemether/lumefantrine (Coartem®) costs about $2.5.

Artemisinin-based combination therapy (ACT), which has become the mainstay for the treatment of uncomplicated malaria, is bedeviled with challenges including the development of resistance and high cost in resource-limited settings. Less than 50 years after its discovery in the 1970s, artemisinin resistance was first reported in 2009. Since then, a good number of reports have continued to surface of the malaria parasite's resistance to artemisinins [4]. There is therefore a great need to discover new, potent therapies for malaria or repurpose existing drugs [5].

Hence, our research was aimed at redesigning quinine using nanotechnology and investigating an alternative route for its administration for the treatment of malaria. To the best of our knowledge, no research investigating the intranasal administration of quinine for the treatment of malaria has been published. However, quinine nasal spray is being investigated for the prevention of infection by the SARS-CoV-2 virus [6]. In this study, a nanosuspension of quinine was formulated to suit intranasal administration.

The nasal cavity is highly vascularized with a relatively large surface area (96 m^2), making it an attractive route for drug administration [7]. Therefore, this route of drug administration can be exploited for drugs such as quinine, which are beset with numerous GIT side effects. The relatively reduced surface area of the nasal cavity compared to the GIT may reduce the rate of saturation of the systemic circulation with quinine, thereby preventing side effects such as cinchonism, which is dependent on the concentration of quinine in circulation [8]. The intranasal route also offers the advantage of ease of application and painless self-administration, just like the oral route. Unlike the oral route, drugs administered through this route can avoid first-pass metabolism, as well as achieve direct brain access through the olfactory region [9]. A previous study targeting intravenously administered quinine to the brain by means of conjugation to Transferrin, aimed at reducing the serum concentration, toxicity and improving efficacy in the treatment of cerebral malaria has been reported [10]. Other side effects that may be avoided by administering quinine through the nasal route include hypotension (which may occur when the drug is given too rapidly, e.g., through intravenous injection), venous thrombosis (this is usual with intravenous administration), pain and sterile abscesses at the site of injection (which happens when intramuscularly administered) [1]. Furthermore, delivery of quinine into the brain through the nasal route in cerebral malaria patients would rapidly eradicate the parasite, averting the initiation of vascular occlusion and disseminated intravascular coagulation and neurological damage.

The rectal route is yet another route which may offer a slower administration of quinine, in addition to avoidance of GIT exposure; however, a negative cultural perception of the rectal route may limit its use [11].

Nanotechnology has been employed innovatively [12] for the revival of drugs and enhancement of their therapeutic potentials, in addition to the reduction of side-effect profiles [13], and enhancement of solubility of poorly water-soluble entities [14]. A good number of antiparasitic drugs such as halofantrine, primaquine, atovaquone as well as quinine [15] have been prepared as nanoformulations in order to improve drug solubility, enhance drug release, and achieve specific drug targeting. The designing of quinine as a nanocapsule was found to increase interaction between quinine and the erythrocyte, resulting in an increase in its *in vivo* efficacy in malaria-infected rats [15]. It is based on this premise that a nanosuspension of quinine was designed to suite intranasal administration.

In this study, the ability of the redesigned and rerouted quinine to treat malaria was investigated by formulating nanosuspension of quinine using different lipids. Characterization of NS was performed, after which QHCl-NS were administered intranasally to mice infected with *Plasmodium berghei* ANKA. The results obtained were compared with orally administered quinine and plain unprocessed quinine solution in water.

2. Materials and Methods

2.1. Materials

QHCl (purity $\geq$ 95%) was obtained from Sigma-Aldrich (England, UK). Softisan® 154 (S154) and Miglyol® 812 N (medium-chain triglycerides (MCT)) were gifts from IOI Oleo GmbH, (Witten, Germany). Compritol® 888 ATO (C888) (Glycerol dibehenate), Compritol® HD 5 ATO (CHD 5) (Behenoyl polyoxyl-8 glycerides), and Transcutol® HP (THP) (Diethylene glycol monoethyl ether) were also gifts from Gattefossé SAS (Saint-Priest Cedex, France). Free samples of Phospholipon® 90H (P90H) were obtained from Lipoid GmbH (Ludwigshafen, Germany). Stearic acid, Poloxamer® 188 (P188) and Tween® 80 (T80) were purchased from Sigma-Aldrich (England, UK). Ultrapure water was sourced from ELGA Purelab Ultra Genetic Water Purification System (UK). All other reagents were of analytical grade and purchased from standard commercial suppliers.

2.2. Methods

2.2.1. Determination of QHCl Solubility in Solid and Liquid Lipids

The solubility of QHCl in solid lipids (stearic acid, S154, CHD 5, and C888) as well as liquid lipids (Miglyol® 812 N and glyceryl monooleate) and the solvent (Transcutol® HP), were determined using a modified version of the shake-flask technique [16]. A 1:2 and 1:3 (w/w) ratio of the drug to lipid were prepared in each case and vortexed for 2 min. For the solid lipids, drugs were added to melted lipid (after 15 min heating at 10 °C above their melting temperatures). The solid and liquid lipids were selected for the formulation based on the transparency of the drug–lipid mixture [16].

2.2.2. Formulation of SRMS Lipid Matrices

The SRMS lipid matrices consisting of 1:2 (30:70) mixtures of P90H and either of the lipids: S154, or CHD 5, or C888, were prepared by fusion method [17]. Each of the lipids and P90H were weighed using an electronic balance (Mettler Toledo-X A204 Delta range), melted together on a hot plate (BioCote® Stuart hotplate stirrer) at 70 °C and stirred with the help of a magnetic bead at 200 rpm. The temperature was reduced to room temperature after melting, and mixing was completed to allow for solidification of lipid matrices. These were gently scraped out of the beakers and stored in air-tight glass bottles.

2.2.3. Preparation of QHCl-NS

QHCl-loaded NS were prepared by the combined ultrasonication technique, first with a cup-horn sonicator, and then a probe sonicator [9]. SRMS lipid matrix (3%) and MCT served as the solid and liquid lipids, respectively, while T80 (2, 3, or 5%) was the surfactant. P188 (1%) and sorbitol (5%) served as the stabilizer and cryoprotectant, respectively. First, melting of the solid lipid was done at 70 °C on a magnetic hot plate (BioCote® Stuart hotplate stirrer); subsequently, QHCl was dissolved in a mixture of THP and MCT and then added to the melted solid lipid matrix and mixed thoroughly at 300 rpm. An aqueous surfactant solution was added to the melted oil phase in drops with the help of a burette [18]. The aqueous surfactant solution contained T80, P188, and sorbitol and was heated to the same temperature as the lipid phase. The mixing of the two phases was carried out briefly on a magnetic stirrer at 70 °C before sonication was performed in a cup-horn sonicator for 30, 60, or 90 min at 100 amp at 43 °C. Additional sonication was achieved with a probe sonicator for 3 min at 70 °C. The same procedure was repeated for NS containing no drug (blanks), which served as placebos. All the NS were divided into two, one part was freeze-dried (Labconco FreeZone 6 Plus, Kansas City, MO, USA) at −80 °C and at a pressure of 0.01 mmHg, while the other portion was used for some characterizations and stability studies. The compositions of QHCl-NS and the cup-horn sonication times of different formulations are stated in Table 1.

Table 1. Quantities of materials used for the formulation of QHCl-NS.

QHCl-NS	Lipid Type	Concentration (%)							Cup-Horn Sonication Time (min)
		QHCl	Solid Lipid	MCT	THP	T80	P188	Sorbitol	
Q1	S154	1.5	3	1.5	1.5	3	1	5	30
Q3	S154	1.5	3	1.5	1.5	3	1	5	90
Q9	S154	1.5	3	1.5	1.5	2	1	5	60
Q10	S154	1.5	3	1.5	1.5	5	1	5	60
Q13	CHD 5	1.5	3	1.5	1.5	3	1	5	60
Q14	CHD 5	1.5	3	1.5	1.5	3	1	5	60
Q15	CHD 5	1.5	3	1.5	1.5	3	1	5	60
Q5	CHD 5	1.5	3	1.5	1.5	2	1	5	30
Q6	CHD 5	1.5	3	1.5	1.5	5	1	5	30
Q7	CHD 5	1.5	3	1.5	1.5	2	1	5	90
Q8	CHD 5	1.5	3	1.5	1.5	5	1	5	90
Q2	C888	1.5	3	1.5	1.5	3	1	5	30
Q4	C888	1.5	3	1.5	1.5	3	1	5	90
Q11	C888	1.5	3	1.5	1.5	2	1	5	60
Q12	C888	1.5	3	1.5	1.5	5	1	5	60

Key: QHCl = Quinine hydrochloride, **S154** = Softisan® 154, **C888** = Compritol® 888 ATO, **CHD 5** = Compritol® HD 5 ATO, **MCT** = Medium-Chain Triglyceride, **THP** = Transcutol® HP, **T80** = Tween® 80, **P188** = Poloxamer® 188.

2.2.4. Characterization of QHCl-NS

Determination of Particle Sizes, Polydispersity Indices, and Zeta (ζ) Potentials of NS

Particle size, polydispersity indices, and zeta potential determination was done using a nanosizer (Malvern-Zetasizer Nano series, Malvern Panalytical, Malvern, United Kingdom) immediately after formulation, as well as 30 and 90 days post formulation. Determination was carried out at 25 °C after dilution, and three replicates were obtained for each measurement. Data are presented as mean ± standard deviation (SD) [19].

Determination of the Morphology of QHCl-NS

(1) Transmission Electron Microscopy (TEM) of QHCl

A transmission electron microscope (JEOL JEM-1400 USA, Inc., Peabody, MA, USA) was used in the determination of the morphology of the QHCl-NS. Appropriate dilution of QHCl-NS (a 100 μL volume of each preparation diluted to 30 mL) using deionized water was done before placing NS on a 300-mesh carbon-coated copper grid (FC300Cu, Formvar/Carbon film on Copper). A drying time of two minutes was allowed to enable NS to adhere to the carbon substrate. Negative staining with 2% uranyl acetate and subsequent air-drying at room temperature for 1 min was carried out before observation was performed using TEM at 300 k magnification and voltage of 120 kV [9].

(2) Cryo-Scanning Electron Microscopy (cryo-SEM) and Field Emission SEM (FESEM) of NS

Morphological characterization was also done using cryo-SEM and field emission SEM. A drop of appropriately diluted QHCl-NS was placed on metal stubs and rapidly frozen with slush nitrogen until −210 °C and sublimated at −90 °C for 90 s. Coating was achieved with palladium under vacuum, and samples were fractured, transferred to the chamber, and examined using FESEM (Philips XL30 FEG ESEM). The observations were done at −150 °C [20], while micrographs were obtained at an accelerated voltage of 5 kV at a pressure of 0.6 mmHg [21]. Morphology of the dry samples (such as excipients, pure drug samples of quinine as well as freeze-dried NS) was carried out using normal FESEM. Samples were placed on double-sided tapes stuck to an aluminum stud. The samples were coated for three minutes with gold using a sputter coater (EMS 7620 Mini Sputter Coater/Glow Discharge System) with deposition control adjusted to 25 mA. Afterwards, the coated samples were loaded into FESEM (Philips XL30 FEG ESEM) and micrographs were obtained at an accelerated voltage of 15 kV, at a pressure of 0.6 mmHg [19].

Fourier Transform Infrared (FTIR) Spectroscopic Analysis

Fourier transform infrared spectroscopy (ATR-FTIR) of the freeze-dried QHCl-NS and excipients was performed on a PerkinElmer Spectrum fitted with a universal ATR sampling accessory and recorded in the range of 4000 to 650 cm^{-1} at ambient temperature. Evaluations were performed using a 16-scan-per-sample cycle and a fixed universal compression force of 80N. Subsequent analyses were carried out using spectrum software [17,22].

Thermal Analysis

Thermal analyses of freeze-dried samples of QHCl-NS, drug, physical mixtures of lipids and drug and other excipients were conducted using a differential scanning calorimeter (DSC) (TA Instruments, DSC-25, U.K.). Samples were weighed into aluminum pans, hermetically sealed, and determinations were performed at a heating rate of 10 °C/min over a temperature range of 10–200 °C under nitrogen purge at a flow rate of 80 cm^3/min. Data analyses, and the determination of melting points and enthalpies were done using TA Rheology Data Analysis software (TA Instruments Trios Version 4.1.33073, UK) [23].

Powder X-ray Diffractometry (XRD)

XRD patterns of drug, excipients, and QHCl-NS were obtained using a diffractometer (SIEMENS/BRUKER D5005, Munich, Germany). The samples were exposed to Cu-Kα radiation (40 kV, 35 mA), and measurements were carried out at room temperature. Scanning was carried out at 2θ from 5° to 90°, for 30 min (for excipients) or 1 h (for NS) [9].

Time-Dependent pH Stability Studies of NS

The pH of QHCl-NS as well as blanks were analysed in triplicate, and at room temperature on day 1, 30, and 90 using a pH meter (pHep® Hanna Instruments, Italy) [24].

Osmolality Determination

The osmolality determinations of NS were done using an Osmometer (Osmomat® 3000, Gonotec, Germany). A 100 µL quantity of each formulation was used, and the average of three determinations was obtained [25].

2.2.5. Solubility Analysis of QHCl in Simulated Nasal Fluid and Alcoholic Buffer

A concentration above the expected solubility level of QHCl (8%) was prepared in simulated nasal fluid (SNF) (8.77 g NaCl, 2.98 g KCl, and 0.59 g $CaCl_2$ per 1000 mL of demineralised water, pH of 6.4) [9]. Suspensions were kept in an incubator/shaker (Multitron Infors HT, Bottmingen, Switzerland) maintained at 37 ± 0.5 °C with continuous stirring at 100 rpm. After 24 h, samples were carefully withdrawn and filtered through 0.45 µm nylon syringe filters, diluted 1000 times, and analysed using HPLC [26].

2.2.6. In Vitro Release Studies of QHCl-NS

Release studies were conducted in SNF. Amounts of freeze-dried QHCl-NS containing 5 mg of the drugs were placed in polycarbonated dialysis membrane (length: 6 cm; pore size: 2.4 nm; molecular weight cut-off: 12, 000–14, 000 Da) which was previously soaked overnight in distilled water prior to the procedure, and immersed in 100 mL of release media [9,17]. The set-up was maintained at 37.0 ± 0.5 °C and stirred at 60 rpm for 24 h. At predetermined time intervals, a 1 mL portion of the dissolution media was withdrawn for HPLC analysis and replaced with 1 mL of fresh medium. The amount of drug released at each time point was calculated with reference to the relevant calibration plot. *In vitro* dissolutions of pure QHCl were also carried out following the same method. Each procedure was performed in triplicate. Kinetic evaluation of release profiles was done by fitting the data into zero-order, first-order, and Higuchi square root models [18]. The Korsmeyer–Peppas equation was also taken into consideration for determination of the release mechanism [27]. For the Korsmeyer–Peppas model, only the data points with less than 60% release were used for model fitting [28].

2.2.7. Ex Vivo Permeation Studies of QHCl-NS

Franz diffusion cells were used for the determination of the permeation efficacy of QHCl-loaded NS across porcine nasal mucosa from one selected formulation (Q9). Diffusion cells with a surface area of 6.6 cm^2 and volume of 10 mL were used. Appropriate volumes of deaerated SNF (pH 6.4) were placed in each of the Franz diffusion cells and allowed to equilibrate at 37.0 ± 0.5 °C for 15 min on a heated magnetic block. Fresh nasal pig tissue was obtained at the local slaughterhouse with prior permission from concerned authorities at the abattoir. Moreover, all animal experimental protocols were carried out in accordance with the guidelines of the Animal Ethics Committee of the University of Nigeria, Nsukka, and EU Directive 2010/63/EU for animal experiments. The nasal mucosa was detached from the septum, connective, and cartilaginous tissues and stored in phosphate-buffered saline (pH 6.4) during transportation. During each procedure, the nasal mucosa was individually placed on the Franz diffusion cells and clamped between the donor and receptor compartments, and sealed with Parafilm® to prevent loss of moisture during the 3 h period. Before sealing was done, amounts of Q9 NS containing 0.5 mg of QHCl were placed on the membrane surface on the donor compartment, and stirring was done at 600 rpm. A gel-loading pipette tip attached to a micropipette was used to withdraw 600 µL of the media from the receptor compartment at predetermined time intervals and replaced with the same volume of fresh SNF [18,29]. Permeations of pure unprocessed QHCl were also carried out to serve as the positive control. The results obtained were used to construct a permeation profile by plotting the amount of drug permeated per unit surface area (mg/cm^2) versus time (min). The steady-state flux (Jss, mg/cm^2 min) was calculated from the slope of the linear portion of the plot using linear regression analysis [9].

The drug apparent permeability coefficient (*Papp*) was calculated according to the equation:

$$Papp = \frac{dQ}{dt} \times \frac{1}{A.C_0.60} \quad (1)$$

where *Papp* is the apparent permeability coefficient (cm/s); dQ/dt is the cumulative amount of drug permeated vs. time per unit area (flux); A is the effective surface area (cm^2); and C_0 is the initial concentration ($\mu g/cm^3$).

2.2.8. In Vivo Pharmacodynamic Studies

The guidelines of the Animal Ethics Committee of the University of Nigeria, Nsukka, and EU Directive 2010/63/EU were observed while conducting the *in vivo* pharmacodynamic studies in mice. CBA/J mice, weighing 18–20 g, were made to develop malaria by intraperitoneal inoculation of 2×10^5 *Plasmodium berghei* ANKA-parasitized erythrocytes from a previously infected mouse [30,31]. Grouping into 5 sets of 6 was done on day seven (7) post infection, as shown in Table 2. Table 2 also shows the treatments administered, doses, as well as routes of administration.

Table 2. *In Vivo* studies design.

Formulation/API	Group	Treatment	Dosing	Route
QHCl	G1	Q9	20 mg/kg at 0 h, 10 mg/kg every 12 h for 4 days	IN
	G2	Plain QHCl solution		
	G3	Q9		Oral
Placebo	G4	Blank NS	20 mg/kg at 0 h, 10 mg/kg every 12 h for 4 days	IN
	G5			Oral

Key: **QHCl** = Quinine hydrochloride, **G1 to G4** = Groups of 6 mice; **Q9** = Q9 (QHCl-NS containing 1.5% *w/v* QHCl, 3% *w/v* S154, 2% *w/v* T80 and sonicated for 60 min); IN = Intranasal.

Mice receiving treatment intranasally were slightly anesthetized using ketamine at a dose of 75 mg/kg. To administer NS intranasally, each mouse was held using the left hand and restrained by anchoring the tail between the small finger and the palm [32,33], after which the mice were held in a supine position with the head elevated. Using the right hand, gel-loading tips attached to a micropipette were used to slowly introduce the right dose of each formulation through the external nares. Dose volumes were adjusted to $\leq$20 µL to avoid suffocation and death [34,35]. Mice were maintained in a standing position until they recovered from the effect of the anaesthesia. On day 13 post infection, each mouse was tail-bled, and a thin blood film was made on a microscope slide. The films were fixed with methanol, stained with gentian violet solution, and examined microscopically to monitor the parasitaemia level. The level of parasitaemia at day 1 post infection was used for comparison.

The percentage of parasitaemia in the blood was calculated using the expression:

$$\text{\% Reduction in Parasitaemia} = 100 - \{\text{No. of parasitized RBC}/(\text{No. of parasitized RBC} + \text{No. of nonparasitized RBC})\} \times 100 \quad (2)$$

Antimalarial activity of formulations and pure drug was determined by using the equation:

$$\text{Activity}(\%) = 100 - \left(\frac{\text{Mean parasitaemia in treated group}}{\text{Mean parasitaemia in control group}}\right) \times 100 \quad (3)$$

RBC (Red blood cell).

2.2.9. Histopathological Studies

After intranasal administration with Q9 NS, mice were sacrificed by cervical dislocation. Histopathological examination of samples collected from the nasal mucosa and lungs was conducted after the samples were fixed with 10% neutral-buffered formalin and dehydrated in graded concentrations of ethanol. Samples were subsequently cleared in xylene and embedded in paraffin wax. Exactly 5-μm-thick sections of the samples were cut and mounted on a glass slide, and later stained with hematoxylin and eosin (H & E). Micrographs of sections were captured using a Moticam Images Plus 2.0 digital camera (Motic China Group Ltd., Fujian, China) attached to a Leica binocular light microscope [36].

2.2.10. Data and Statistical Analysis

Results were expressed as mean ± standard deviation. For group comparisons, statistically significant differences were determined at $p < 0.05$ using one-way analysis of variance. Statistical analyses were done using GraphPad Prism version 8.2.0 (Prisma, Graphpad Software, La Jolla, CA, USA) [37].

3. Results and Discussion

3.1. Solubility of QHCl in Solid Lipids and Liquid Lipids

Since the solubility of any drug (especially hydrophilic drugs such as QHCl) is crucial in any lipid-based drug delivery system, the solubility of QHCl was analysed in selected solid and liquid lipids. Of all the solid lipids analysed, CHD 5 was a better solvent for QHCl, with stearic acid showing the least solubility (Table 3). Solubilization of QHCl in stearic acid was only possible after heating at 80 °C for up to 15 min. The Compritol® lipids (CHD 5 and C888) and S154 were selected for the preparation of NS, because QHCl was better solubilized in them and at lower temperatures. On the other hand, after the solubility test, QHCl was significantly dissolved in all the liquid lipids tested, with THP exhibiting the highest solubility. Hence, THP and Miglyol® 812 N were selected as the cosolvent and liquid lipid, respectively. Transcutol® HP, the highest purity grade of diethylene glycol monoethyl ether, has been known to exhibit good solubility for poorly soluble drugs such as risperidone [38] due to the presence of an ether and an alcohol functional groups in its molecule [39]. It is therefore used as cosolvent and surfactant. THP was also selected for use in this intranasal formulation due to its skin permeation enhancement property [40]. Despite its high solubility, it could not be used alone as the solvent, because it is toxic at high concentrations [41].

Table 3. Solubility of QHCl in solid and liquid lipids.

Solid Lipids	**Stearic Acid**	**Softisan® 154**	**Compritol® HD 5 ATO**	**Compritol® 888 ATO**
QHCl Solubility	+	++	+++	++
Liquid Lipids		Glyceryl monooleate	Transcutol® HP	
QHCl Solubility	+++	+++	++++	

Key: + = Sparingly soluble; ++ = Slightly soluble; +++ = Soluble; ++++ = Freely soluble, **QHCl** = Quinine hydrochloride.

3.2. Mean Particle Size and Particle Distribution Indices Analyses of QHCl-NS

Even though CHD 5 produced a higher solubility of QHCl, lipid matrices made with C888 and S154 were also used in the formulation of NS. In addition, three concentrations of surfactant and different sonication times were employed to determine the effect of varying these process parameters on average particle sizes and PDI of NS. All QHCl-NS had mean particles in the nanometer size range (68.60 ± 0.86 to 300.80 ± 10.11 nm) (Table 4; Figure 1).

Table 4. Particle size (24 h, 30 and 90 days), PDI and Zeta potential of QHCl-NS.

QHCl-NS	24 h			30 Days		90 Days	
	Particle Size (nm)	PDI	Zeta Potential	Particle Size (nm)	PDI	Particle Size (nm)	PDI
Q1	194.76 ± 4.495	0.441 ± 0.006	6.72 ± 0.259	199.27 ± 0.4225	0.433 ± 0.011	192.1 ± 2.371	0.427 ± 0.013
Q3	118.4 ± 0.7216	0.426 ± 0.017	3.24 ± 0.304	143.4 ± 0.6202	0.406 ± 0.2646	177.0 ± 3.704	0.237 ± 0.016
Q9	117.5 ± 1.53	0.282 ± 0.004	6.95 ± 0.416	112.2 ± 1.715	0.279 ± 0.004	113.7 ± 0.7550	0.279 ± 0.007
Q10	80.9 ± 1.57	0.465 ± 0.015	4.74 ± 0.371	68.15 ± 0.8786	0.456 ± 0.003	60.85 ± 0.5901	0.442 ± 0.013
Q13	92.9 ± 2.765	0.564 ± 0.032	4.81 ± 0.067	97.51 ± 0.767	0.431 ± 0.029	90.58 ± 0.747	0.315 ± 0.003
Q14	109.4 ± 0.814	0.599 ± 0.011	4.92 ± 0.096	115.2 ± 0.7092	0.443 ± 0.005	130.2 ± 0.4173	0.211 ± 0.011
Q15	120.5 ± 4.828	0.574 ± 0.027	4.28 ± 0.180	97.67 ± 1.640	0.425 ± 0.015	104.2 ± 0.7024	0.280 ± 0.003
Q5	83.52 ± 0.676	0.467 ± 0.008	6.19 ± 0.396	86.92 ± 0.999	0.319 ± 0.020	92.82 ± 0.5046	0.255 ± 0.010
Q6	150 ± 1.595	0.475 ± 0.007	5.36 ± 0.106	120.5 ± 0.808	0.371 ± 0.006	130.8 ± 0.625	0.268 ± 0.005
Q7	68.6 ± 0.861	0.491 ± 0.003	5.93 ± 0.18	79.9 ± 2.87	0.261 ± 0.010	87.31 ± 1.160	0.225 ± 0.005
Q8	90.36 ± 0.520	0.445 ± 0.003	4.36 ± 0.076	90.50 ± 2.859	0.307 ± 0.028	97.51 ± 0.197	0.307 ± 0.028
Q2	300.8 ± 10.11	0.603 ± 0.029	4.23 ± 0.294	206 ± 27.97	0.596 ± 0.146	186.9 ± 16.12	0.557 ± 0.073
Q4	121.5 ± 39.62	0.397 ± 0.132	2.31 ± 0.061	121.5 ± 39.62	0.397 ± 0.132	143.3 ± 1.739	0.441 ± 0.012
Q11	119.4 ± 0.945	0.511 ± 0.008	0.738 ± 0.138	109.1 ± 1.649	0.488 ± 0.006	113.2 ± 4.192	0.475 ± 0.017
Q12	118.6 ± 1.093	0.492 ± 0.022	2.16 ± 0.659	103.5 ± 1.580	0.434 ± 0.007	111.8 ± 4.925	0.318 ± 0.037

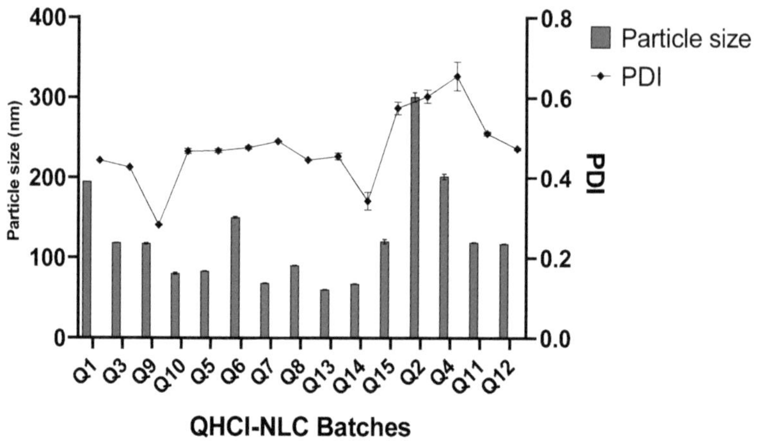

Figure 1. Particle size and PDI of QHCl-NS within 24 h.

In all the lipid matrices used, increasing sonication time from 30 to 90 min did better at reducing the particle size than increasing the surfactant concentration from 2 to 5%. In the case of QHCl-NS prepared with CHD 5 and C888, increasing the concentration of the surfactant resulted instead in an increase in particle size (compare Q5 and Q7, and Q6 and Q8, Figure 2). Hence, it can be concluded that it is not in all cases that increasing the concentration of the surfactant will result in a reduction in particle sizes [42]. Increasing the surfactant concentration may produce smaller particle sizes than increasing sonication time, especially when the concentration of the liquid lipid is high [9]. NS formulated with S154 produced more predictable results in terms of the particle sizes of QHCl-NS. Increasing sonication or the surfactant resulted in a reduction of particle sizes, even though increasing sonication time still yielded a smaller particle size better than using a higher concentration of surfactant (Q1 compared with Q3, and Q9 compared with 10; Table 4).

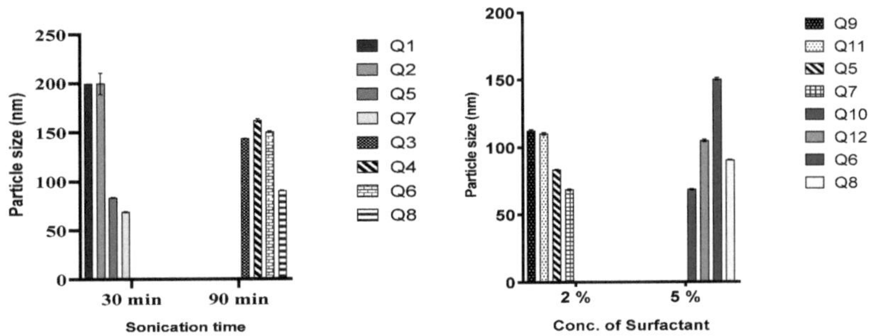

Figure 2. Effect of duration of sonication and concentration of T80 on particle size of QHCl-NS.

The differences observed in the particle sizes of the QHCl-NS made from the three lipid matrices can be attributed to their constituents and characteristics; S154 is hydrogenated palm oil made up of C14–C18 fatty acids and has a melting point of approximately 55 °C [43]; while C888 is docosanoic acid, monoester with glycerine, having a melting point of 69–74 °C. In addition to the constituents of C888, CHD 5 also contains poly(ethylene)glycol and melts at a lower temperature (56–63 °C) [44,45].

The PDIs of QHCl-NS formulations were in the range of 0.282 ± 0.004 to 0.603 ± 0.029 immediately after formulation, and the PDI of QHCl-NS prepared with C888 (Q2, Q4, and Q11) were all significantly higher ($p < 0.05$) than NS prepared with CHD 5, and especially S154 (Table 4). It was observed that PDI increased with a reduction in mean particle size. Thus, smaller particles had higher PDI than larger particles (compare Q9 with Q7 and Q10 in Table 4).

The nanosizes obtained may be beneficial to the usefulness of QHCl, since drugs in such sizes can access the brain through the olfactory region of the nasal cavity. It can also improve permeability through the nasal mucosa. Deductions made from the results on the influence of sonication of duration and concentration of surfactant show that these parameters significantly affect the final size of lipid nanoparticles. Therefore, preformulation studies to monitor average particle size while varying these parameters must be performed in order to determine the optimal conditions that would yield monodispersed, nano-sized lipid particles.

3.2.1. Stability Studies of QHCl-NS

The QHCl-NS formulated were very stable after 90 days. Only 6 formulations out of the 12 increased in size (negligibly) after 1 month. In some cases, a reduction in size and PDI was recorded (Table 4). This has been reported by other researchers [9,46,47] and is caused by a loss of solubilized water situated within the core of the nanoformulation. It may still be due to the ultrasonic energy used for size reduction, which is known to cause the collision of smaller oil droplets to form large ones [48,49], a phenomenon termed "sonication induced aggregate formation" [16,50]. Removal of the ultrasonic energy during storage may have led to a stable system resulting in a reduction in particle size and PDI. The high stability of the QHCl-NS formulation may be due to the surfactant (T80) and stabilizer use (P188) during formulation.

3.2.2. Surface Charge (Zeta (ζ) Potential) of NS

The low zeta potential values recorded for QHCl-NS (0.738 ± 0.138 to 6.720 ± 0.259 mV) are attributed to the use of non-ionic surfactants (T80 and P188) (Tables 4 and S1) [51].

The zeta potential of blank formulations was higher than that of the drug containing NS (Table S1). All QHCl-NS were positively charged within the first 24 h after preparation; however, some C888-containing formulations became slightly negatively charged after 30 and 90 days of storage. The low zeta potential did not cause an increase in sizes of nanoparticles in storage. This stability in sizes in storage may be due to the presence of P188, which is known to cause steric stabilization [52,53].

Three QHCl-NS formulations (Q7, Q9, and Q12) representing the three lipid matrices used (CHD 5, S154, and C888, respectively) were selected for further characterizations based on their acceptable particle sizes, PDI (highlighted on Table 4), and high stability during storage.

3.3. *Time-Dependent pH Stability Studies and Osmolality of QHCl-NS*

The pH of QHCl-NS formulations (5.03–5.55) were within the acceptable pH range for nasal formulations (pH of 4.5 to 6.5) [54], and remained stable for 90 days while being stored at 8 ± 2 °C (Figure 3). Osmolality of the formulations ranged from 422.3 ± 12.3 to 517.0 ± 21.7 mOsmol/kg (Table 5), which is within the range of most marketed nasal products (300–700 mOsmol/kg) [55]. Therefore, nasal irritation due to pH disparity or tonicity is unlikely to occur during administration.

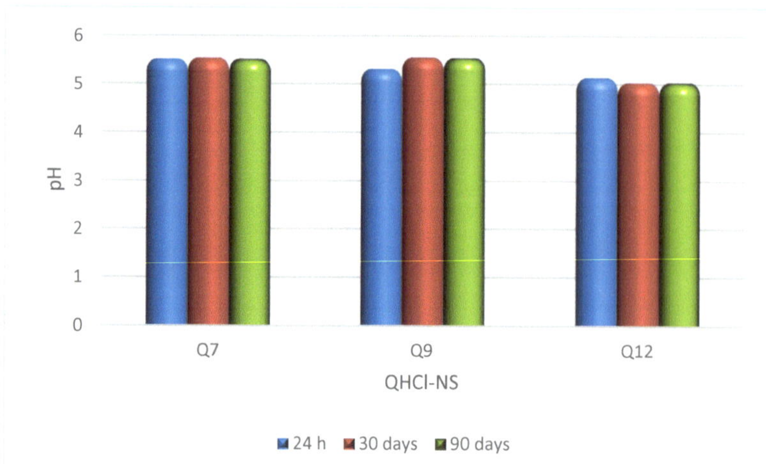

Figure 3. pH stability study of QHCl-NS.

Table 5. Osmolality, *ex vivo* permeation results and drug release mechanism and kinetics of QHCl-NS.

Batches	Osmolality (mOsmol/kg)	Flux (μg/cm²min)	Permeation Coefficient (cm/s)	Zero Order	First Order	Higuchi	Korsmeyer–Peppas	
				r^2	r^2	r^2	r^2	n
Q7	422.3 ± 12.3	-	-	0.7807	0.9161	0.9939	0.9756	0.5055
Q9	492.7 ± 17.9	320.710	2.18×10^{-2}	0.7466	0.8885	0.9298	0.9993	0.8869
Q12	517.0 ± 21.7	-	-	0.8603	0.9611	0.9710	1	0.7730
Plain Solution of QHCl	-	56.973	3.87×10^{-4}	-	-	-	-	-

3.4. Morphology of QHCl-NS

The micrographs obtained from both TEM and SEM revealed round and oval-shaped particles (Figure 4A–D). During our experiments, we observed that cryo-SEM was a better method for visualizing NS compared to SEM (Figure 4C,D). Cryo-SEM was able to reveal more distinct particles (Figure 4D). The images obtained from SEM were clumped together, making visualization of individual particles difficult (Figure 4C). This difference may have been caused by lyophilization, since the samples used in SEM were lyophilized solid samples, while NS formulations in liquid form were used for cryo-SEM analysis. The micrographs of excipients and other NS formulations are presented in Figure S2.

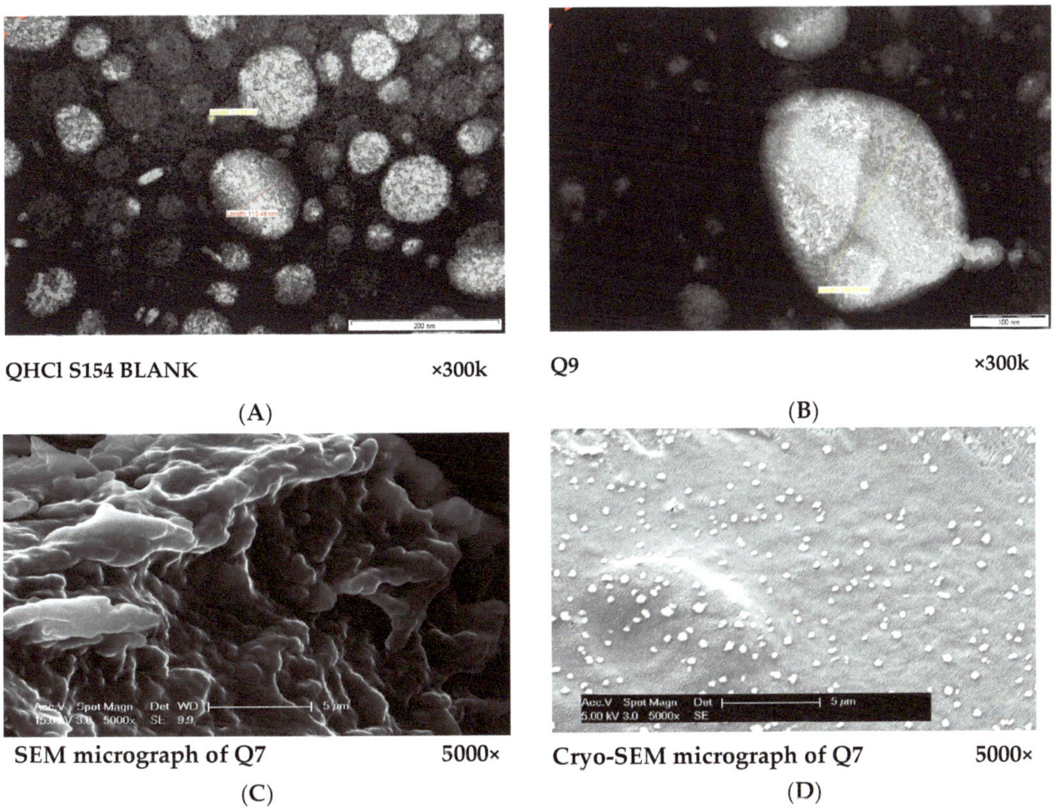

Figure 4. (**A**,**B**): TEM micrographs of S154 BLANK for QHCl-NS formulations (**A**) and Q9 (containing 1.5% *w*/*v* QHCl, 3% *w*/*v* S154, 2% *w*/*v* T80 and sonicated for 60 min). (**C**,**D**): Comparison between SEM (**C**) and cryo-SEM (**D**) micrographs of batch Q7 (containing 1.5% *w*/*v* QHCl, 3% *w*/*v* CHD 5, 2% *w*/*v* T80 and sonicated for 90 min).

3.5. FTIR Spectroscopic Analysis

A Fourier transform infrared spectrophotometer was used to determine possible interactions between the constituents of the lipid matrix (Softisan® 154, Compritol® HD 5 ATO and Compritol® 888 ATO, and Phospholipon® 90H) and between the lipid matrix and drug. The lipids (S154, CHD 5, and C888) were compatible with the P90H, as revealed in the FTIR spectra of the lipid matrices (Figure S2). The same principal peaks were evident in the FTIR spectra of the individual lipids and phospholipid as in the spectra of the lipid matrices. This suggests that the use of heat in the fusion of P90H and the lipids did not result in a chemical interaction.

In addition, FTIR results confirmed the lack of chemical interaction, and the compatibility of QHCl with the lipid matrices used as well as other excipients. Apart from the broadening and shortening of some principal peaks in the FTIR spectra of the selected formulations (Figure S2), no other difference was noticed. This difference may be attributed to hydrogen bond interactions as well as the presence of other excipients [27]. Hence, the designing of QHCl as a nanosuspension did not alter the chemistry of the drug, and it is expected not to lose its antimalaria activity.

3.6. Crystalline State of QHCl-NS

DSC analysis was conducted for unprocessed QHCl, excipients, and physical mixtures of the drug and lipids. All the solid lipids were crystalline in nature (sharp melting endothermic peaks at 57.68, 72.53, and 59.56 °C, enthalpies of 94.675, 116.25, and 117.83 J/g for pure S154, C888, and CHD5, respectively (Table S2)). P90H also showed an endothermic peak (122.30 °C). Fusion of the solid lipids with P90H during the formation of the SRMS caused reductions in melting points and enthalpies (Table S2). A complete disappearance of the melting peak due to P90H with the fusion of C888 and P90H implied that P90H was completely molecularly dispersed or amorphous in C888 (Figure S3) [56]. On the other hand, the inclusion of the liquid lipid (MCT) also caused further depression in the melting points of all solid lipid matrices, and so did the addition of Transcutol® HP (Table S2). The reduction of the melting points of solid lipids in the presence of liquid lipids and drugs has been previously reported by Garcia-Fuentes et al. [57] and Hu et al. [58]. Le-Jiao and coauthors [59] also reported a decrease in enthalpies of nanostructured lipid carriers with the addition of MCT. This is defined as a eutectic behaviour [56] and suggests a disordered lattice that can accommodate more drug molecules [9].

The melting point and enthalpy of unprocessed QHCl (117.29 °C) reduced when formulated as NS (57.11, 56.29, and 63.63 °C, for Q7, Q9, and Q12, respectively), suggesting amorphization (Figure 5). However, this was further investigated using X-ray diffractometry, because it may be erroneous to conclude on the crystal nature of a drug using DSC when it is available in low concentrations (less than 10%) [9].

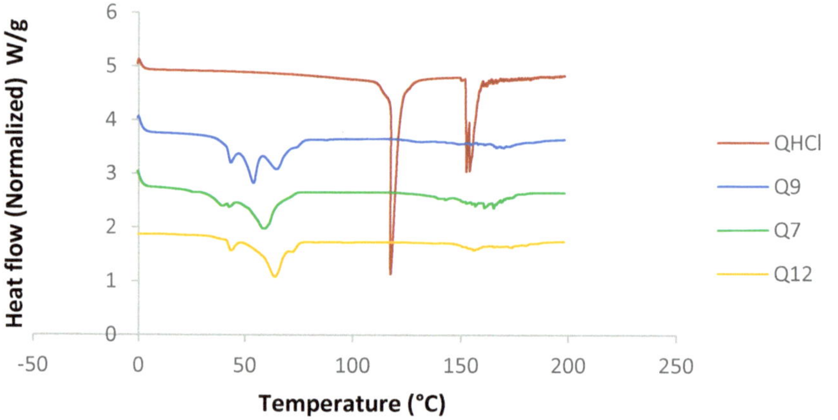

Figure 5. DSC thermogram of selected batches from QHCl-NS.

3.7. Powder X-ray Diffractometry of NS

X-ray diffractograms of QHCl-NS confirmed a reduction in crystallinity, since there were disappearances of some peaks in the diffractogram of the NS, unlike in the unprocessed drug (Figure 6A–D). For instance, some peaks in unprocessed QHCl at 2θ of 10° to 20° (Figure 6A) were not seen in Q7 (Figure 6B), while peaks at 2θ of 13.4° and 24.0° in pure QHCl were lost in Q9 (Figure 6C). Also, so many sharp peaks were not visible in Q12 (Figure 6D) compared to the pure quinine hydrochloride. For example, sharp distinct peaks at 2θ of 9.2°, 14.6°, 17.9°, and 23.9° to 48.9° in pure quinine sample were not detected in Q12 (Figure 6D). Amorphous halos were also very visible on the XRD patterns of the NS; an indication of the presence of some amorphous form of the drug. It can therefore be concluded that QHCl were in microcrystalline or semi-crystalline form in the NS [60,61].

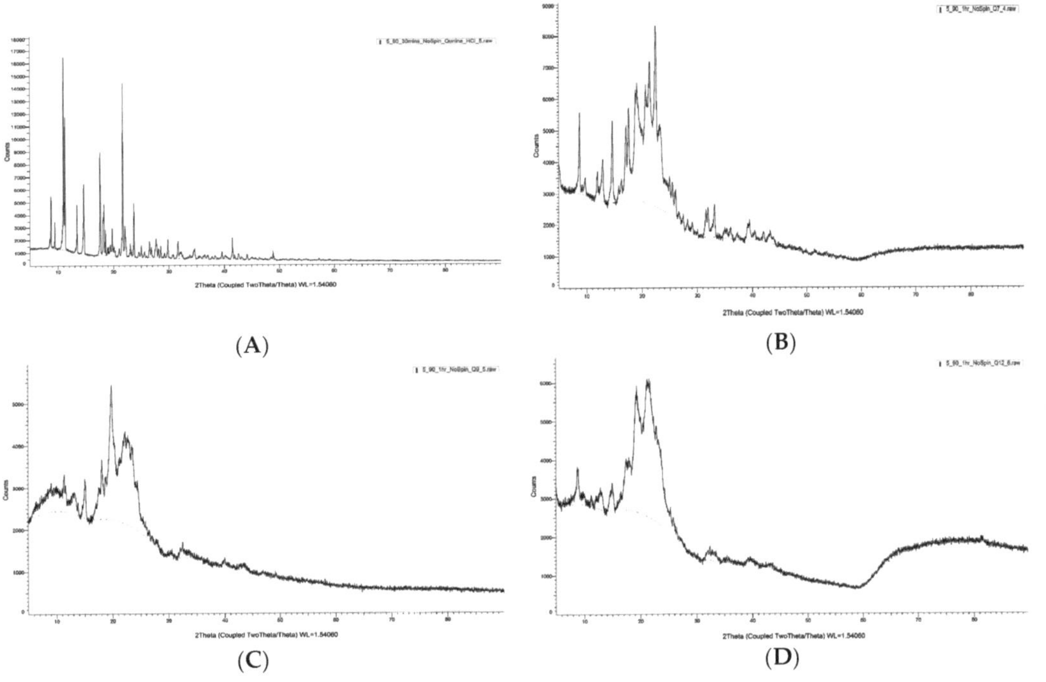

Figure 6. XRD pattern of pure sample of (**A**) QHCl, (**B**) Q7, (**C**) Q9, (**D**) Q12.

3.8. *In Vitro Release Studies of QHCl-NS*

3.8.1. *In Vitro* Release of QHCl-NS

In Vitro drug release was conducted in SNF in order to mimic the nasal environment. The solubility of QHCl in SNF was determined to be 26.49 mg/mL. The sink condition was maintained throughout the procedure by using a volume that will accommodate a solubilized drug without saturation occurring, as well as by replenishing with fresh media after each withdrawal of samples for analyses. Neither the concentration of surfactant nor particle size significantly influenced drug release among QHCl-loaded formulations (Figure 7). This may have been due to the high solubility of QHCl in the release media (26.49 mg/mL). Although pure unprocessed QHCl released slightly more drug than all the selected QHCl formulations (Q7, Q9, Q12), after the 6th hour, all formulations released above 80% of their drug content, with Q9 (formulated with S154) NS showing the fastest rate of release. The faster rate of release of pure unprocessed API compared to drugs in nanoformulations has been reported by other researchers [28] and may be attributed to time required for the release media to bypass the lipid carrier system wherein the drug is encapsulated or dispersed.

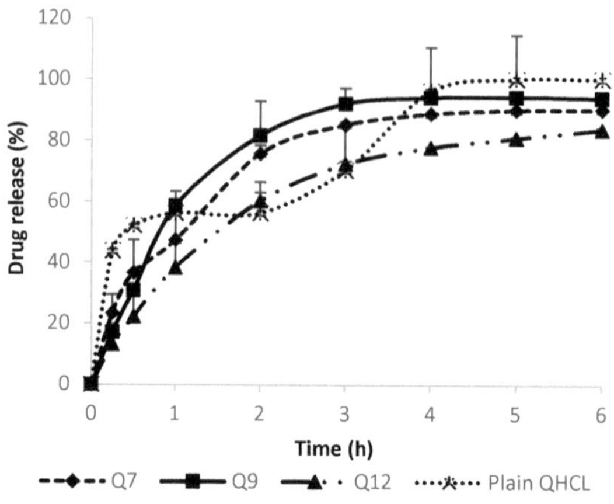

Figure 7. *In Vitro* release of QHCl from NS in simulated nasal fluid.

3.8.2. Evaluation of Drug Release Mechanism and Kinetics of QHCl-NS

Zero-order, first-order and Higuchi mathematical models were used to evaluate the kinetics of drug release from NS. The model that gave the highest correlation coefficient value was considered the best fit for the release data being analysed [27,62]. Drug release from all QHCl-NS followed Higuchi release kinetics (Table 5). The mechanism of drug release was determined using the Korsmeyer–Peppas model (Table 5). The "n" value for Q7 was 0.5055, indicating a Fickian diffusion-controlled mechanism. Diffusion of quinine from Q9 and Q12 was by anomalous diffusion, implying that the release of quinine from these NS formulations was by both diffusion and erosion of the lipid matrix, since the diffusional release exponent was greater than 0.5 but less than 1 (Table 5).

3.9. *Ex Vivo Permeation Analysis of QHC1-NS*

The *ex vivo* permeation analysis and the *in vivo* antimalarial study were conducted using only Q9, because this formulation exhibited a higher *in vitro* release than the other selected formulations (Q7 and Q12). The results obtained were compared with the unprocessed QHCl to determine the effect of formulating quinine as NS. The flux and permeation coefficient of Q9 (320.71 $\mu g/cm^2 min$ and 2.18×10^{-2} cm/s, respectively) were significantly ($p < 0.05$) higher than that of the unprocessed pure sample of QHCl (56.97 $\mu g/cm^2 min$ and 3.87×10^{-4} cm/s, respectively) (Table 5). A 5-fold and 56-fold increase in the flux and permeation coefficient, respectively, were observed. This implies that the rate of absorption as well as ease of drug permeation through porcine nasal mucosa was impressively enhanced by formulating QHCl as NS. A similar outcome has been earlier reported for artesunate NLC and gentamicin lipid-based microsuspension [9,25]. The lipophilicity of NS improved the permeation of QHCl, a hydrophilic drug, through excised porcine nasal mucosa. This also implies that the reformulation of QHCl can improve its permeation through lipid bilayers.

3.10. *In Vivo Antimalarial Studies of QHCl-NS*

Antimalarial investigation showed that there was a significant ($p < 0.05$) reduction of parasitaemia achieved with both intranasal and oral administration of QHCl-NS compared with the placebo. The reductions in parasitaemia caused by Q9 administered through the intranasal and oral routes were 51.16% and 52.12%, respectively (Figure 8). Interestingly, reductions in parasitaemia observed in both routes of administration were similar and not significantly different, though that of the oral route was higher. This is an indication that malaria treatment with quinine through the intranasal route is a possibility. The activity of the drug when given through the nasal route (56.26%) was higher than when administered through the oral route (54.22%) (Figure 9). The administration of unprocessed quinine solution exhibited a significantly ($p < 0.05$) higher reduction in parasitaemia and activity. This varies from the results obtained in the *ex vivo* permeation studies. However, for the intranasal administration, an NS formulation of quinine will still be preferrable, because such a formulation being mucoadhesive will be better retained in the nasal region [63]. A plain solution of the drug may easily be cleared in the nasal cavity, as well as run off to the oral region of the mouth or throat, causing the patient to experience an unpleasant taste.

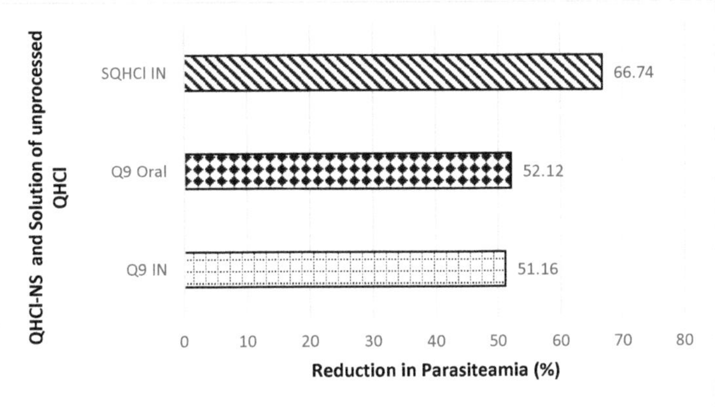

Figure 8. Percentage reduction in parasitaemia after IN and oral treatment with Q9 NS formulation and plain solution of QHCl (SQHCl).

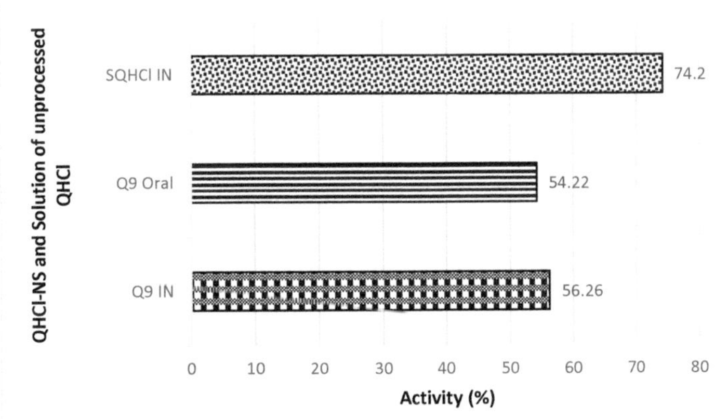

Figure 9. *In Vivo* antimalarial activity after IN and oral treatment with Q9 NS formulation and plain solution of QHCl (SQHCl).

3.11. Histopathological Studies

The histopathological analysis of the lungs and the nasal mucosa showed a clear difference between the treated and the untreated groups. While there was evidence of inflammation and congestion of red blood cells in the lungs and nasal mucosa of the mice in the untreated group (Figures 10A and 11A), none was observed in the lungs and nasal mucosa of mice treated intranasally with Q9 (Figures 10B and 11B,C). The observed congestion of pulmonary vessels in the mice treated intranasally with placebo is indicative of severe malaria [64]. No sign of injury was seen in the group treated with Q9 through the intranasal route (Figures 10B and 11B,C). However, the pseudostratified columnar ciliated epithelium and nonciliated goblet cells covered by thin mucus membrane were intact in both groups and undamaged by intranasal administration of formulation and placebo. This implies that intranasal administration of NS formulation of QHCl for the treatment of malaria caused no damage to the nasal mucosa.

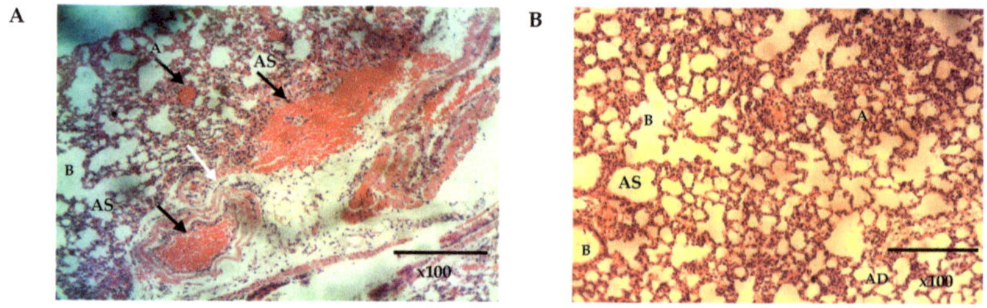

Figure 10. The histomicrograph of the lungs of mice treated with placebo (**A**) and Q9 (**B**) H & E ×100, showing normal respiratory epithelium of the smaller bronchioles (**B**) (arrows: ciliated simple columnar epithelium). Note also the normal Alveolar duct (AD), Alveolar sac (AS), and Alveoli (**A**) in A and B, and the sequestration of parasitized red blood cells (black arrows) and congested pulmonary vessels (white arrows) in (**A**). (For interpretation of the references to colour in this figure legend, the reader is referred to the web version of this article).

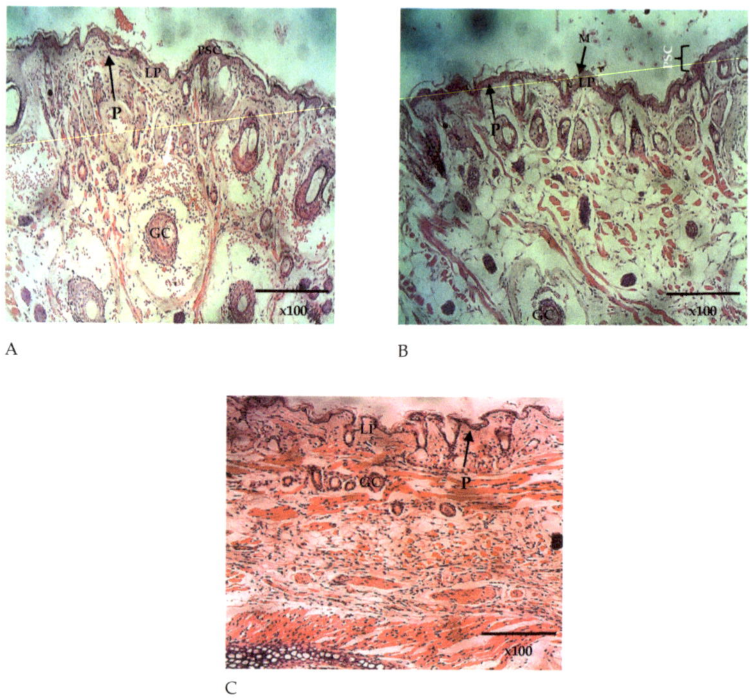

Figure 11. The histomicrograph of the nasal epithelium of mice treated intranasally with placebo (A) and Q9 (**B**,**C**) H & E ×100, showing the nasal epithelium (epidermis; P) composed of pseudostratified columnar ciliated epithelium (PSC) and covered by mucus membrane ((**C**); MM). Note also the goblet cells (GC) and the lamina propria (LP). H & E ×100 (**A**–**C**). The sequestration of parasitized red blood cells (white arrows).

4. Conclusions

Intranasal administration of QHCl-NS successfully treated malaria in mice infected with *Plasmodium berghei* ANKA, and can therefore serve as an alternative to other routes of administration of QHCl for the treatment of malaria. It can be beneficial in preventing the GIT-associated side effects caused by QHCl and can be helpful to patients (especially pregnant woman) who may be having the side effect of vomiting due to morning sickness and malaria. The relatively reduced surface area of the nasal cavity compared to the GIT may reduce the rate of saturation of the systemic circulation with quinine, thereby preventing cinchonism, which is dependent on the concentration of quinine in circulation. The intranasal route also offers the advantage of ease of application, painless self-administration, avoidance of first-pass metabolism, and direct access to the brain in the case of cerebral malaria. The results of the *ex vivo* studies revealed the potential of the nanosuspension of QHCl improving permeation through the nasal mucosa. Its bioadhesive nature compared to unprocessed QHCl solution will increase the residence time of the drug in the nasal cavity and retard flow of the bitter-tasting drug to the oral cavity.

Further studies to compare parenteral administration of QHCl to intranasal administration, as well as quantification of the amount of drug in the brain relative to systemic circulation after intranasal administration, will be conducted.

Supplementary Materials: The following supporting information can be downloaded at: https://www.mdpi.com/article/10.3390/pr11061811/s1: Table S1: zeta potentials of blank-NS formulations; Figure S1: SEM micrographs of drug and some excipients; Figure S2: FTIR Spectra of Excipients and QHCl-NS formulation; Table S2: DSC analysis results of plain drugs, and lipids and lipid matrices; Figure S3: DSC thermogram of lipid matrix made with C888 and P90H.

Author Contributions: Conceptualization, C.P.A., P.A.A., J.N.R.-O., C.M. and A.A.A.; data curation, T.C.U.; formal analysis, C.P.A., J.N.R.-O., C.M. and A.A.A.; funding acquisition, P.A.A., C.M., A.A.A., M.A.M. and K.C.O.; investigation, T.C.U. and K.C.O.; methodology, C.P.A., T.C.U., J.D.O., L.O.U., P.O.N., C.M., A.A.A. and K.C.O.; project administration, C.P.A., C.M. and M.A.M.; resources, K.C.O.; validation, J.N.R.-O. and J.O.O.; visualization, O.C.E.; writing—original draft, C.P.A. and T.C.U.; writing—review and editing, C.P.A., O.C.E., A.L.O., A.C.E., C.S.N., S.W.U., J.D.O., P.O.N., J.N.R.-O. and C.M. All authors have read and agreed to the published version of the manuscript.

Funding: The conducting of the research work was funded by the Commonwealth Scholarshipa Commission in the United Kingdom and 2021 African–German Network of Excellence in Science-Programme Advocating Women in Science (AGNES-PAWS) Grant for Junior Researchers, supported by the Federal Ministry of Education and Research (BMBF) and the Alexander von Humboldt Foundation (AvH). The funders provided financial support but did not influence the study design, analysis, and interpretation of data generated.

Institutional Review Board Statement: The study was conducted in accordance with the National Code of Conduct for Animal Research Ethics (NCARE). The animal study protocol was approved by the Animal Research Ethics Committee of the University of Nigeria, Nsukka (7 February 2019).

Informed Consent Statement: Not applicable.

Data Availability Statement: Not applicable.

Acknowledgments: We wish to thank the management of the University of Nigeria, Nsukka, and the University of Birmingham, where this research was conducted. Our appreciation also goes to Lipoid GmbH and IOI Oleo GmbH for the gift of Phospholipon® 90H, Softisan® 154, Compritol® 888 ATO, Compritol® HD5 ATO, Transcutol® HP, Transcutol® P, and Miglyol® 812 N. Our sincere appreciation also goes to Offor Ejike, Nwabueze Harrison, Ubachukwu Ugochukwu, Ugwu Chinedu, Fredrick Ugwuoke, Gábor Drávavölgyi, Hanouf Bafhaid, Ali Al Amri, Sarah Lastakchi, Justyna Hofmanová, and Sandya Rajesh for assisting in the laboratory procedures and the animal studies.

Conflicts of Interest: The authors declare no conflict of interest. The funders had no role in the design of the study; in the collection, analyses, or interpretation of data; in the writing of the manuscript, or in the decision to publish the results.

References

1. Achan, J.; Talisuna, A.O.; Erhart, A.; Yeka, A.; Tibenderana, J.K.; Baliraine, F.N.; Rosenthal, P.J.; Alessandro, U.D. Quinine, an Old Anti-Malarial Drug in a Modern World: Role in the Treatment of Malaria. *Malar. J.* **2011**, *10*, 144. [CrossRef] [PubMed]
2. World Health Organization. *Guideline for the Treatment of Malaria*, 3rd ed.; World Health Organization: Geneva, Switzerland, 2015. [CrossRef]
3. Fairhurst, R.M.; Wellems, T.E. Malaria (Plasmodium Species). In *Mandell Douglas, and Bennett's Principles and Practice of Infectious Diseases*; Elsevier: Amsterdam, The Netherlands, 2019; pp. 3070–3090.
4. Fairhurst, R.M.; Dondrop, A.M. Artemisinin-Resistant Plasmodium Falciparum Malaria. *Microbiol. Spectr.* **2016**, *4*, 4-3. [CrossRef] [PubMed]
5. Uzondu, S.; Echezona, A.; Nwagwu, C.; Onugwu, A.; Ugorji, L.; Agbo, C.; Kenechukwu, F.; Ogbonna, J.; Akpa, P.; Nnamani, P.; et al. *Combating Antimalarial Drug Resistance: Recent Advances and Future Perspectives*; IntechOpen: London, UK, 2022; pp. 1–16. [CrossRef]
6. U.S. National Library of Medicine. GLS-1200 Topical Nasal Spray to Prevent SARS-CoV-2 Infection (COVID-19). NCT04408183. Available online: https://clinicaltrials.gov/ct2/show/NCT04408183 (accessed on 31 August 2022).
7. Gizurarson, S. Anatomical and Histological Factors Affecting Intranasal Drug and Vaccine Delivery. *Curr. Drug Deliv.* **2012**, *9*, 566–582. [CrossRef] [PubMed]
8. Quinine. In *Meyler's Side Effects of Drugs*; Aronson, J.K. (Ed.) Elsevier: Oxford, UK, 2016; pp. 27–35. [CrossRef]
9. Agbo, C.P.; Ugwuanyi, T.C.; Ugwuoke, W.I.; McConville, C.; Attama, A.A.; Ofokansi, K.C. Intranasal Artesunate-Loaded Nanostructured Lipid Carriers: A Convenient Alternative to Parenteral Formulations for the Treatment of Severe and Cerebral Malaria. *J. Control. Release* **2021**, *334*, 224–236. [CrossRef]
10. Gupta, Y.; Jain, A.; Jain, S.K. Transferrin-conjugated Solid Lipid Nanoparticles for Enhanced Delivery of Quinine Dihydrochloride to the Brain. *J. Pharm. Pharmacol.* **2007**, *59*, 935–940. [CrossRef]
11. Marijon, A.; Bonnot, G.; Fourier, A.; Bringer, C.; Lavoignat, A.; Gagnieu, M.-C.; Bienvenu, A.-L.; Picot, S. Efficacy of Intranasal Administration of Artesunate in Experimental Cerebral Malaria. *Malar. J.* **2014**, *13*, 501. [CrossRef]
12. Torrino, E.; De Marco, I.; Reverchon, E. Organic Nanoparticles Recovery in Supercritical Antisolvent Precipitation. *J. Supercrit. Fluids* **2010**, *55*, 300–306. [CrossRef]
13. Mumuni, M.A.; Frankline, K.C.; Ugwu, C.E.; Musiliu, A.O.; Agboke, A.A.; Agbo, P.; Ossai, E.C.; Ofomata, A.C.; Youngson, D.C.; Omeje, C.E.; et al. Development and Evaluation of Artemether-Loaded Microspheres Delivery System for Oral Application in Malaria Treatment. *Trop. J. Nat. Prod. Res.* **2021**, *5*, 2030–2036.
14. Wang, C.; Yan, T.; Yan, T.; Wang, Z. Fabrication of Hesperetin/Hydroxypropyl-B-Cyclodextrin Complex Nanoparticles for Enhancement of Bioactivity Using Supercritical Antisolvent Technology. *J. Mol. Struct.* **2023**, *1279*, 134947. [CrossRef]
15. Haas, S.E.; Bettoni, C.C.; de Oliveira, L.K.; Guterres, S.S.; Costa, T.D. Nanoencapsulation Increases Quinine Antimalarial Efficacy against *Plasmodium berghei In Vivo*. *Int. J. Antimicrob. Agents* **2009**, *34*, 156–161. [CrossRef]
16. Shah, B.; Khunt, D.; Bhatt, H.; Misra, M.; Padh, H. Application of Quality by Design Approach for Intranasal Delivery of Rivastigmine Loaded Solid Lipid Nanoparticles: Effect on Formulation and Characterization Parameters. *Eur. J. Pharm. Sci.* **2015**, *78*, 54–66. [CrossRef] [PubMed]
17. Agbo, C.; Umeyor, C.; Kenechukwu, F.; Ogbonna, J.; Chime, S.; Lovelyn, C.; Agubata, O.; Ofokansi, K.; Attama, A. Formulation Design, *in vitro* Characterizations and Anti-Malarial Investigations of Artemether and Lumefantrine-Entrapped Solid Lipid Microparticles. *Drug Dev. Ind. Pharm.* **2016**, *42*, 1708–1721. [CrossRef] [PubMed]
18. Jain, K.; Sood, S.; Gowthamarajan, K. Optimization of Artemether-Loaded NLC for Intranasal Delivery Using Central Composite Design. *Drug Deliv.* **2015**, *22*, 940–954. [CrossRef] [PubMed]
19. Gaba, B.; Fazil, M.; Khan, S.; Ali, A.; Baboota, S.; Ali, J. Nanostructured Lipid Carrier System for Topical Delivery of Terbinafine Hydrochloride. *Bull. Fac. Pharm. Cairo Univ.* **2015**, *53*, 147–159. [CrossRef]
20. Tichota, D.M.; Silva, A.C. Design, Characterization, and Clinical Evaluation of Argan Oil Nanostructured Lipid Carriers to Improve Skin Hydration. *Int. J. Nanomed.* **2014**, *20*, 3855–3864.
21. Pretorius, E. Influence of Acceleration Voltage on Scanning Electron Microscopy of Human Blood Platelets. *Microsc. Res. Technol.* **2010**, *73*, 225–228. [CrossRef]
22. Jiang, Y.; Meng, X.; Wu, Z.; Qi, X. Modified Chitosan Thermosensitive Hydrogel Enables Sustained and Efficient Anti-Tumor Therapy via Intratumoral Injection. *Carbohydr. Polym.* **2016**, *144*, 245–253. [CrossRef]
23. Mallappa, P.; Prabirkumar, S.; Panchakshari, A. Taste Masked Quinine Sulphate Loaded Solid Lipid Nanoparticles for Flexible Pediatric Dosing. *Indian J. Pharm. Educ. Res.* **2014**, *48*, 93–99. [CrossRef]
24. Momoh, M.A.; Franklin, K.C.; Agbo, C.P.; Ugwu, C.E.; Adedokun, M.O.; Anthony, O.C.; Chidozie, O.E.; Okorie, A.N. Microemulsion-Based Approach for Oral Delivery of Insulin: Formulation Design and Characterization. *Heliyon* **2020**, *6*, e03650. [CrossRef]
25. Onugwu, A.L.; Agbo, C.P.; Nwagwu, C.S.; Uzondu, E.; Echezona, A.C.; Dike, J.; Ogbonna, N.; Akpa, P.A.; Momoh, M.A.; Nnamani, P.O.; et al. Development of Lipid-Based Microsuspensions for Improved Ophthalmic Delivery of Gentamicin Sulphate. *Ther. Deliv.* **2021**, *12*, 671–683. [CrossRef]
26. FDA. *Dissolution Methods Database*; FDA: Washington, DC, USA, 2019.

27. Ümİt, G.; Melİke, Ü.; Gülgün, Y.; Ecem Fatma, K.; Aydoğmuş, Z. Formulation and Characterization of Solid Lipid Nanoparticles, Nanostructured Lipid Carriers and Nanoemulsion of Lornoxicam for Transdermal Delivery. *Acta Pharm.* **2015**, *65*, 771–791. [CrossRef]
28. Weng, J.; Tong, H.H.Y.; Chow, S.F. *In vitro* Release Study of the Polymeric Drug Nanoparticles: Development and Validation of a Novel Method. *Pharmaceutics* **2020**, *12*, 732. [CrossRef]
29. Ng, S.F.; Rouse, J.; Sanderson, D.; Eccleston, G. A Comparative Study of Transmembrane Diffusion and Permeation of Ibuprofen across Synthetic Membranes Using Franz Diffusion Cells. *Pharmaceutics* **2010**, *2*, 209–223. [CrossRef]
30. Clemmer, L.; Martins, Y.C.; Zanini, G.M.; Frangos, J.A.; Carvalho, L.J.M. Artemether and Artesunate Show the Highest Efficacies in Rescuing Mice with Late-Stage Cerebral Malaria and Rapidly Decrease Leukocyte Accumulation in the Brain. *Antimicrob. Agents Chemother.* **2011**, *55*, 1383–1390. [CrossRef] [PubMed]
31. Craig, A.G.; Grau, G.E.; Janse, C.; Kazura, J.W.; Milner, D.; Barnwell, J.W.; Turner, G.; Langhorne, J.; on behalf of the participants of the Hinxton Retreat meeting on "Animal Models for Research on Severe Malaria". The Role of Animal Models for Research on Severe Malaria. *PLoS Pathog.* **2012**, *8*, e1002401. [CrossRef]
32. Shimizu, S. Routes of Administration. In *The Laboratory Mouse (Handbook of Experimental Animals)—USP*; Elsevier: Tsukuba, Japan, 2004; pp. 527–541.
33. Simmons, M.L.; Brick, J.O. *The Laboratory Mouse*; Hollaender, A., Ed.; Prentice-Hall Inc.: Englewood Cliffs, NJ, USA, 1970.
34. Shen, X.; Lagergård, T.; Yang, Y.; Lindblad, M.; Fredriksson, M.; Holmgren, J.A.N.; Mmun, I.N.I. Group B Streptococcus Capsular Polysaccharide-Cholera Toxin B Subunit Conjugate Vaccines Prepared by Different Methods for Intranasal Immunization. *Infect. Immun.* **2001**, *69*, 297–306. [CrossRef]
35. Shen, X.; Lagergård, T.; Yang, Y.; Lindblad, M.; Fredriksson, M.; Holmgren, J.A.N. Systemic and Mucosal Immune Responses in Mice after Mucosal Immunization with Group B Streptococcus Type III Capsular Polysaccharide-Cholera Toxin B Subunit Conjugate Vaccine. *Infect. Immun.* **2000**, *68*, 5749–5755. [CrossRef] [PubMed]
36. Ashoori, Y.; Mohkam, M.; Heidari, R.; Abootalebi, S.N.; Mousavi, S.M.; Hashemi, S.A.; Golkar, N.; Gholami, A. Development and *In vivo* Characterization of Probiotic Lysate- Treated Chitosan Nanogel as a Novel Biocompatible Formulation for Wound Healing. *Biomed Res. Int.* **2020**, *2020*, 8868618. [CrossRef]
37. Gratieri, T.; Martins, G.; Melani, E.; Hugo, V.; Freitas, O.; De Fonseca, R.; Lopez, V. A Poloxamer/Chitosan in Situ Forming Gel with Prolonged Retention Time for Ocular Delivery. *Eur. J. Pharm. Biopharm.* **2010**, *75*, 186–193. [CrossRef] [PubMed]
38. Khames, A. Investigation of the Effect of Solubility Increase at the Main Absorption Site on Bioavailability of BCS Class II Drug (Risperidone) Using Liquisolid Technique. *Drug Deliv.* **2017**, *24*, 328–338. [CrossRef]
39. Gattefossé. *Transcuto$^{l®}$ P For Efficient Drug Solubilization and Skin Penetration*; Gattefossé: Saint-Priesr Cedex, France, 2020; pp. 1–24. Available online: https://www.gattefosse.com/pharmaceuticals-products/transcutol-p (accessed on 29 April 2023).
40. Salimi, M.; Fouladi, A. Effect of the Various Penetration Enhancers on the *In vitro* Skin Permeation of Meloxicam through Whole Rat Skin. *Eur. J. Bio. Pharm. Sci* **2015**, *2*, 1282–1291.
41. Sullivan, D.W.; Gad, S.C.; Julien, M. A Review of the Nonclinical Safety of Transcutol Ò, a Highly Purified Form of Diethylene Glycol Monoethyl Ether (DEGEE) Used as a Pharmaceutical Excipient. *Food Chem. Toxicol.* **2014**, *72*, 40–50. [CrossRef]
42. Zirak, M.B.; Pezeshki, A. Effect of Surfactant Concentration on the Particle Size, Stability and Potential Zeta of Beta Carotene Nano Lipid Carrier. *Int. J. Curr. Microbiol. Appl. Sci.* **2015**, *4*, 924–932.
43. Softisan. Available online: www.warnergraham.com/images/SoftisanHardFatsProdIn.pdf (accessed on 10 February 2022).
44. Available online: https://pubchem.ncbi.nlm.nih.gov/compound/Glyceryl-behenate (accessed on 10 February 2022).
45. Gycerol Dibehenate. *European Pharmacopeia*; Conseil de l'Europe: Strasbourg, French, 2007; pp. 2110–2111.
46. Cirri, M.; Mennini, N.; Maestrelli, F.; Mura, P.; Ghelardini, C.; Di Cesare Mannelli, L. Development and *In Vivo* Evaluation of an Innovative "Hydrochlorothiazide-in Cyclodextrins in Solid Lipid Nanoparticles" Formulation with Sustained Release and Enhanced Oral Bioavailability for Potential Hypertension Treatment in Pediatrics. *Int. J. Pharm.* **2017**, *521*, 73–83. [CrossRef]
47. Radomska-Soukharev, A.; Muller, R.H. Chemical Stability of Lipid Excipients in SLN-Production of Test Formulations, Characterization and Short-Term Stability. *Pharmazie* **2006**, *61*, 425–430.
48. Tan, S.F.; Masoumi, H.R.F.; Karjiban, R.A.; Stanslas, J.; Kirby, B.P.; Basri, M.; Basri, H.B. Ultrasonic Emulsification of Parenteral Valproic Acid-Loaded Nanoemulsion with Response Surface Methodology and Evaluation of Its Stability. *Ultrason. Sonochem.* **2016**, *29*, 299–308. [CrossRef] [PubMed]
49. Tang, S.Y.; Shridharan, P.; Sivakumar, M. Impact of Process Parameters in the Generation of Novel Aspirin Nanoemulsions—Comparative Studies between Ultrasound Cavitation and Microfluidizer. *Ultrason. Sonochem.* **2013**, *20*, 485–497. [CrossRef]
50. Aoki, M.; Ring, T.A.; Haggerty, J.S. Analysis and Modeling of the Ultrasonic Dispersion Technique. *Adv. Ceram. Mater.* **1987**, *2*, 209–212. [CrossRef]
51. Özdemir, S.; Çelik, B.; Üner, M. Properties and Therapeutic Potential of Solid Lipid Nanoparticles and Nanostructured Lipid Carriers as Promising Colloidal Drug Delivery Systems. In *Materials for Biomedical Engineering*; Elsevier: Amsterdam, The Netherlands, 2019; pp. 451–499. [CrossRef]
52. Moghimi, S.M.; Hunter, A.C. Poloxamers and Poloxamines in Nanoparticle Engineering and Experimental Medicine. *Trends Biotechnol.* **2000**, *18*, 2958–2964. [CrossRef] [PubMed]
53. Han, F.; Li, S.; Yin, R.; Liu, H.; Xu, L. Effect of Surfactants on the Formation and Characterization of a New Type of Collodial Drug Delivery System: Nanostructured Lipid Carriers. *Colloids Surfaces A Physicochem. Eng. Asp.* **2008**, *315*, 210–216. [CrossRef]

54. Appasaheb, P.S. A Review on Intranasal Drug Delivery System. *J. Adv. Pharm. Edu. Res.* **2013**, *3*, 333–346.
55. Thorat, S. Formulation and Product Development of Nasal Spray: An Overview. *Sch. J. Appl. Med. Sci.* **2016**, *4*, 2976–2985. [CrossRef]
56. Bunjes, H.; Unruh, T. Characterization of Lipid Nanoparticles by Differential Scanning Calorimetry, X-ray and Neutron Scattering. *Adv. Drug Deliv. Rev.* **2007**, *59*, 379–402. [CrossRef] [PubMed]
57. Garcia-Fuentes, M.; Alonso, M.J.; Torres, D. Design and Characterization of a New Drug Nanocarrier Made from Solid–Liquid Lipid Mixtures. *J. Colloid Interface Sci.* **2005**, *285*, 590–598. [CrossRef] [PubMed]
58. Hu, F.Q.; Jiang, S.-P.; Du, Y.-Z.; Yuan, H.; Ye, Y.Q.; Zeng, S. Preparation and Characterization of Stearic Acid Nanostructured Lipid Carriers by Solvent Diffusion Method in an Aqueous System. *Colloids Surf. B Biointerfaces* **2005**, *45*, 167–173. [CrossRef]
59. Jia, L.-J.; Zhang, D.-R.; Li, Z.-Y.; Feng, F.-F.; Wang, Y.-C.; Dai, W.-T.; Duan, C.-X.; Zhang, Q. Preparation and Characterization of Silybin-Loaded Nanostructured Lipid Carriers. *Drug Deliv.* **2010**, *17*, 11–18. [CrossRef]
60. Boyer, R.F. *Transitions and Relaxations in Amorphous and Semicrystalline Organic Polymers and Copolymers. Encyclopedia of Polymer Science and Technology*; John Wiley & Sons, Inc.: New York, NY, USA, 1977; pp. 745–839.
61. Boutonnet-Fagegaltier, N.; Menegotto, J.; Lamure, A.; Duplaa, H.; Caron, A.; Lacabanne, C.; Bauer, M. Molecular Mobility Study of Amorphous and Crystalline Phases of a Pharmaceutical Product by Thermally Stimulated Current Spectrometry. *J. Pharm. Sci.* **2002**, *91*, 1548–1560. [CrossRef]
62. Brandl, F.; Kastner, F.; Gschwind, R.M.; Blunk, T.; Teßmar, J.; Göpferich, A. Release Kinetics. *J. Control. Release* **2009**, *142*, 221–228. [CrossRef]
63. Marx, D.; Williams, G.; Birkhoff, M. Intranasal Drug Administration—An Attractive Delivery Route for Some Drugs. In *Drug Discovery and Development*; IntechOpen: London, UK, 2015; pp. 1–23. [CrossRef]
64. Basir, R.; Rahiman, S.F.; Hasballah, K.; Chong, W.; Talib, H.; Yam, M.; Jabbarzare, M.; Tie, T.; Othman, F.; Moklas, M.; et al. Plasmodium Berghei ANKA Infection in ICR Mice as a Model of Cerebral Malaria. *Iran. J. Parasitol.* **2012**, *7*, 62–74. [PubMed]

processes MDPI

Article

Formulation and Characterization of a Novel Palm-Oil-Based α-Mangostin Nano-Emulsion (PO-AMNE) as an Antimicrobial Endodontic Irrigant: An In Vitro Study

Omer Sheriff Sultan [1,*], Haresh Kumar A/L Kantilal [2], Khoo Suan Phaik [1], Hira Choudhury [3] and Fabian Davamani [4,*]

1 School of Dentistry, International Medical University, Kuala Lumpur 57000, Malaysia
2 School of Medicine, International Medical University, Kuala Lumpur 57000, Malaysia
3 School of Pharmacy, International Medical University, Kuala Lumpur 57000, Malaysia
4 School of Health Sciences, International Medical University, Kuala Lumpur 57000, Malaysia
* Correspondence: omersheriff@imu.edu.my (O.S.S.); fabian_davamani@imu.edu.my (F.D.); Tel.: +60-173987864 (O.S.S.)

Abstract: Aim: To formulate and characterize a palm-oil-in-water-based α-Mangostin nano-emulsion (PO-AMNE) endodontic irrigant, in order to evaluate its antibacterial efficacy against *Enterococcus faecalis*, *Staphylococcus epidermidis*, and *Candida albicans* biofilms, as well as its capacity to remove smear layer. Methods: The solubility of α-Mangostin in various oils was determined and selected, surfactants and co-surfactants were used for the nano-emulsion trial. PO-AMNE was prepared and optimized. The MIC was performed, and the antimicrobial efficacy was estimated against biofilms. The optimized 0.2% PO-AMNE irrigant antimicrobial efficacy in a tooth model was done using colony-forming units. The treated teeth were processed by scanning electron microscopic examination for debris and smear layer removal. An Alamar Blue assay was used to evaluate cell viability. The optimization of the PO-AMNE irrigant was performed using Box–Behnken statistical design. Results: The optimized 0.2% PO-AMNE irrigant was found to have a particle size of 340.9 nm with 0.246 PDI of the dispersed droplets, and a zeta potential (mV) of -27.2 ± 0.7 mV. The MIC values showed that 0.2% PO-AMNE (1.22 ± 0.02) were comparable to 2% CHX (1.33 ± 0.01), and 3.25% NaOCl (2.2 ± 0.09) had the least inhibition for *E. faecalis*. NaOCl (3.25%) showed the maximum inhibition of *S. epidermidis* (0.26 ± 0.05), whereas 0.2% PO-AMNE (1.25 ± 0.0) was comparable to 2% CHX (1.86 ± 0.07). For *C. albicans*, 2% CHX (8.12 ± 0.12) showed the least inhibition as compared to 0.2% PO-AMNE (1.23 ± 0.02) and 3.25% NaOCl (0.59 ± 0.02). The 0.2% PO-AMNE irrigant was then evaluated for its antimicrobial efficacy against the three biofilms, using colony-forming units. The 0.2% PO-AMNE was comparable to both 3.25% NaOCl and 2% CHX in inhibiting the growth of biofilms. The 0.2% PO-AMNE and 17% EDTA eliminated the smear layer with the lowest mean scores ($p < 0.001$). Finally, 0.2% PO-AMNE was shown to be biocompatible when compared to 17% EDTA, 3.25% NaOCl, and 2% CHX in immortalized oral keratinocyte cells. Conclusion: Overall, the formulated 0.2% PO-AMNE irrigant was an effective antimicrobial and biocompatible which could combat endodontic-infection-related polymicrobial biofilms.

Keywords: palm oil; α-Mangostin; nano-emulsion; antimicrobial; endodontic irrigant

Citation: Sultan, O.S.; Kantilal, H.K.A/L.; Phaik, K.S.; Choudhury, H.; Davamani, F. Formulation and Characterization of a Novel Palm-Oil-Based α-Mangostin Nano-Emulsion (PO-AMNE) as an Antimicrobial Endodontic Irrigant: An In Vitro Study. *Processes* **2023**, *11*, 798. https://doi.org/10.3390/pr11030798

Academic Editors: Ibrahim M. Abu-Reidah and Paolo Trucillo

Received: 2 February 2023
Revised: 10 February 2023
Accepted: 26 February 2023
Published: 7 March 2023

1. Introduction

Medicinal plants are an inexhaustible source of novel bioactive compounds which have a promising future in medicine [1,2]. Mangosteen (*Garcinia mangostana Linn*) is a fruit that grows in Asian countries, such as Malaysia, Myanmar, Thailand, the Philippines, Sri Lanka, and India. α-Mangostin, a natural xanthone produced from the pericarp of mangosteen, has been documented for its distinct pharmacological activities [3]. This compound has been shown to possess antibacterial [4], antifungal [5], and antiparasitic properties [6]. Multi-species biofilms cause infections of the root canal system in teeth and periradicular tissues, and microbial persistence appears to have a major role in the endodontic treatment outcome [7]. In cases of failed endodontic therapy and canals with persistent infections, *Enterococcus faecalis* (*E. faecalis*) and yeast, primarily *Candida albicans* (*C. albicans*), have been repeatedly identified. [8,9]. Murad et al. (2014) reported that *Staphylococcus epidermidis* (*S. epidermidis*) plays an important role in persistent endodontic infections [10]. Various studies have evaluated the effects of α-Mangostin on *E. faecalis*, *S. epidermidis*, and *C. albicans*. Flavonoids derived from the mangosteen pericarp were found to be effective against *E. faecalis* bacterial biofilm [11].

The instrumentation during the shaping of canals leads to organic and inorganic remnants on the canal, leading to an uneven and granular layer called the smear layer [12]. This layer obstructs the dentinal tubules by reducing dentinal permeability, delaying the action of topical medications and irrigants, and preventing close contact of the filling material with the dentinal walls [13]. As a result, there is an increased risk of bacterial infection and failure [14]. Removing the smear layer leads to better irrigant interaction with canal walls, resulting in improved cleaning and closer interfacial proximity between root canal fillings and dentin. Citric acid has been commonly employed to remove the smear layer, which has strong chemical stability, is inexpensive [15], and has excellent results [16,17].

Root canal irrigants have been used as an adjunct to enhance the antibacterial effect of cleaning and shaping in endodontic treatment for many years. Because of its dissolving action on pulp tissue and antibacterial activity, sodium hypochlorite (NaOCl) is the most commonly used irrigant in the treatment of infected root canals. However, when injected into periapical tissues, it has a cytotoxic effect, leaves a bad odor and taste, has a corrosive potential, and may cause allergic reactions [18]. Chlorhexidine (CHX) is a cationic bis-biguanide and has been used as an irrigating solution during root canal treatment due to its broad antibacterial activity. However, the use of CHX as a root canal irrigant is limited because it has no tissue solvent activity and some patients may have allergic reactions to it [19]; additionally, it can discolor teeth [20]. Previous research has shown that both NaOCl and CHX were highly cytotoxic to human periodontal ligament (PDL) cells, by inhibiting mitochondrial activity [21]. As a result, there is a growing need for irrigants with effective antimicrobial properties that are also biocompatible with oral tissues.

Nano-emulsions are colloidal systems that are made up of two immiscible phases, and are translucent, with droplet sizes ranging from 50 to 500 nm. Surfactants, or a mixture of surfactants and co-surfactants, are used to kinetically stabilize these systems and reduce the droplet size of the nano-emulsion [22,23]. Controlling medication release and providing a wide range of therapeutic agents are two main advantages of nano-emulsions [24]. Furthermore, they outperform macroemulsions in terms of surface area and free energy, and coalescence, flocculation, creaming, and sedimentation are also avoided. Nano-emulsions can be made with lower emulsifying agent concentrations, lowering surfactant-related toxicity [25].

The antimicrobial efficacy of α-Mangostin nano-emulsion was evaluated against *E. faecalis*, *S. epidermidis*, and *C. albicans* biofilms, as these species cause the most prevalent root canal infections. *E. faecalis* is a microorganism that is detected in persistent and asymptomatic endodontic infections, which is quite prevalent in such infections ranging from 24% to 77% [26]. Palm oil is a medium-chain fatty-acid-rich source of lauric acid (LA). Hess et al., 2015 [27] and Hinton and Ingram (2006) [28] found that LA inhibited the formation of *E. faecalis* biofilms, while Krishnapriya et al. (2019) reported that LA was effective against *C. albicans* [29]. *S. epidermidis* is also a Gram-positive facultative anaerobic bacterium that infects the root canal and stays in a benign relationship with the host. Moreover, this is an opportunistic pathogen, as it invades the host defenses like antimicrobial peptides (AMPs) on human skin and penetrates the epithelial barrier [30].

Our work aims to formulate and characterize an α-Mangostin oil-in-water nano-emulsion-based endodontic irrigant, targeting antimicrobial efficacy against *E. faecalis*, *S. epidermidis*, and *C. albicans* biofilms with smear layer removal capability. In addition, it will be tested for biocompatibility in immortalized oral kerantinocyte OKF-6 cells.

2. Materials and Methods

The α-Mangostin (purity > 95%) was procured from Chengdu Biopurify Phytochemicals (China). The procurement of palm oil, olive oil, and avocado was done from Country Farms Sdn Bhd (Malaysia). The macadamia oil was procured from Macadamia Nut Oil, Malaysia, almond oil from IN-SCENT (USA), and primrose oil from iHerb (Malaysia). The surfactants Tween 80, Tween 20, glycerol, and Span 80 were obtained from Merck, Sigma-Aldrich, St. Louis, USA. The other surfactants, Lipophille, Labrafac™ PG, Labrafil CS, and Peceol, were gifted by Gattefosse, Saint Priest, France. The Maisine and Transcutol HP co-surfactants were procured from Gattefosse, Saint Priest, France. The dialysis membrane (12 kDa cut-off) was purchased from HiMedia Laboratories Mumbai, India. A low-speed diamond saw was obtained from Buehler (Isomet, Buehler Ltd., Lake Bluff, IL, USA). A Gates Glidden drill was obtained from Mani®, Utsunoniya, Tochigi, Japan and the low-speed handpiece was obtained from Kavo, Charlotte, NC, USA). The K-files and ProTaper Universal rotary system were obtained from Dentsply, Maillefer, Ballaigues, Switzerland.

2.1. Development of PO-AMNE: Formulation and Characterization

Various oils, surfactants, and co-surfactants were employed to evaluate the solubility of α-Mangostin during the selection of components for the development of PO-AMNE.

2.1.1. Selection of Suitable Internal Phase (Oil): Screening of Oils for α-Mangostin Nano-Emulsion Development

α-Mangostin was added to the oils every 6–12 h, until the additive was no longer visibly soluble. To establish equilibrium, the mixture was maintained for 72 h at 37 °C in a shaking water bath incubator, at 100 rpm. α-Mangostin was added every 24 h, until it could no longer be visually dissolved. After 72 h, the samples were centrifuged for 15 min to produce the residue, which was diluted with methanol (Solution A) and filtered through a 0.22 mm membrane filter [31]. α-Mangostin concentration was determined using the UV spectrophotometric technique, at a wavelength of 248 nm.

2.1.2. Selection of Surfactants and Co-Surfactants

α-Mangostin was added to the surfactants and co-surfactants every 6–12 h, until the additive was no longer visibly soluble. The absorbance of the diluted sample was then measured at 248 nm. Then, to determine the final surfactant and co-surfactant, 1 mL of each, with a combination of palm oil, was added dropwise, followed by 1 min of vortexing. This was performed until the mixture turned turbid. Each sample was performed in triplicate. For the nano-emulsion test, the chosen surfactant and co-surfactant were used [31].

2.1.3. Preparation of PO-AMNE

The PO-AMNE was prepared by high-pressure homogenization for 3 cycles, alongside an aqueous titration method, based on the different ratios of the components as suggested by the software. The weighed quantity of the drug was added to the oil and stirred until complete solubilization was achieved. The mixtures of surfactants, acid, and water were added into the respective tubes in specific ratios, followed by the dropwise addition of an aqueous phase with continuous vortex-mixing, to form a clear, transparent, and homogeneous PO-AMNE. Finally, the clear, transparent, or slightly whitish PO-AMNEs were evaluated for the droplet size, polydispersity index (PDI), and zeta potential, using Zetasizer (Nano-ZS90, Malvern Instruments, Worcestershire, UK).

2.2. Thermodynamic Stability Testing of α-Mangostin Nano-Emulsion

2.2.1. Centrifugation Test

All formulations were centrifuged at 5000 rpm for 30 min and checked for phase separation, creaming, and/or cracking. The homogenous stable formulations were chosen for further research after triple testing [31].

2.2.2. Heating–Cooling Cycle

To determine the temperature effect formulation stability, formulations were kept at 40 °C and 4 °C for 48 h, in three cycles. After this, formulations were observed for any instability (phase separation, turbidity, etc.) [31].

2.2.3. Freeze–Thaw Cycle

Three freeze–thaw cycles at −20 °C and 25 °C, for 48 h each, were used to assess thermodynamically stable materials. Creaming, cracking, and phase separation was examined after the freeze–thaw cycles. To identify any drastic change in the droplet size, PDI, and surface charge, these parameters of all the stable formulations were compared [31].

2.3. Box–Behnken Statistical Design: Optimization of the α-Mangostin Nano-Emulsion (PO-AMNE)

According to the literature, oil and surfactant percentages play a significant influence in achieving the ideal droplet size and polydispersity index in nano-emulsion formulation [31,32]. In addition, the cosolvent may have affected the physical properties of nano-emulsions [33]. Therefore, in this optimization process, the percentage of oil (palm oil), percentage of Smix (Span 80: Tween 80::1:2), and percentage of cosolvent in nano-emulsion formulation were regarded as independent variables, and their effects on droplet size and polydispersity index of the nano-emulsion were investigated using a three-factor, three-level Box–Behnken statistical design (Design Expert®, version 12; State-Ease Inc., Minneapolis, MN, USA).

Three levels of the three independent variables were determined and included in the software [33] based on our preliminary studies and literature data. The droplet size and polydispersity index of each formulation were analyzed based on the software's recommendation of 17 batches, containing a combination of different levels of independent variables (Table 1). In addition, based on the experimental data of the droplet size and the PDI of the suggested 17 formulations, the software was used to obtain the software-recommended optimal formulation. Analysis of variance (ANOVA) was used to analyze the interaction of three independent variables with their three-level dependent variables.

In addition, the effect of the interaction of the three independent variables on droplet size and PDI was illustrated using software-generated perturbation plots, experimental versus predicted plot contour plots, and 3D surface plots.

$$Y = b0 + b1A + b2B + b3C + b12AB + b13AC + b23BC + b11A2 + b22B2 + b33C2 \quad (1)$$

where Y represents dependent dent variables, droplet size, and PDI, whereas b0 is the intercept, and the coefficient for the respective model, where the terms are represented by b1, b2, b3, b12, b13, b23, b11, b22, and b33 [34].

2.3.1. Measurement of Droplet Size, Polydispersity Index, and Zeta Potential

Using Zetasizer, the mean droplet size and degree of droplet size distribution were analyzed. Using the same instrument, the surface charge of the formulations was measured. As stated previously, thermodynamic stability studies were conducted on the developed PO-AMNE.

Table 1. Box–Behnken Design: three independent variables and their three levels.

Batch	Level of Independent Variables		
	A	B	C
F1	−1	0	−1
F2	1	0	1
F3	0	0	0
F4	1	0	−1
F5	0	0	0
F6	0	1	−1
F7	0	0	0
F8	−1	1	0
F9	1	1	0
F10	−1	−1	0
F11	0	−1	1
F12	0	0	0
F13	1	−1	0
F14	0	0	0
F15	−1	0	1
F16	0	1	1
F17	0	−1	−1
Independent variable	Levels		
	Low (−1)	Medium (0)	High (1)
A = Oil (% *v/v*)	5	7.5	10
B = Smix (% *v/v*) Smix :: Span 80: tween 80 :: 1:2	9	15	21
C = Glycerol (% *v/v*)	10	15	20
Dependent variables			
Y1 = Droplet size (nm)			

2.3.2. Preparation of Optimized Palm-Oil-Based α-Mangostin Nano-Emulsion (PO-AMNE)

α-Mangostin-loaded oil-in-water nano-emulsions (NEs) were formulated with the same proportions of palm oil, Smix, 10% citric acid, and aqueous phases as the optimized formulation. The low energy emulsification technique was utilized to formulate the drug-loaded NE formulation, in which 20 mg α-Mangostin was dissolved in the oil phase, followed by the addition of Smix, acid, and aqueous phase to yield 10 mL of the NE formulation (0.2% PO-AMNE).

2.3.3. Transmission Electron Microscopy (TEM)

Transmission electron microscopic evaluation was performed to determine the structure and morphology of the formulated PO-AMNE. A drop of the 100-fold diluted sample was placed on a 300-mesh carbon-coated copper grid for analysis. Finally, 2% phosphotungstic acid was used to negatively stain the droplets, which were then examined using a 100 kV microscope [35].

2.4. Antimicrobial Studies

2.4.1. Microorganisms

E. faecalis, *S. epidermidis*, and *C. albicans* in the biofilms formed in the microtiter dish were quantified directly by counting the microbial cells adhering to the surface. For the identification of any metabolic alterations of three test organisms, *E. faecalis*, *S. epidermidis*, and *C. albicans*, an analytical profile index (API) identification scheme had been used [36,37].

2.4.2. Determination of Minimum Inhibitory Concentration (MIC)

For the present study, MIC was performed for *E. faecalis*, *S. epidermidis*, and *C. albicans*. The concentration of α-Mangostin nano-emulsion that has been used for the study is 0.2% of α-Mangostin [38].

2.4.3. Micro Dilution Method

Here, the MIC of PO-AMNE against *E. faecalis*, *S. epidermidis*, and *C. albicans* chloride (TTC) was used as an indicator [39].

2.4.4. Biofilm Assay (Microtiter Plate)

Bacteria are incubated at the "U"-shaped bottom of 96-well microtiter plates, in order to observe the adherence of bacteria to an abiotic surface. Cells that are adhered to the wells get stained for visualization [40].

2.5. Ex Vivo Experiment—Antimicrobial Efficacy—0.2% PO-AMNE Irrigant in aTooth Model Using Colony-Forming Units (CFU)

2.5.1. Tooth Sample Preparation

One hundred sixty freshly extracted caries-free single-rooted mandibular premolars extracted for orthodontic treatment were collected from patients with their informed consent in accordance with a protocol, reviewed and approved by the Institutional Review Board of International Medical University, Joint Committee on Research and Ethics for the research project ID: PMHS I-2018 (01). In this research, the experiment model was a modified version of the Haapasalo and Orstavik tooth model that included bovine teeth. This model was used because it offered a more accurate simulation of clinical circumstances for evaluating the antimicrobial efficacy of endodontic irrigants in dentinal tubules. To evaluate the CFU, one hundred and sixty dentin blocks were prepared. These teeth specimens of 8 mm width were obtained by removing the crowns (2–3 mm from the cement–enamel junction), and 3 to 5 mm of the apical portion of the root, using a low-speed diamond saw (Isomet, Buehler Ltd., Lake Bluff, IL, USA). The inner diameter of the root canals of each sample was standardized using a Gates Glidden size 3 (Mani®, Utsunoniya, Tochigi, Japan) drill at 300 rpm with a low-speed handpiece (Kavo, Charlotte, NC, USA) [41].

2.5.2. Antimicrobial Assessment Using Colony-Forming Units (CFU)

The CFU measurement technique was used to estimate the antimicrobial efficacy of a nano-formulation irrigant in tooth models. The *E. faecalis*, *S. epidermidis*, and *C. albicans* were grown overnight in a BHI broth [42]. Each dentin block was submerged in microcentrifuge tubes containing BHIA (Brain Heart Infusion Agar) broth that had been pre-sterilized. To contaminate the dentin block, 50 μL of the overnight culture of *E. faecalis*, *S. epidermidis*, and *C. albicans* cultures were placed in each microcentrifuge tube. On BHIA (Brain Heart Infusion Agar) plates, 5 μL of the broth from the incubated dentin blocks was subcultured to determine the purity of the culture. The samples of teeth were irrigated. The dentin blocks were randomly divided into four groups of 10 specimens for each organism (*E. faecalis*, *S. epidermidis*, and *C. albicans*): group 1: 0.2% PO-AMNE, group 2: 2% CHX (CHX), group 3: 3.25% NaOCl, and group 4: saline (control). After irrigation, bacterial growth was evaluated by harvesting dentin chips at 200 μm and 400 μm with Gates Glidden drills No. 4 and No. 5 (Mani®, Utsunoniya, Tochigi, Japan), respectively [43].

2.6. Smear Layer Removal in Tooth Model

To perform this experiment, 30 extracted human single-rooted premolars were selected for this study. The samples were randomly divided into 3 groups ($n = 10$), where treatments were as follows: group 1: 0.2% PO-AMNE, group 2: 17% EDTA and group 3: saline.

Access cavities were prepared and working lengths were determined by deducting 1 mm from the length at which the file was barely visible to the naked eye at the apical foramen. Using the ProTaper Universal rotary system (Dentsply, Maillefer, Ballaigues, Switzerland), canals were prepared. Each canal was finished with a #F3 apical preparation. Each solution (30 mL) was used to determine the effects of experimental and control solutions as a final rinse on the surface of root canals after instrumentation, for a total duration of one minute. A 30-gauge side-vented needle (Dentsply Tulsa Dental, Tulsa, OK, USA) was passively placed within the middle third of the root canals to deliver the irrigating solution. The teeth were then divided longitudinally into two halves. The samples were then scheduled for scanning electron microscopic (SEM) evaluation [44].

2.6.1. Selection of Representative SEM Sections

After the central beam of the SEM had been directed to the center of the object by the SEM operator at 10× magnification, the magnification was increased to 200× and subsequently, 1000×, respectively, and the canal wall region appearing on the screen was photographed [45].

2.6.2. SEM Analysis and Scoring

The smear layer score systems, introduced by Hülsmann et al., [45] were followed to score the smear layer (Table 2). These scoring systems were applied to the coronal, middle, and apical thirds of the canal [45].

Table 2. Scoring criteria for smear layer evaluation.

Score	Criteria
1	Clean root canal wall and only a few small debris particles
2	A few small agglomerations of debris
3	Many agglomerations of debris covering less than 50% of the root canal wall
4	More than 50% of the root canal walls were covered with debris
5	Complete or nearly complete root canal wall coverage with debris

The results were then dichotomized into "clean canal wall", which included scores 1 and 2, and "smear layer present", which included scores 3, 4, and 5.

2.7. Biocompatibility Test Using an Alamar Blue Assay on Immortalized Oral Kerantinocyte OKF-6 Cells

The Alamar Blue assay was used to determine the viability of immortalized oral keratinocyte OKF-6 cells. The cells were cultured in keratinocyte serum-free medium (SFM, Thermo Fischer, Waltham, MA, USA) and seeded at a density of 1.8 × 104 cells per well in a 96-well black plate. The cells were allowed to attach overnight. The cells were then treated with group 1: 20 mgs/mL α-Mangostin, group 2: 0.2% PO-AMNE, group 3: 3.25% NaOCl, group 4: 2% CHX, and group 5: 17% EDTA. The cells were then treated for 1 min, 5 min, 10 min, 30 min, and 1 h. Later, the medium in each well was changed to a standard culture medium. Then, 10 μL of Alamar Blue (Invitrogen, Carlsbad, CA, USA) was added to each well, followed by a 4 h incubation at 37 °C and 5% carbon dioxide. Furthermore, the fluorescence of each well was measured using the FLx800 fluorescence microplate reader at 570 nm excitation and 590 nm emission wavelengths (Bio-Tek, Winooski, VT, USA) [46,47].

2.8. Statistical Analysis

The data obtained were entered manually in Microsoft Excel Data, and the data were cleaned. IBM SPSS Statistics for Windows software, Version 29.0. (IBM Corp: Armonk, NY, USA)) was used for data analysis, to represent the data graphically and provide statistical analysis. The mean and standard deviation was analyzed using descriptive statistics. The normality of the data was assessed by using the Kolmogorov–Smirnov test. A one-way analysis of variance (ANOVA), with a Dunnett's test, was used to investigate significant differences between the four groups of data that followed (differences between the four groups). To do the pair-wise comparison, a Tukey's post-hoc test was used. The significance level was set at $p < 0.05$.

3. Results

3.1. Formulation and Characterization of PO-AMNE

Selection of Oils, Surfactants, and Co-Surfactant for PO-AMNE Development

The selection of suitable components for the development of NE was determined in different components.

The solubility profiles of α-Mangostin in various components are depicted in Table 3, where the highest solubility of α-Mangostin was obtained in palm oil (29.579 ± 0.101 mg/mL), Tween 80 (70.804 ± 0.102 mg/mL), Span 80 (69.236 ± 0.032 mg/mL), and glycerol (212. 412 ± 0.07 mg/mL). As a result, palm oil 90, Tween 80, Span 80, and glycerol were chosen as the oil, surfactant, and co-surfactant for the next phase of NE development.

Table 3. Solubility profiles of α-Mangostin in different oils surfactants and co-surfactants.

Oils	α-Mangostin Solubility * (mg/mL)
Palm oil	29.579 ± 0.101
Olive	27.288 ± 0.006
Avocado	27.38 ± 0.010
Macadamia	27.182 ± 0.011
Almond	27.885 ± 0.045
Primrose	26.807 ± 0.008
Surfactants and Co-surfactants	**α-MangostinSolubility (mg/mL) ***
Tween 80	70.804 ± 0.102
Tween 20	64.31 ± 0.0588
Lipophille	34.71 ± 0.03
Labrafac PG	54.632 ± 0.014
Labra CS	64.104 ± 0.074
Peceol	39.2422 ± 0.25
Maisine	34.51 ± 0.045
Glycerol	212. 412 ± 0.07
Transcutol HP	418.457 ± 0.122
Transcutol HP	417.755 ± 0.076
Span 80	69.236 ± 0.032

* The data are presented in this table as the mean ± SD, $n = 3$.

3.2. Measurement of Droplet Size, Polydispersity Index, and Zeta Potential

3.2.1. Droplet Size

Among the various physical properties of nano-emulsion, droplet size is one of the most crucial parameters because it affects the formulation aesthetic appeal and stability [48]. To lessen the experimental burden, we included droplet size as one of the dependent variables in the optimization procedure. Table 4 displays the results of a statistical analysis of the effect of the interaction between the percentage of oil, Smix, and glycerol (cosolvent) on the droplet size of blank nano-emulsion.

Table 4. Analysis of variance data for droplet size.

Source	F-Ratio	*p*-Value
Model	79.96	<0.0001
A	568.47	<0.0001
B	46.82	0.0002
C	0.0808	0.7845
AB	16.76	0.0046
AC	7.79	0.0269
BC	9.46	0.0179
A^2	48.02	0.0002
B^2	15.36	0.0058
C^2	8.53	0.0223
Residual		
Lack of Fit	2.68	0.1823
Pure Error		

The model F-value of 79.96 and *p*-value of <0.05 represented the significance of the best fit quadratic model, and the representative *p*-values ($p < 0.05$) for the model terms A, B, AB, AC, BC, A2, B2, and C2 have the statistical significance influence on droplet size of the developed nano-emulsion.

The F-value of 2.68 and $p > 0.05$ for the lack of fit indicated the insignificance of the lack of fit relative to the pure error. Non-significance of the lack of fit is good as it represents the model to be fit. An adequate precision (signal-to-noise ratio) value of 28.764 indicated an adequate signal for the suggested quadratic model. Hence, this model could be used to navigate the design space. A polynomial equation on the responses of three independent variables on globule size was generated in the fitted model (Equation (2)), where Y1 represents the droplet size of the formulation. Coefficient values for the respective model terms help to identify the relative impact of the independent variables on globule size.

Together with increasing palm oil and glycerol %, the positive coefficient of model terms A (+230.71) and C (+2.75) represented that the droplet size of the nano-emulsion was also increased; although, the effect of glycerol content on the droplet size of the formulation was insignificant, as represented in Table 4 ($p > 0.05$). As per the literature, a significant effect of glycerol content was reported when the % of glycerol was more than 20% [49]. In our experiment, the percentage was within 20%, which might the reason for the insignificant effect of glycerol content on droplet size. Alternatively, a negative coefficient value of −66.21 for the model term indicated increasing droplet size, with increasing % of Smix. The effect of oil content and Smix content on droplet size were in agreement with the literature data. The negative coefficient value (−56.02) of model term AB (Equation (2)) indicated decreasing in droplet size with increasing the combined effect of model terms A and B. Similarly, model term C was also associated with a negative coefficient, which represents a decrease in droplet size, with increasing combined interaction of model terms A and C. Combined effects of model terms AB and AC on droplet size were significant, which was evident in their respective *p*-values; whereas, a positive coefficient (+38.20) was associated with the model term AC, which represents the combined effect of model terms A and C as inverse with globule size.

$$Y1 = +428.76 + 230.71 \times A - 66.21 \times B + 2.75 \times C + 92.43 \times A2 - 56.02 \times A \times B + 38.20 \times A \times C - 52.27 \times B2 - 42.10 \times B \times C + 38.96 \times C2 \quad (2)$$

Moreover, proportional increases in droplet size with increasing oil % are evident with a steep positive slope in the perturbation plot (Figure 1A) associated with increasing model term A, which is in agreement with the positive coefficient value (+230.71) of the model term A in Equation (2). Alternatively, the negative slope in the perturbation plot (Figure 1A) associated with increasing model term B indicated decreasing droplet size with increasing Smix %, which is follows the negative slope (−66.21) of the model term B in Equation (2). The similar effect of model terms A and B are also evident in the 3D surface plot (Figure 1B). Increasing the surfactant % might provide a layer for the adsorption of the oil and water phase and reduce the interfacial tension between the dispersible droplets and the external aqueous phase, which leads to a decrease in droplet size [50]. Predicted and experimental globule size data are close together, which can be seen in the predicted vs. actual plot (Figure 1C).

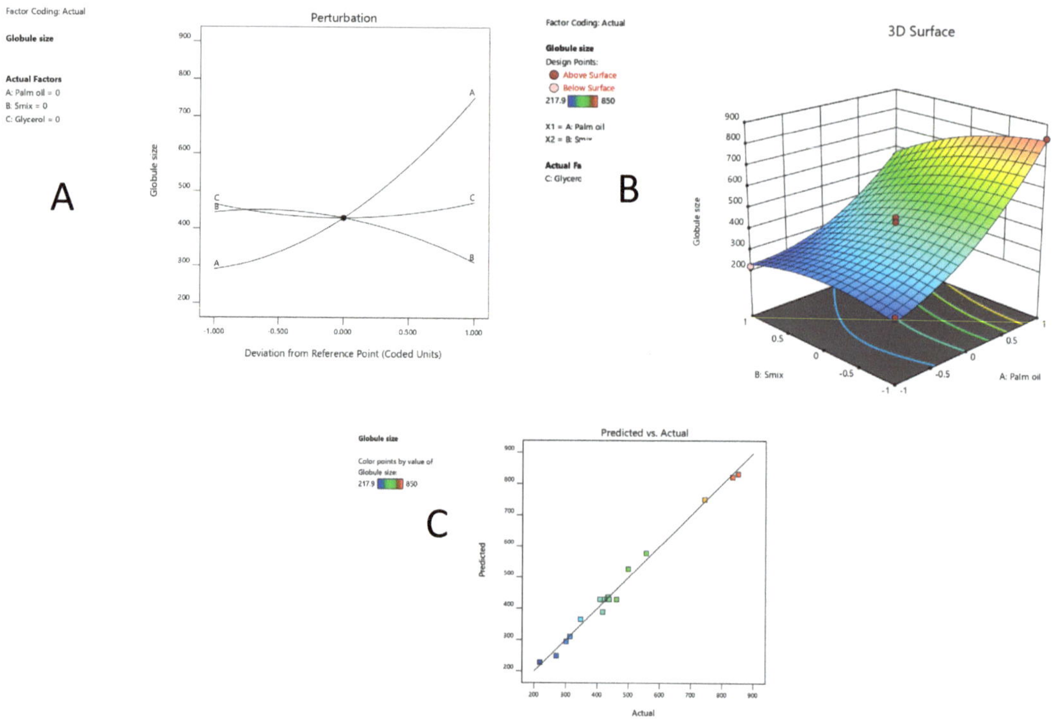

Figure 1. Effect of interaction of independent variables on the droplet size of nano-emulsion formulation: (**A**) Perturbation plot, (**B**) 3D surface plot, (**C**) predicted vs. actual graph.

3.2.2. Effect of the Independent Variable on Polydispersity Index

The PDI measures the homogeneity of dispersed oil droplets. Additionally, it ensures the formulation stability. Low PDI values signify the high kinetic stability of nano-emulsion formulation, while high PDI values signify low formulation stability [51]. Uniform droplets with a narrow size distribution are represented by low PDI values, which signify high kinetic stability.

The statistical data from the quadratic model, of the interaction of three independent variables on PDI, is mentioned in Table 5. The model F-value of 15.12 and p-value of 0.0008 ($p < 0.05$) indicated model significance. Further, model terms A, B, C, A^2, and B^2 are found to have a significant effect on the PDI of the droplet size range of the formulation, as the respective $p < 0.05$. The F-value of 3.57 for the lack of fit term and respective p-value ($p > 0.05$) represented that the lack of fit is not significant relative to the pure error. An insignificant lack of fit is good, as we want the model to fit. Additionally, an adequate precision value of 14.113, more than 4, indicated a sufficient signal. Hence, this model could be used to navigate the design space.

Table 5. Analysis of variance data for PDI.

Source	F-Ratio	p-Value
Model	15.12	0.0008
A	19.96	0.0029
B	70.92	<0.0001
C	7.34	0.0303
AB	1.72	0.2306
AC	0.3818	0.5562
BC	4.54	0.0705
A^2	17.68	0.0040
B^2	11.64	0.0113
C^2	0.9561	0.3607
Residual		
Lack of Fit	3.57	0.1254
Pure Error		

The closeness of actual PDI and predicted PDI is evident in the predicted vs. actual plots (Figure 2C). The quadratic model generated a polynomial equation on the effect of three independent variables on the PDI of the formulation, presented in (Equation (3)), where Y2 represents the PDI of the formulation. Coefficient values corresponding to the respective model terms indicate the relative impact of the independent variables on the PDI of the nano-emulsion.

$$Y2 = +0.2396 + 0.0511 \times A - 0.0964 \times B + 0.0310 \times C + 0.0663 \times A2 - 0.0212\text{terms} - 0.0100 \times A \times C + 0.0538 \times B2 - 0.0345 \times B \times C - 0.0154 \times C2 \quad (3)$$

The positive coefficient of the model terms A and C, along with a significant p-value ($p < 0.05$) (Table 5 and Equation (3)), indicated the significant increase in PDI with increasing oil and glycerol % in nano-emulsion; whereas the negative coefficient value associated with the model term B indicated that there was a decrease in PDI with increasing surfactant content. The perturbation plot and 3D surface plot on the effect of % of oil, Smix, and glycerol on PDI of the formulation, are shown in Figure 2A,B, respectively, where an increasing PDI with increasing oil content, and contrarily, the decreasing PDI with increasing Smix, is evident. The obtained results are in agreement with the respective negative and positive coefficient values in Equation (3). Additionally, the highest coefficient value associated with the model term B was also reflected with a sharp negative decline in the perturbation plot with model term B. The positive coefficient value of A was reflected in the positive slope associated with model term A in the perturbation plot, whereas the least coefficient value associated with model term C in Equation (3) was also related to model term C in Table 5 and the smallest slope in the perturbation plot, per the insignificant p-value (Figure 2C).

Increasing the amount of surfactant may result in the successful coating of oil droplets, which may result in a low PDI value. In addition, as the number of oil droplets increases, there may be insufficient surfactant to effectively coat the dispersed droplets in nanoemulsion formulation, leading to the coalescence of incompletely coated dispersed droplets. Our findings are consistent with the published literature [51].

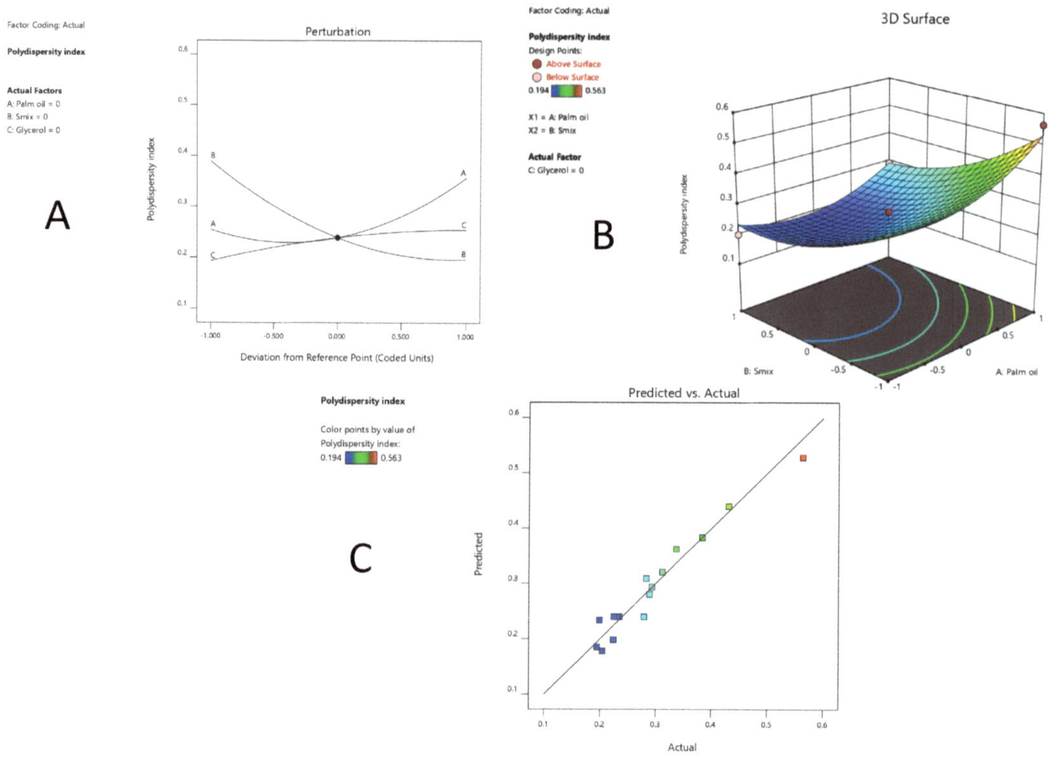

Figure 2. Effect of independent variables on PDI: (**A**) Perturbation graph of the droplet size, (**B**) 3D surface plot, (**C**) predicted vs. actual graph.

3.3. Optimization and Characterization of 0.2% PO-AMNE

α-Mangostin was loaded into the optimized NE formulation by dissolving the drug to obtain a final concentration of 0.2% mg/mL (20 mg/mL) in the oil phase, and formulated as described above, incorporating desired percentages of Smix and the aqueous phase. The globule size of the drug-loaded NE was determined as 340.9 nm, whereas the PDI 0.246 and surface charge were recorded as 27.2 ± 0.7 mV (Figure 3). α-Mangostin was loaded into the optimized NE formulation by dissolving the drug to achieve a final concentration of 0.2% mg/mL (20 mg/mL) in the oil phase and formulating with the desired proportions of Smix and the aqueous phase, as described above. The globule size of the drug-loaded NE was measured to be 340.9 nm, while the PDI 0.246 and surface charge were measured to be −27.2 ± 0.7 mV, respectively (Figure 3b).

The 0.2% PO-AMNE was formulated using concentrations of 0.2% α-Mangostin solution, and the formulation was then characterized based on the change in globule size, PDI, and surface charge. TEM analysis of the 0.2% PO-AMNE formulation revealed the coated droplets' spherical morphology (Figure 3A). Figure 3B histogram indicates that the particles in the PO-AMNE formulation have a smaller range of sizes, which is consistent with Table 6 PDI value (0.246) and the morphological observation in the TEM image (Figure 3B). In addition, the uniformity of the droplet size (340.9 nm) in the absence of crystalline AM confirmed that the drug was completely entrapped within the oil phase of the formulation.

Table 6. Physical characterization of 0.2% PO-AMNE.

Nano-Formulation	Average Particle Size (nm)	Polydispersity Index (PDI)	Zeta Potential (mV)
0.2% PO-AMNE	340.9 nm	PDI 0.246	−27.2 ± 0.7 mV

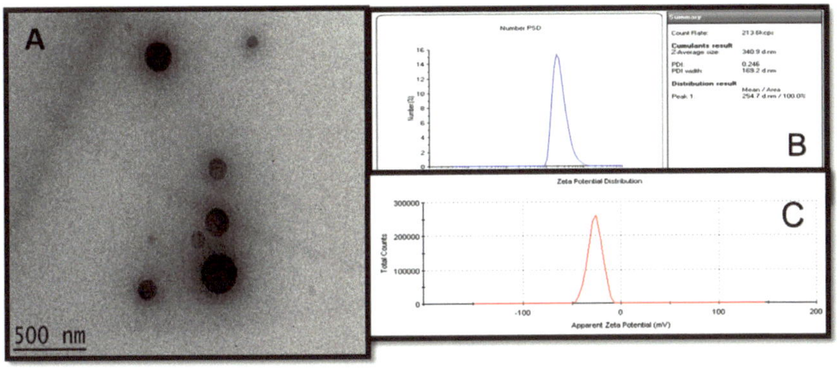

Figure 3. Representation of morphology of the optimized (0.2%) PO-AMNE (**A**), particle size distribution (**B**), and zeta potential (**C**).

3.4. *Ex Vivo Antimicrobial Studies*

3.4.1. Enumeration and Identification of *Enterococcus Faecalis*, *Staphylococcus epidermidis*, and *Candida albicans*

For the identification of any metabolic alterations for three test organisms, *E. faecalis*, *S. epidermidis*, and *C. albicans*, an analytical profile index (API) identification scheme was used (Figure 4A–C).

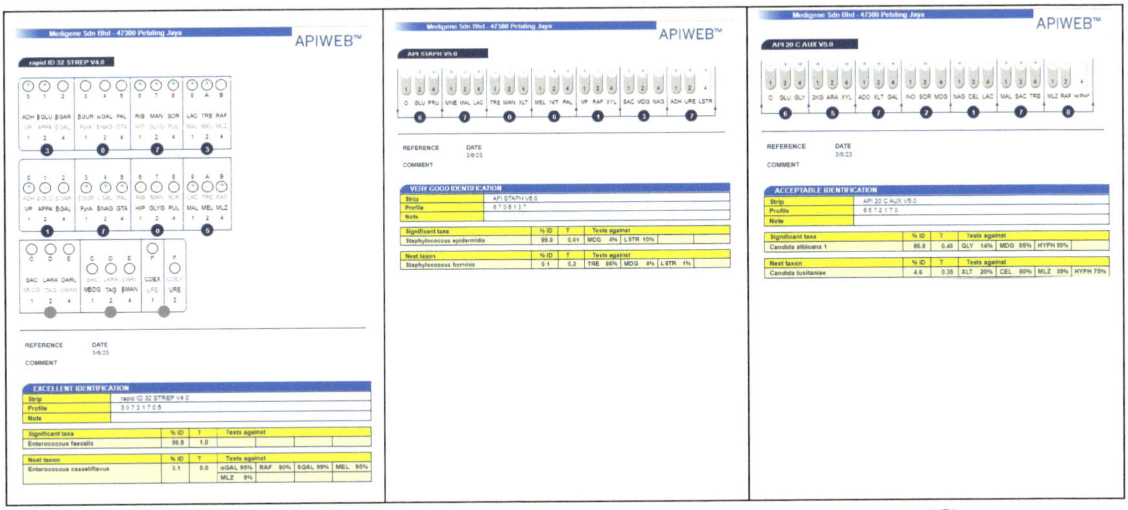

(**A**) (**B**) (**C**)

Figure 4. (**A**) API identification of *E. faecalis;* (**B**) API identification of *S. epidermidis;* (**C**) API identification of *C. albicans*.

3.4.2. Determination of Minimum Inhibitory Concentration (MIC)

α-Mangostin is comparable to both 3.25% NaOCl and 2% CHX in inhibiting the growth of tested microbes. The MIC values showed that 0.2% α-Mangostin (1.22 ± 0.02) was comparable to 2% CHX (1.33 ± 0.01) and 3.25% NaOCl (2.2 ± 0.09) and had the least inhibition of *E. faecalis*. NaOCl (3.25%) showed the maximum inhibition of *S. epidermidis* (0.26 ± 0.05), whereas 0.2% α-Mangostin (1.25 ± 0.0) was comparable to 2% CHX (1.86 ± 0.07). For *C. albicans*, 2% CHX (8.12 ± 0.12) showed the least inhibition, as compared to 0.2% α-Mangostin (1.23 ± 0.02) and 3.25% NaOCl (0.59 ± 0.02), as shown in Figure 5.

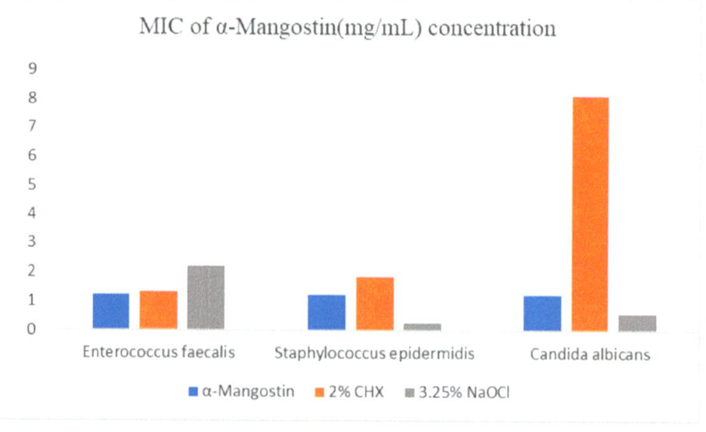

Figure 5. MIC of α-Mangostin(mg/mL) concentration against *E. faecalis*, *S. epidermidis*, and *C. albicans*.

3.5. Ex Vivo Experiment: Preparation of Human Teeth Specimens to Check the Antimicrobial Efficacy of PO-AMNE Irrigant in a Tooth Model Using Colony-Forming Units (CFU)

3.5.1. Antimicrobial Assessment Using CFU

A reduction in the number of CFUs was statistically significant in all groups compared to the control group ($p < 0.05$).

Results for *E. faecalis* at 200 μm

The mean CFU of 0.2% PO-AMNE was 4.63 ± 0.26, 4.68 ± 0.35 for 2% CHX, 4.87 ± 0.15 for 3.25% NaOCl, and 6.05 ± 0.04 for saline, as shown in Table 7. In the presence of nanoformulation, the CFU formation of *E. faecalis* was lower compared to either 2% CHX or 3.25% NaOCl at a depth of 200 μm, showing its potent antimicrobial effect.

Table 7. Mean and standard deviation of irrigants against *E. faecalis* at 200 μm.

Irrigants	N	Mean	Std. Deviation
0.2% PO-AMNE	10	4.63972392	0.266228615
2% CHX	10	4.68676222	0.350818870
3.25% NaOCl	10	4.87883986	0.158879880
Saline	10	6.05010073	0.043594317
Total	40	5.06385668	0.625931661

Overall, 0.2% PO-AMNE had statistical significance when compared to 2% CHX against *E. faecalis*. The 95% confidence interval was −0.2455 to 0.3396, with a standard error of 0.139. The *p*-value was found to be <0.0001. Similarly, the 0.2% PO-AMNE irrigant had statistical significance when compared to 3.25% NaOCL. The 95% confidence interval was observed to be 0.0331 to 0.4451, with a *p*-value of <0.0001.

Results for *E. faecalis* at 400 μm

The mean CFU of 0.2% PO-AMNE was 5.20 ± 0.09, 5.52 ± 0.07 for 2% CHX, 5.52 ± 0.27 for 3.25% NaOCl and 6.14 ± 0.01 for saline, as shown in Table 8. In the presence of nanoformulation, the CFU formation of *E. faecalis* was lower when compared to either CHX or NaOCl at a depth of 400 μm, showing its potent antimicrobial effect.

Table 8. Mean and standard deviation of irrigants against *E. faecalis* at 400 μm.

Irrigants	N	Mean	Std. Deviation
0.2% PO-AMNE	10	5.20722816	0.098434833
2% CHX	10	5.52280850	0.074578498
3.25% NaOCl	10	5.52824399	0.027822768
Saline	10	6.14335492	0.016507675
Total	40	5.60040889	0.349093792

Overall, 0.2% PO-AMNE had statistical significance when compared to 2% CHX against *E. faecalis*. The 95% confidence interval was −0.3976 to −0.2335, with a standard error of 0.039. The *p*-value was found to be <0.0001. Similarly, the 0.2% PO-AMNE irrigant had statistical significance when compared to 3.25% NaOCL. The 95% confidence interval was observed to be 0.2531 to 0.3890, with a *p*-value of <0.0001.

Results for *S. epidermidis* at 200 μm

The mean CFU of 0.2% PO-AMNE was 3.77 ± 0.37, 4.44 ± 0.34 for 2%, 4.78 ± 0.86 for 3.25% NaOCl and 6.07 ± 0.01 for saline, as shown in Table 9. In the presence of nanoformulation, the CFU of *S. epidermidis* was lower when compared to either CHX or NaOCl at a depth of 200 μm, showing its potent antimicrobial effect.

Table 9. Mean and standard deviation of irrigants against *S. epidermidis* at 200 μm.

Irrigants	N	Mean	Std. Deviation
0.2% PO-AMNE	10	3.77083657	0.379532943
2% CHX	10	4.44137166	0.342868124
3.25% NaOCl	10	4.78653050	0.086264246
Saline	10	6.07355352	0.013985945
Total	40	4.76807306	0.884068563

Overall, 0.2% PO-AMNE had statistical significance when compared to 2% CHX against *E. faecalis*. The 95% Confidence interval was 0.3307 to 1.0103, with a standard error of 0.162. The *p*-value was found to be <0.0001. Similarly, the 0.2% PO-AMNE irrigant had statistical significance when compared to 3.25% NaOCL. The 95% confidence interval was observed to be 0.7571 to 1.2743, with a *p*-value of <0.0001.

Results for *S. epidermidis* at 400 μm

The mean CFU of 0.2% PO-AMNE was 4.40 ± 0.17, 4.54 ± 0.14 for 2% CHX, 5.00 ± 0.29 for 3.25% NaOCl and 6.13 + 0.01 for saline, as shown in Table 10. In the presence of nanoformulation, the CFU formation of *S. epidermidis* was lower when compared to either CHX or NaOCl at a depth of 400 μm, showing its potent antimicrobial effect.

Table 10. Mean and standard deviation of irrigants against *S. epidermidis* at 400 μm.

Irrigants	N	Mean	Std. Deviation
0.2% PO-AMNE	10	0.171520772	0.054239631
2% CHX	10	0.469422690	0.148444489
3.25% NaOCl	10	0.094514831	0.029888214
Saline	10	0.016537682	0.005229674
Total	40	0.731541254	0.115666828

Overall, 0.2% PO-AMNE had statistical significance when compared to 2% CHX against *S. epidermidis*. The 95% confidence interval was −0.1951 to 0.4690, with a standard error of 0.158. The *p-value* was found to be <0.0001. Similarly, the 0.2% PO-AMNE irrigant had statistical significance when compared to 3.25% NaOCL. The 95% confidence interval was observed to be 0.4658 to 0.7260, with a *p-value* of <0.0001.

Results for *C. albicans* at 200 μm

The mean CFU of 0.2% PO-AMNE was 2.83 + 0.24, 3.14 ± 0.18 for 2%CHX, 3.34 ± 0.24 for 3.25% NaOCl and 5.09 ± 0.02 for saline, as shown in Table 11. In the presence of nano-formulation, the CFU formation of *C. albicans* was lower when compared to either CHX or NaOCl at a depth of 200 μm, showing its potent antimicrobial effect.

Table 11. Mean and standard deviation of irrigants against *C. albicans* at 200 μm.

Irrigants	N	Mean	Std. Deviation
0.2% PO-AMNE	10	2.83659430	0.247612845
2% CHX	10	3.14848172	0.181043992
3.25% NaOCl	10	3.34090591	0.240291718
Saline	10	5.09852677	0.020267240
Total	40	3.60612717	0.910935697

Overall, 0.2% PO-AMNE had statistical significance when compared to 2% CHX against *C. albicans*. The 95% confidence interval was 0.1081 to 0.5157, with a standard error of 0.097. The *p*-value was found to be <0.0001. Similarly, the 0.2% PO-AMNE irrigant had statistical significance when compared to 3.25% NaOCL. The 95% confidence interval was observed to be 0.2751 to 0.7335, with a *p*-value of <0.0001.

Results for *C. albicans* at 400 μm

The mean CFU of 0.2% PO-AMNE was 2.95 ± 0.15, 3.11 ± 0.14 for 2% CHX, 3.33 ± 0.11 for 3.25% NaOCl and 5.14 ± 0.02 for saline, as shown in Table 12. In the presence of nano-formulation, the CFU formation of *C. albicans* was lower when compared to either CHX or NaOCl at a depth of 400 μm, showing its potent antimicrobial effect.

Overall, 0.2% PO-AMNE had statistical significance when compared to 2% CHX against *C. albicans*. The 95% confidence interval was 0.0143 to 0.2965, with a standard error of 0.067. The *p*-value was found to be <0.0001. Similarly, the 0.2% PO-AMNE irrigant had statistical significance when compared to 3.25% NaOCL. The 95% confidence interval was observed to be 0.2559 to 0.4978, with a *p*-value of <0.0001.

Table 12. Mean and standard deviation of irrigants against *C. albicans* at 400 μm.

Irrigants	N	Mean	Std. Deviation
0.2% PO-AMNE	10	2.95885666	0.150876977
2% CHX	10	3.11424553	0.149477061
3.25% NaOCl	10	3.33568568	0.101826728
Saline	10	5.14021700	0.025604369
Total	40	3.63725122	0.896448891

3.6. Smear Layer Removal in Tooth Models

The results obtained from this study are summarized in Tables 13–15, and show the SEM images (Figure 6) of the tested irrigants. Group 1: 0.2% PO-AMNE and group 2: 17% EDTA exhibited better efficacy in removing the smear layer without altering the normal dentinal structures with the lowest mean scores ($p < 0.001$), followed by group 3: saline. For all the treated groups, there was no statistically significant difference ($p < 0.05$) between the scores at each root third (cervical, middle, apical).

Table 13. Group 1: Debris score with 0.2% PO-AMNE.

Coronal Third					Middle Third					Apical Third				
Clean		Debris Present			Clean		Debris Present			Clean		Debris Present		
1 *	2	3	4	5	1	2	3	4	5	1	2	3	4	5
0	8/10 +	2/10 +	0	0	1/10 +	8/10 +	0	0	0	0/10 +	8/10 +	1/10 +	1/10 +	0
8/10 + (80%) *+		2/10 + (20%) *+			9/10 + (30%) *+		0/10 + (70%) *+			8/10 + (60%) *+		2/10 + (100%) *+		

* Debris scores (Hülsmann et al., 1997 [45]). + Number of canals presented with a given score. *+ Dichotomized scores: scores 1 to 2 (clean canal wall) versus 3 to 5 (debris present).

Table 14. Group 2: Debris score with 17% EDTA.

Coronal Third					Middle Third					APICAL THIRD				
Clean		Debris Present			Clean		Debris Present			Clean		Debris Present		
1 *	2	3	4	5	1	2	3	4	5	1	2	3	4	5
0	8/10 +	2/10 +	0	0	0	4/10 +	6/10 +	0	0	1/0 +	4/10 +	2/10 +	1/0 +	0
8/10 + (100%) *+		2/10 + (100%) *+			4/10 + (100%) *+		6/10 + (100%) *+			5/10 + (100%) *+		3/10 + (100%) *+		

* Debris scores (Hülsmann et al., 1997 [45]). + Number of canals presented with a given score. *+ Dichotomized scores: scores 1 to 2 (clean canal wall) versus 3 to 5 (debris present).

Table 15. Group 2: Debris score with saline.

Coronal Third					Middle Third					Apical Third				
Clean		Debris Present			Clean		Debris Present			Clean		Debris Present		
1 *	2	3	4	5	1	2	3	4	5	1	2	3	4	5
0	0	0	0	10/10 +	0	0	0	0	10/10 +	0	0	0	0	10/10 +
0/10 + (100%) *+		10/10 + (100%) *+			0/10 + (100%) *+		10/10 + (100%) *+			0/10 + (100%) *+		10/10 + (100%) *+		

* Debris scores (Hülsmann et al., 1997 [45]). + Number of canals presented with a given score. *+ Dichotomized scores: scores 1 to 2 (clean canal wall) versus 3 to 5 (debris present).

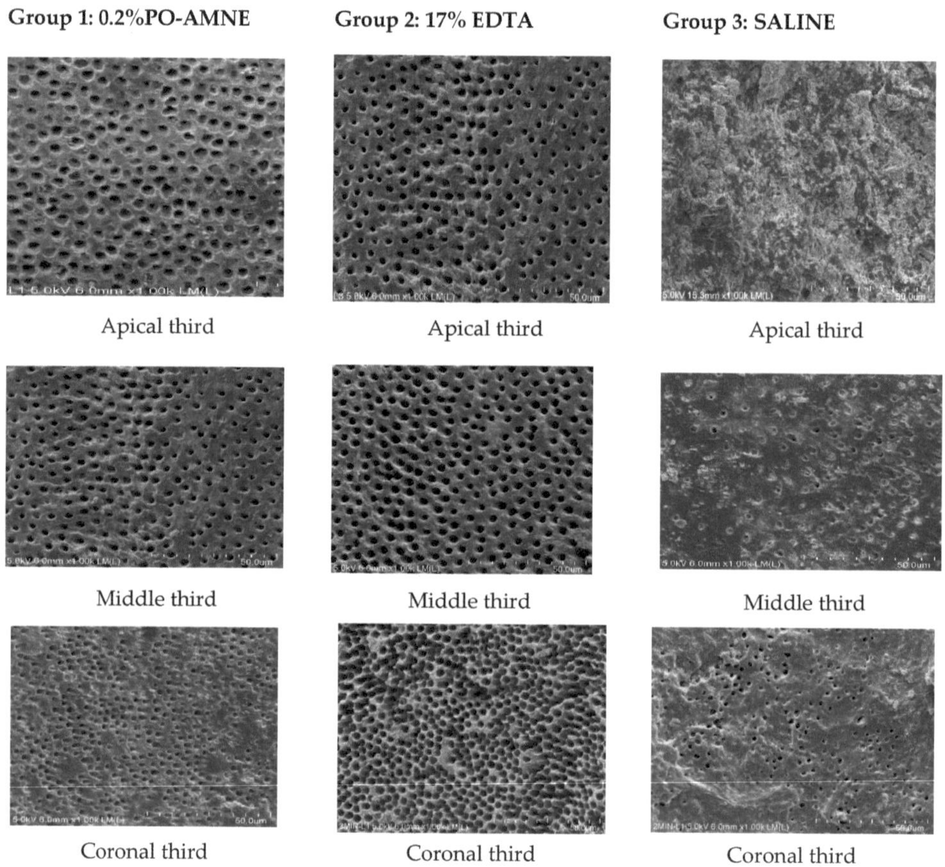

Figure 6. SEM images of all groups.

3.6.1. Debris Scores for 0.2% PO-AMNE

In the coronal third, 8 out of 10 (80%) samples were scored as 2, respectively, representing a clean dentin surface. In the middle third, 9 out of 10 (90%) were scored 1 or 2, respectively, representing a clean dentin surface, and 0 out of 10 (60%) were scored as 3, 4, and 5. In the apical third, 8 out of 10 (80%) samples were scored as 2, representing a clean dentin surface, and 2 out of 10 (20%) samples were scored as 3 and 4, respectively, representing debris present on the dentin surface, as shown in Table 13.

3.6.2. Debris Scores for 17% EDTA

In the coronal third, 8 out of 10 (80%) and 2 out of 10 (20%) samples were scored as either 1 or 2, respectively, representing a clean dentin surface. In the middle third, 4 out of 10 (40%) were scored as 2, representing a clean dentin surface, and 6 out of 10 (60%) were scored as 3, respectively, representing debris present on the dentin surface. In the apical third, 5 out of 10 (50%) samples were scored as either 1 or 2, respectively, representing a clean dentin surface, and 5 out of 10 (50%) samples were scored as either 3, 4 or 5, respectively, representing debris present on the dentin surface, as shown in Table 14.

3.6.3. Debris Score for Saline

In the coronal third, no samples were characterized with scores of 1 or 2, and 10 out of 10 (100%) samples were scored as 5, representing debris present on the dentin surface.

In the middle third, no samples were characterized with scores of 1 or 2, and 10 out of 10 (100%) samples were scored as 5, representing debris present. In the apical third, 0 out of 10 (100%) samples were scored as either 1 nor 2, and 10 out of 10 canals (100%) were scored as 5, representing debris present on the dentin surface as shown in Table 15.

Group 1 (0.2%) PO-AMNE and group 2 (17% EDTA) exhibited better efficacy in removing the smear layer without altering the normal dentinal structures with the lowest mean scores ($p < 0.001$), followed by group 3 (saline). There was no statistically significant difference ($p < 0.05$) between the scores at each root third (cervical, middle, apical) for group 1 and group 2, though the apical third scores were less than the other root thirds.

3.7. Biocompatibility Test Using Alamar Blue Assay on Immortalized Oral Kerantinocytes OKF-6 Cells

The proliferation percentage of the OKF-6 cells in the presence of various irrigants is presented in Figure 7. The percentage of OKF-6 cells was calculated at 1 min, 5 min,10 min, 30 min, and 60 min. The results show that group 1 (20 mg α-Mangostin) was comparable to group 2 (0.2% PO-AMNE) after 60 min, with cell viability percentages of 35.48% and 32.51%, whereas there was a decrease in the cell viability percentages for group 3 (3.25% NaOCl), group 4 (CHX), and group 5 (17% EDTA) proliferation percentages of 6.47%, 4.30%, and 4.15%, respectively.

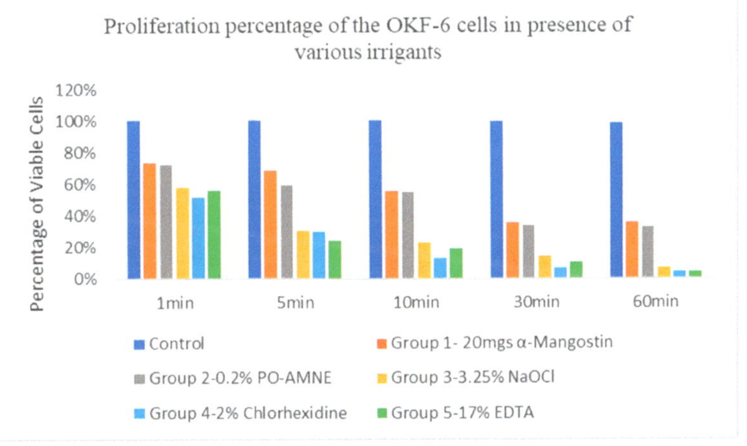

Figure 7. Proliferation percentage of OKF-6 cells in the presence of various irrigants.

4. Discussion

For many years, root canal irrigants have been utilized as an adjuvant to improve the antibacterial efficacy of endodontic cleaning and shaping. Because of its pulp-dissolving and antibacterial properties, NaOCl is the most used irrigant for treating infected root canals. However, when injected into the periapical tissues, it has a cytotoxic effect, leaves a bad odor and taste, has a corrosive potential, and may cause allergic reactions. Due to its extensive antibacterial action, the cationic bis-biguanide CHX has been employed as an irrigating solution during root canal therapy. However, the use of CHX as a root canal irrigant is limited, because it has no tissue solvent activity and some patients may have allergic reactions to it; additionally, it can discolor teeth [52]. A previous study has shown that both NaOCl and CHX impair the mitochondrial activity of human periodontal ligament (PDL) cells, making them extremely cytotoxic [53]. As a result, there is a growing need for irrigants with effective antimicrobial properties, that are also biocompatible with oral tissues.

The filling of the dentinal walls and organic remnants present in the canal leaves an uneven and granular smear layer during root canal instrumentation, called the smear layer [54]. This layer obstructs the dentinal tubules by reducing dentinal permeability, delaying the action of topical medications and irrigants, and preventing close contact of the filling material with the dentinal walls [55]. Consequently, there is a greater chance of bacterial infection and failure. The choice of commercially available chelating agents used in the study was EDTA and citric acid for smear layer removal, as there are ample data in the literature showing the effectiveness of these agents for the elimination of the smear layer from root canals. Citric acid is a weak acid that removes the smear layer, has strong chemical stability, is inexpensive, and has excellent results [56].

The selection of *E. faecalis*, *C. albicans*, and *S. epidermidis* biofilms was based on scientific data demonstrating that these pathogens are the most often isolated bacterial species in endodontic patients. *E. faecalis* in dentinal tubules is 1000 times more resistant to phagocytosis, antibodies, and antimicrobials than non-biofilm-producing organisms. Additional virulence features have been observed in *E. faecalis* clinical isolates from asymptomatic, chronic endodontic infections of the root canal and oral cavity [57]. *C. albicans* has been regularly identified as the species most often recovered from failing root canals requiring retreatment. Waltimo et al. proposed fungi as microorganisms resistant to endodontic therapy in apical periodontitis and proved that *C. albicans* species need incubation with a saturated calcium hydroxide solution for 16 h [58]. Peciuliene et al. [59] identified fungi as resistant microorganisms in the obturated root canals of teeth diagnosed with chronic apical periodontitis. Egan et al. conducted a study to determine the association between the presence of fungi in the saliva and the root canals of teeth in patients with apical periodontitis. They discovered that fungi were 13.8 times more prevalent in the root canals when they were also present in the patient's saliva [60].

The Gram-positive bacterium, *S. epidermidis*, was frequently found to be the cause of endodontic flare-ups and secondary infections. However, in many instances, the infection has persisted despite multiple attempts to alleviate symptoms (such as intracanal medication placement and antimicrobials) [61].

Our study aimed to formulate an oil-based nano-emulsion containing a mixture of α-Mangostin, palm oil, Tween® 80, Span® 80, glycerol, and 10%, citric acid and investigate its effectiveness at inhibiting *E. faecalis*, *S. epidermis*, and *C. albicans* biofilms, as well as to assess its biocompatibility using immortalized oral keratinocyte (OKF-6) cells.

In our study, the solubility profiles of α-Mangostin in various oils, such as olive oil, macadamia oil, almond oil, and primrose oil, were evaluated. The highest solubility of AM was obtained with palm oil (29.57 mg/mL), Tween® 80 (70.80 mg/mL), Span® 80 (69.23 mg/mL), and glycerol (212.41 mg/mL). As a result, palm oil 90, Tween® 80, Span® 80, and glycerol were the oils, surfactants, and co-surfactants of choice for the next phase of NE development. Based on the Box–Behnken statistical design, the composition of nano-emulsion 0.2% PO-AMNE had an average particle size of 340.9 nm. This was in agreement with studies conducted by Marcel et al., who developed a limonene nano-emulsion formed by a high-pressure homogenizer. The average particle size in their study varied from 264.50 nm to 434.20 nm [62]. The zeta potential is used for predicting dispersion stability, and its value depends on the physicochemical property of the drug, polymer, vehicle, presence of electrolytes, and adsorption. The zeta potential of 0.2% PO-AMNE in this study was -27.2 + 0.7 mV. It is in accordance with a study done by Đorđević et al., who concluded that a zeta potential of ±30 mV is sufficient to ensure the physical stability of NE [63]. Ahmad et al. concluded that all their nano-formulations had a zeta potential of −28.5 to −40.2 mV. The formulations having a zeta potential charge value greater than −30 mV indicate that nano-emulsion formulations are stable [64]. The PDI represents the consistency of nano-emulsion droplet size. The greater the value, the less uniform the nano-emulsion droplet size will be. It is the ratio of the standard deviation to the mean droplet size. A PDI of 0.08 or less indicates a monodispersed sample, while 0.08 to 0.70 represents the middle range of PDI [65]. Our research revealed a PDI of 0.246, which showed the consistency of nano-emulsion droplet sizes.

The MIC values in our study showed that 0.2% α-Mangostin (1.22 ± 0.02) was comparable to 2% CHX (1.33 ± 0.01) and 3.25% NaOCl (2.2 ± 0.09) for *E. faecalis*. These results were consistent with the study by Kaomongkolgit et al., which showed that α-Mangostin is effective against *E. faecalis* [66]. In our study, 3.25% NaOCl showed maximum inhibition of *S. epidermidis* (0.26 ± 0.05), whereas 0.2% α-Mangostin (1.25 ± 0.00) was comparable to 2% CHX (1.86 ± 0.07), in accordance with Sivaranjani et al. [67]. For *C. albicans*, 2% CHX (8.12 ± 0.12) showed the least inhibition as compared to 0.2% α-Mangostin (1.23 ± 0.02) and 3.25%NaOCl (0.59 ± 0.02), as per the results reported by Kaomongkolgit et al., who determined that α-Mangostin's potent antifungal activity and low toxicity made it a viable drug for the treatment of oral candidiasis [68]. The irrigant 0.2% PO-AMNE was bactericidal and fungicidal against *E. faecalis*, *S. epidermidis*, and *C. albicans* biofilms. According to a study by Leelapornpisid et al., the planktonic inhibitory concentrations of α-Mangostin were bactericidal on *E. faecalis* at 2–4 mg/L, and candidal on *C. albicans* at 1000 mg/L [69]. We discovered that 0.2% PO-AMNE was both bactericidal and fungicidal against *E. faecalis*, *S. epidermidis*, and *C. albicans* biofilms at 0.625 mg/L, which is lower than previously published literature [69]. Our results are in agreement with the study by Nourzadeh et al., where the mean CFU of 0.2% PO-AMNE (4.63 ± 0.26) was lower than for 2% CHX (4.68 ± 0.35), 3.25% NaOCl (4.87 ± 0.15) and saline (6.05 ± 0.04) at a depth of 200 μm for *E. faecalis*. The same findings were observed at a depth of 400 μm, where the mean CFU of 0.2% PO-AMNE (5.20 ± 0.09) was lower than 2% CHX (5.52 ± 0.07), 3.25% NaOCl (5.52 ± 0.27) and saline (6.14 ± 0.01) [70]. Overall, 0.2% PO-AMNE had a statistically significant difference when compared to 2% CHX and 3.25% NaOCl against *E. faecalis* at 200 μm and 400 μm.

The mean CFU in our study for *S. epidermidis*, at a depth of 200 μm, was 3.77 ± 0.37 for 0.2% PO-AMNE, 4.44 ± 0.34 for 2% CHX, 4.78 ± 0.86 for 3.25% NaOCl and 6.07 ± 0.01 for saline. At a depth of 400 μm, the mean CFU of *S. epidermidis* was 4.40 ± 0.17 for 0.2% PO-AMNE, 4.54 ± 0.14 for 2% CHX, 5.00 ± 0.29 for 3.25% NaOCl and 6.13 ± 0.01 for saline. Overall, 0.2% PO-AMNE had statistical significance when compared to 2% CHX and 3.25% NaOCl against *S. epidermidis* at 200 μm and 400 μm. The CFU formation at a depth of 200 μm and 400 μm for *S. epidermidis* in the presence of 0.2% PO-AMNE was lower when compared to either 2% CHX or 3.25% NaOCl, showing its potent antimicrobial effect. The overall results for *S. epidermidis* at 200 and 400 were in agreement with the study done by Schmidt et al., who studied the effects of CHX, povidone–iodine, and triple antibacterial solution against *S. epidermidis* biofilms [71].

In our study, the mean CFU of 0.2% PO-AMNE (2.83 ± 0.24) was lower than 2% CHX (3.14 ± 0.18), 3.25% NaOCl (3.34 ± 0.24) and saline (5.09 ± 0.02), at a depth of 200 μm for *C. albicans*. The same results were observed at a depth of 400 μm, where the mean CFU of 0.2% PO-AMNE (2.95 ± 0.15) was lower than 2% CHX (3.11 ± 0.14), 3.25% NaOCl (3.33 ± 0.11) and saline (5.14 ± 0.02). This shows that 0.2% PO-AMNE has a potent anti-fungal effect. This is in accordance with the study conducted by Vianna et al., 2004, who evaluated the in vitro antifungal activity of CHX and NaOCl. They concluded that CHX eliminated *C. albicans* within 15 s [72]. Radcliffe et al. tested the effects of various concentrations of NaOCl on *C. albicans* and found that all concentrations of NaOCl reduced CFU below the limit of detection within 10 s [73].

In our study, the SEM analysis showed that none of the tested root canal irrigants could completely remove the smear layer. According to our study, 0.2% PO-AMNE and 17% EDTA exhibited better efficacy in removing the smear layer without altering the normal dentinal structures with the lowest mean scores ($p < 0.001$), followed by saline. There was no statistically significant difference ($p < 0.05$) between the scores at each root third (cervical, middle, apical) for all groups. This is per the findings by Sakinah et al., who concluded that the results of SEM photomicrograph assessment showed a little debris on the surface of the root canal walls and plenty of opened dentin tubules on the canal walls irrigated with extract of mangosteen peel [74]. Charlie et al. reported similar results, where they concluded that EDTA might not penetrate the narrow apical region of teeth [75].

The oral keratinocyte cells were used for the study, as they act as the major barrier to physical, microbial, and chemical agents that may cause local cell injury [76]. They participate in the proinflammatory process by producing cytokines constitutively, or in response to a number of stimuli. Therefore, oral keratinocytes may contribute to the management of oral infections through an inflammatory process involving several interleukins, including IL-13 and IL-18 [77]. Oral mucosa keratinocytes are the primary source of IL-13, an inflammatory cytokine that regulates the production of IL-6 and IL-8 [78]. In our investigation, the proportion of OKF-6 cells proliferating in the presence of different irrigants was determined at 1 min, 5 min, 10 min, 30 min, and 60 min. According to our findings, 0.2% PO-AMNE had a better cell viability percentage of 32.51% after 60 min compared to 3.25% NaOCl, CHX, and 17% EDTA proliferation percentages, which decreased to 6.47%, 4.30%, and 4.15%, respectively. This is consistent with the studies done by Abate et al. and Ngawhirunpat et al., who found that *G. mangostana* extracts are not cytotoxic to human keratinocytes [79,80].

5. Conclusions

α-Mangostin from *Garcinia mangostana Linn* has been explored by many researchers and it has been observed that it is not just effective against bacteria, but also other microbes, such as fungi and mycobacteria. The optimization of the palm-oil-based α-Mangostin nanoemulsion (PO-AMNE) irrigant was performed using a Box–Behnken statistical design. The formulated 0.2% PO-AMNE endodontic irrigant had an overall significant antimicrobial effectiveness against *E. faecalis*, *S. epidermidis*, and *C. albicans* biofilms. 0.2%PO-AMNE endodontic irrigant was further evaluated for the smear layer removal and was found to be comparable with 17% EDTA. Finally, 0.2% PO-AMNE was found to be biocompatible with immortalized oral keratinocyte OKF-6 cells. Overall, the antimicrobial efficacy and safety of the formulated 0.2% PO-AMNE endodontic irrigant has the potential to combat polymicrobial biofilms related to endodontic infections.

Author Contributions: Conceptualization, O.S.S.; methodology, O.S.S.; software, O.S.S.; validation, H.K.A/L.K., F.D.; formal analysis; investigation, H.K.A/L.K., O.S.S.; resources, F.D. and H.C.; data curation, H.C.; writing—original draft preparation, O.S.S. and K.S.P.; writing—review and editing, O.S.S. and K.S.P.; visualization, O.S.S.; H.K.A/L.K., F.D. and K.S.P.; project administration, F.D.; funding acquisition, O.S.S. Authorship is limited to those who have contributed substantially to the work reported. All authors have read and agreed to the published version of the manuscript.

Funding: This research was funded by the Institutional Review Board of International Medical University, Joint Committee on Research and Ethics under the research project ID- PMHS I-2018 (01).

Institutional Review Board Statement: This study required ethical approval.

Informed Consent Statement: Not applicable.

Data Availability Statement: Not Applicable.

Acknowledgments: The authors thank the labs at the International Medical University (IMU), and MIMOS Research Center for the research experiments and analysis. This work was supported in part by the International Medical University (IMU).

Conflicts of Interest: The authors declare no conflict of interest.

References

1. Tapsell, L.C.; Hemphill, I.; Cobiac, L.; Patch, C.S.; Sullivan, D.R.; Fenech, M.; Roodenrys, S.; Keogh, J.B.; Clifton, P.M.; Williams, P.G.; et al. Health benefits of herbs and spices: The past, the present, the future. *Med. J. Aust.* **2006**, *185*, S1–S24. [CrossRef] [PubMed]
2. Triggiani, V.; Resta, F.; Guastamacchia, E.; Sabbà, C.; Licchelli, B.; Ghiyasaldin, S.; Tafaro, E. Role of antioxidants, essential fatty acids, carnitine, vitamins, phytochemicals and trace elements in the treatment of diabetes mellitus and its chronic complications. *Endocr. Metab. Immune Disord. Drug Targets* **2006**, *6*, 77–93. [CrossRef] [PubMed]
3. Schmidt, K.; Estes, C.; McLaren, A.; Spangehl, M.J. Chlorhexidine Antiseptic Irrigation Eradicates Staphylococcus epidermidis From Biofilm: An In Vitro Study. *Clin. Orthop. Relat. Res.* **2018**, *476*, 648–653. [CrossRef] [PubMed]

4. Sakagami, Y.; Iinuma, M.; Piyasena, K.G.N.P.; Dharmaratne, H.R.W. Antibacterial activity of α-Mangostin against vancomycin-resistant *Enterococci* (VRE) and synergism with antibiotics. *Phytomedicine* **2005**, *12*, 203–208. [CrossRef]
5. Kaomongkolgit, R.; Jamdee, K.; Chaisomboon, N. Antifungal activity of alpha-mangostin against Candida albicans. *J. Oral. Sci.* **2009**, *51*, 401–406. [CrossRef]
6. Larson, R.T.; Lorch, J.M.; Pridgeon, J.W.; Becnel, J.J.; Clark, G.G.; Lan, Q. The biological activity of alpha-mangostin, a larvicidal botanic mosquito sterol carrier protein-2 inhibitor. *J. Med. Entomol.* **2010**, *47*, 249–257.
7. Fabricius, L.; Dahlén, G.; Öhman, A.E.; Möller, A.J. Predominant indigenous oral bacteria isolated from infected root canals after varied times of closure. *Scand. J. Dent. Res.* **1982**, *90*, 134–144. [CrossRef]
8. Love, R.M. *Enterococcus faecalis*: A mechanism for its role in endodontic failure. *Int. Endod. J.* **2001**, *34*, 399–405. [CrossRef]
9. Poptani, B.; Sharaff, M.; Archana, G.; Parekh, V. Detection of Enterococcus faecalis and Candida albicans in previously root-filled teeth in a population of Gujarat with polymerase chain reaction. *Contemp. Clin. Dent.* **2013**, *4*, 62–66. [CrossRef]
10. Murad, C.F.; Sassone, L.M.; Faveri, M.; Hirata, R., Jr.; Figueiredo, L.; Feres, M. Microbial diversity in persistent root canal infections investigated by checkerboard DNA-DNA hybridization. *J. Endod.* **2014**, *40*, 899–906. [CrossRef]
11. Dalhar, H.; Latief, M.; Ketut, S.; Dian, A.W. Effectiveness of flavonoid from mangosteen pericarp (*Garcinia mangostana* L.) as *Enterococcus faecalis* antibiofilm. *Conserv. Dent. J.* **2017**, *7*, 18–22.
12. McComb, D.; Smith, D.C. A preliminary scanning electron microscopic study of root canals after endodontic procedures. *J. Endod.* **1975**, *1*, 238–242. [CrossRef]
13. Hülsmann, M. Effects of mechanical instrumentation and chemical irrigation on the root canal dentin and surrounding tissues. *Endod. Top.* **2013**, *9*, 55–86. [CrossRef]
14. Sen, B.H.; Wesselink, P.R.; Türkün, M. The smear layer: A phenomenon in root canal therapy. *Int. Endod. J.* **1995**, *28*, 141–148. [CrossRef]
15. Di Lenarda, R.; Cadenaro, M.; Sbaizero, O. Effectiveness of 1 mol L-1 citric acid and 15% EDTA irrigation on smear layer removal. *Int. Endod. J.* **2000**, *33*, 46–52. [CrossRef]
16. Haznedaroglu, F. Efficacy of various concentrations of citric acid at different pH values for smear layer removal. *Oral. Surg. Oral. Med. Oral. Pathol. Oral. Radiol. Endod.* **2003**, *96*, 340–344. [CrossRef] [PubMed]
17. Dioguardi, M.; Gioia, G.D.; Illuzzi, G.; Laneve, E.; Cocco, A.; Troiano, G. Endodontic irrigants: Different methods to improve efficacy and related problems. *Eur. J. Dent.* **2018**, *12*, 459–466. [CrossRef]
18. Mohammadi, Z.; Asgary, S. A comparative study of antifungal activity of endodontic irrigants. *Iran Endod. J.* **2015**, *10*, 144–147. [PubMed]
19. Mohammadi, Z. Chlorhexidine gluconate, its properties and applications in endodontics. *Iran Endod. J.* **2008**, *2*, 113–125. [PubMed]
20. Yamashita, J.C.; Tanomaru Filho, M.; Leonardo, M.R.; Rossi, M.A.; Silva, L.A. Scanning electron microscopic study of the cleaning ability of chlorhexidine as a root-canal irrigant. *Int. Endod. J.* **2003**, *36*, 391–394. [CrossRef]
21. Chang, Y.C.; Huang, F.M.; Tai, K.W.; Chou, M.Y. The effect of NaOCl and CHX on cultured human periodontal ligament cells. *Oral Surg. Oral Med. Oral Pathol. Oral Radiol. Endod.* **2001**, *92*, 446–450. [CrossRef]
22. Narang, J.K.; Narang, R. Emerging role of nanoemulsions in oral health management. *Int. J. Pharm. Investig.* **2017**, *7*, 1–3. [CrossRef]
23. Kotta, S.; Khan, A.W.; Pramod, K.; Ansari, S.H.; Sharma, R.K.; Ali, J. Exploring oral nanoemulsions for bioavailability enhancement of poorly water-soluble drugs. *Expert Opin. Drug Deliv.* **2012**, *9*, 585–598. [CrossRef]
24. Sabjan, K.B.; Munawar, S.M.; Rajendiran, D.; Vinoji, S.K.; Kasinathan, K. Nanoemulsion as Oral Drug Delivery-A Review. *Curr. Drug Res. Rev.* **2020**, *12*, 4–15. [CrossRef]
25. Liu, Q.; Huang, H.; Chen, H.; Lin, J.; Wang, Q. Food-Grade Nanoemulsions: Preparation, Stability and Application in Encapsulation of Bioactive Compounds. *Molecules* **2019**, *21*, 4242. [CrossRef]
26. Jhajharia, K.; Parolia, A.; Shetty, K.V.; Mehta, L.K. Biofilm in endodontics: A review. *J. Int. Soc. Prev. Community Dent.* **2015**, *5*, 1–12. [CrossRef]
27. Hess, D.J.; Henry-Stanley, M.J.; Wells, C.L. The natural surfactant glycerol monolaurate significantly reduces development of *Staphylococcus aureus* and *Enterococcus faecalis* biofilms. *Surg Infect.* **2015**, *16*, 538–542. [CrossRef]
28. Hinton, A., Jr.; Ingram, K.D. Antimicrobial activity of potassium hydroxide and lauric acid against microorganisms associated with poultry processing. *J. Food Prot.* **2006**, *69*, 1611–1615. [CrossRef]
29. Devan, K.; Peedikayil, F.C.; Chandru, T.P.; Kottayi, S.; Dhanesh, N.; Suresh, K.R. Antimicrobial efficacy of medium chain fatty acids as root canal irrigants: An in vitro study. *J. Indian Soc. Pedod. Prev. Dent.* **2019**, *37*, 258–264. [CrossRef]
30. Ghonmode, W.N.; Balsaraf, O.D.; Tambe, V.H.; Saujanya, K.P.; Patil, A.K.; Kakde, D.D. Comparison of the antibacterial efficiency of neem leaf extracts, grape seed extracts and 3% sodium hypochlorite against *E. feacalis*-An in vitro study. *J. Int. Oral Health* **2013**, *5*, 61–66.
31. Gorain, B.; Choudhury, H.; Kundu, A.; Sarkar, L.; Karmakar, S.; Jaisankar, P.; Pal, T.K. Nanoemulsion strategy for olmesartan medoxomil improves oral absorption and extended antihypertensive activity in hypertensive rats. *Colloids Surf. B Biointerfaces* **2014**, *115*, 286–294. [CrossRef] [PubMed]
32. Choudhury, H.; Zakaria, N.F.B.; Tilang, P.A.B.; Tzeyung, A.S.; Pandey, M.; Chatterjee, B.; Alhakamy, N.A.; Bhattamishra, S.K.; Kesharwani, P.; Gorain, B.; et al. Formulation development and evaluation of rotigotine mucoadhesive nanoemulsion for intranasal delivery. *J. Drug Deliv. Sci. Technol.* **2019**, *54*, 101301. [CrossRef]

33. Chong, W.T.; Tan, C.P.; Cheah, Y.K.; Lajis, A.F.B.; Habi Mat Dian, N.L.; Kanagaratnam, S.; Lai, O.M. Optimization of process parameters in preparation of tocotrienol-rich red palm oil-based nanoemulsion stabilized by Tween80-Span 80 using response surface methodology. *PLoS ONE* **2018**, *13*, e0202771. [CrossRef] [PubMed]
34. Kumbhar, S.A.; Kokare, C.R.; Shrivastava, B.; Gorain, B.; Choudhury, H. Preparation, characterization, and optimization of asenapine maleate mucoadhesive nanoemulsion using Box-Behnken design: In vitro and in vivo studies for brain targeting. *Int. J. Pharm.* **2020**, *30*, 119499. [CrossRef]
35. Akrawi, S.H.; Gorain, B.; Nair, A.B.; Choudhury, H.; Pandey, M.; Shah, J.N.; Venugopala, K.N. Development and Optimization of Naringenin-Loaded Chitosan-Coated Nanoemulsion for Topical Therapy in Wound Healing. *Pharmaceutics* **2020**, *20*, 893. [CrossRef]
36. Hofherr, L.; Votava, H.; Blazevic, D.J. Comparison of three methods for identifying nonfermenting gram-negative rods. *Can. J. Microbiol.* **1978**, *24*, 1140–1144. [CrossRef]
37. Otto, L.A.; Blachman, U. Nonfermentative bacilli: Evaluation of three systems for identification. *J. Clin. Microbiol.* **1979**, *10*, 147–154. [CrossRef]
38. Owuama, C.I. Determination of minimum inhibitory concentration (MIC) and minimum bactericidal concentration (MBC) using a novel dilution tube method. *Afr. J. Microbiol. Res.* **2017**, *11*, 977–980.
39. Wu, G.; Yang, Q.; Long, M.; Guo, L.; Li, B.; Meng, Y.; Zhang, A.; Wang, H.; Liu, S.; Zou, L. Evaluation of agar dilution and broth microdilution methods to determine the disinfectant susceptibility. *J. Antibiot.* **2015**, *68*, 661–665. [CrossRef]
40. Coffey, B.M.; Anderson, G.G. Biofilm formation in the 96-well microtiter plate. In *Pseudomonas Methods and Protocols*; Humana Press: New York, NY, USA, 2014; pp. 631–641.
41. Haapasalo, M.; Ørstavik, D. In vitro infection and disinfection of dentinal tubules. *J. Dent. Res.* **1987**, *66*, 1375–1379. [CrossRef]
42. Galán, C.; Ariatti, A.; Bonini, M.; Clot, B.; Crouzy, B.; Dahl, A.; FernandezGonzález, D.; Frenguelli, G.; Gehrig, R.; Isard, S.; et al. Recommended terminology for aerobiological studies. *Aerobiologia* **2017**, *33*, 293–295. [CrossRef]
43. Chum, J.D.; Lim, D.J.Z.; Sheriff, S.O.; Pulikkotil, S.J.; Suresh, A.; Davamani, F. *In vitro* evaluation of octenidine as an antimicrobial agent against *Staphylococcus epidermidis* in disinfecting the root canal system. *Restor. Dent. Endod.* **2019**, *8*, e8.
44. Andrabi, S.M.; Kumar, A.; Tewari, R.K.; Mishra, S.K.; Iftekhar, H. An In Vitro SEM Study on the Effectiveness of Smear Layer Removal of Four Different Irrigations. *Iran Endod. J.* **2012**, *7*, 171–176.
45. Hülsmann, M.; Rümmelin, C.; Schäfers, F. Root canal cleanliness after preparation with different endodontic handpieces and hand instruments: A comparative SEM investigation. *J. Endod.* **1997**, *23*, 301–306. [CrossRef]
46. Matt, C.; Jennifer, S.C. *Guide to Research Techniques in Neuroscience*, 2nd ed.; Elsevier Publication: Amsterdam, The Netherlands, 2010.
47. Ngo, Y.X.; Haga, K.; Suzuki, A.; Kato, H.; Yanagisawa, H.; Izumi, K.; Sada, A. Isolation and Culture of Primary Oral Keratinocytes from the Adult Mouse Palate. *J. Vis. Exp.* **2021**, *24*, 175.
48. Saberi, A.H.; Fang, Y.; McClements, D.J. Effect of glycerol on formation, stability, and properties of vitamin-E enriched nanoemulsions produced using spontaneous emulsification. *J. Colloid Interface Sci.* **2013**, *411*, 105–113. [CrossRef]
49. Posocco, P.; Perazzo, A.; Preziosi, V.; Laurini, E.; Priclab, S.; Guidocde, S.; Posocco, P. Interfacial tension of oil/water emulsions with mixed non-ionic surfactants: Comparison between experiments and molecular simulations. *RSC Adv.* **2016**, *6*, 4723–4729. [CrossRef]
50. Algahtani, M.S.; Ahmad, M.Z.; Ahmad, J. Investigation of Factors Influencing Formation of Nanoemulsion by Spontaneous Emulsification: Impact on Droplet Size, Polydispersity Index, and Stability. *Bioengineering* **2022**, *9*, 384. [CrossRef]
51. Sarheed, O.; Dibi, M.; Ramesh, K.V.R.N.S. Studies on the Effect of Oil and Surfactant on the Formation of Alginate-Based O/W Lidocaine Nanocarriers Using Nanoemulsion Template. *Pharmaceutics* **2020**, *17*, 1223. [CrossRef]
52. Vasconcelos, B.C.de.; Luna-Cruz, S.M.; De-Deus, G.; Moraes, I.G.D.; Maniglia-Ferreira, C.; Gurgel-Filho, E.D. Cleaning ability of chlorhexidine gel and sodium hypochlorite associated or not with EDTA as root canal irrigants: A scanning electron microscopy study. *J. Appl. Oral Sci.* **2007**, *15*, 387–391. [CrossRef]
53. Karkehabadi, H.; Yousefifakhr, H.; Zadsirjan, S. Cytotoxicity of Endodontic Irrigants on Human Periodontal Ligament Cells. *Iranian Endodontic Journal.* **2018**, *13*, 390–394. [PubMed]
54. Scelza, M.Z.; de Noronha, F.; da Silva, L.E.; Maurício, M.; Gallito, M.A.; Scelza, P. Effect of Citric Acid and Ethylenediaminetetraacetic Acid on the Surface Morphology of Young and Old Root Dentin. *Iran Endod. J.* **2016**, *11*, 188–191. [PubMed]
55. Bystrom, A.; Sundqvist, G. The antibacterial action of NaOCl and EDTA in 60 cases of endodontic therapy. *Int. Endod. J.* **1985**, *18*, 35–40. [CrossRef] [PubMed]
56. Sedgley, C.; Buck, G.; Appelbe, O. Prevalence of Enterococcus faecalis at multiple oral sites in endodontic patients using culture and PCR. *J. Endod.* **2006**, *32*, 104–109. [CrossRef] [PubMed]
57. Stuart, C.H.; Schwartz, S.A.; Beeson, T.J.; Owatz, C.B. *Enterococcus faecalis*: Its role in root canal treatment failure and current concepts in retreatment. *J. Endod.* **2006**, *32*, 93–98. [CrossRef]
58. Waltimo, T.M.T.; Siren, E.K.; Torkko, H.L.; Olsen, I.; Haapasalo, M.P. Fungi in therapy-resistant apical periodontitis. *Int. Endod. J.* **1997**, *30*, 96–101. [CrossRef]
59. Peciuliene, V.; Reynaud, A.H.; Balciuniene, I.; Haapasalo, M.P.P. Isolation of yeasts and enteric bacteria in root-filled teeth with chronic apical periodontitis. *Int. Endod. J.* **2001**, *34*, 429–434. [CrossRef]

60. Egan, M.W.; Spart, D.A.; Ng, Y.L.; Lam, J.M.; Moles, D.R.; Gulabivala, K. Prevalence of yeast in saliva and root canals of teeth associated with apical periodontitis. *Int. Endod. J.* **2002**, *35*, 321–329. [CrossRef]
61. Röhner, E.; Jacob, B.; Böhle, S.; Rohe, S.; Löffler, B.; Matziolis, G.; Zippelius, T. NaOCl is more effective than CHX for eradication of bacterial biofilm of *staphylococci* and *Pseudomonas aeruginosa*. *Knee Surg. Sport. Traumatol. Arthrosc.* **2020**, *28*, 3912–3918. [CrossRef]
62. Hidajat, M.J.; Jo, W.; Kim, H.; Noh, J. Effective Droplet Size Reduction and Excellent Stability of Limonene Nanoemulsion Formed by High-Pressure Homogenizer. *Colloids Interfaces* **2020**, *4*, 5. [CrossRef]
63. Đorđević, S.M.; Cekić, N.D.; Savić, M.M.; Isailović, T.M.; Ranđelović, D.V.; Marković, B.D.; Savić, S.R.; Timić Stamenić, T.; Daniels, R.; Savić, S.D. Parenteral nanoemulsions as promising carriers for brain delivery of risperidone: Design, characterization and in vivo pharmacokinetic evaluation. *Int. J. Pharm.* **2015**, *493*, 40–54. [CrossRef]
64. Eid, A.; Elmarzugi, N.; El Enshasy, H. Preparation and evaluation of olive oil nanoemulsion using sucrose monoester. *Int. J. Pharm. Pharm. Sci.* **2013**, *5*, 434–440.
65. Danaei, M.; Dehghankhold, M.; Ataei, S.; Hasanzadeh, D.F.; Javanmard, R.; Dokhani, A.; Khorasani, S.; Mozafari, M.R. Impact of Particle Size and Polydispersity Index on the Clinical Applications of Lipidic Nanocarrier Systems. *Pharmaceutics* **2018**, *18*, 57. [CrossRef]
66. Kaomongkolgit, R.; Jamdee, K.; Pumklin, J.; Pavasant, P. Laboratory evaluation of the antibacterial and cytotoxic effect of alpha-mangostin when used as a root canal irrigant. *Indian J. Dent.* **2013**, *4*, 12–28. [CrossRef]
67. Sivaranjani, M.; Prakash, M.; Gowrishankar, S.; Rathna, J.; Pandian, S.K.; Ravi, A.V. In vitro activity of alpha-mangostin in killing and eradicating Staphylococcus epidermidis RP62A biofilms. *Appl. Microbiol. Biotechnol.* **2017**, *101*, 3349–3359. [CrossRef]
68. Ibrahim, M.Y.; Hashim, N.M.; Mariod, A.A.; Mohan, S.; Abdulla, M.A.; Abdelwahab, S.I.; Arbab, I.A. α-Mangostin from Garcinia mangostana Linn: An updated review of its pharmacological properties. *Arab. J. Chem.* **2016**, *9*, 31–329. [CrossRef]
69. Leelapornpisid, W. Efficacy of alpha-mangostin for antimicrobial activity against endodontopathogenic microorganisms in a multi-species bacterial-fungal biofilm model. *Arch. Oral Biol.* **2022**, *133*, 105304. [CrossRef]
70. Nourzadeh, M.; Amini, A.; Fakoor, F.; Raoof, M.; Sharififar, F. Comparative Antimicrobial Efficacy of *Eucalyptus Galbie* and *Myrtus Communis L.* Extracts, Chlorhexidine and Sodium Hypochlorite against *Enterococcus Faecalis*. *Iran Endod. J.* **2017**, *12*, 205–210.
71. Smith, D.C.; Maiman, R.; Schwechter, E.M.; Kim, S.J.; Hirsh, D.M. Optimal irrigation and debridement of infected total joint implants with chlorhexidine gluconate. *J. Arthroplasty.* **2015**, *30*, 1820–1822. [CrossRef]
72. Vianna, M.E.; Gomes, B.P.; Berber, V.B.; Zaia, A.A.; Ferraz, C.C.; de Souza-Filho, F.J. In vitro evaluation of the antimicrobial activity of chlorhexidine and sodium hypochlorite. *Oral Surg. Oral Med. Oral Pathol. Oral Radiol. Endod.* **2004**, *97*, 79–84. [CrossRef]
73. Radcliffe, C.E.; Potouridou, L.; Qureshi, R.; Habahbeh, N.; Qualtrough, A.; Worthington, H.; Drucker, D.B. Antimicrobial activity of varying concentrations of sodium hypochlorite on the endodontic microorganisms *Actinomyces israelii, A. naeslundii, Candida albicans* and *Enterococcus faecalis*. *Int. Endod. J.* **2004**, *37*, 438–446. [CrossRef] [PubMed]
74. Sakinah, A.; Setyowati, L. The cleanliness differences of root canal irrigated with 0.002% saponin of mangosteen peel extract and 2.5% NaOCl. *Dent. J.* **2015**, *48*, 104–107. [CrossRef]
75. Charlie, K.M.; Kuttappa, M.A.; George, L.; Manoj, K.V.; Joseph, B.; John, N.K. A Scanning Electron Microscope Evaluation of Smear Layer Removal and Antimicrobial Action of Mixture of Tetracycline, Acid and Detergent, Sodium Hypochlorite, Ethylenediaminetetraacetic Acid, and Chlorhexidine Gluconate: An In vitro Study. *J. Int. Soc. Prev. Community Dent.* **2018**, *8*, 62–69.
76. Niyonsaba, F.; Ushio, H.; Nakano, N.; Ng, W.; Sayama, K.; Hashimoto, K.; Nagaoka, I.; Okumura, K.; Ogawa, H. Antimicrobial peptides human beta-defensins stimulate epidermal keratinocyte migration, proliferation and production of proinflammatory cytokines and chemokines. *J. Investig. Dermatol.* **2007**, *27*, 594–604. [CrossRef] [PubMed]
77. Grone, A. Keratinocytes and cytokines. *Vet. Immunol. Immunopathol.* **2002**, *6*, 1–12. [CrossRef]
78. Sfakianakis, A.; Barr, C.E.; Kreutzer, D.L. Localization of the chemokine interleukin-8 and interleukin-8 receptors in human gingiva and cultured gingival keratinocytes. *J. Periodontal. Res.* **2002**, *37*, 154–160. [CrossRef]
79. Abate, M.; Pagano, C.; Masullo, M.; Citro, M.; Pisanti, S.; Piacente, S.; Bifulco, M. Mangostanin, A Xanthone Derived from *Garcinia mangostana* Fruit, Exerts Protective and Reparative Effects on Oxidative Damage in Human Keratinocytes. *Pharmaceuticals* **2022**, *15*, 84. [CrossRef]
80. Ngawhirunpat, T.; Opanasopi, P.; Sukma, M.; Sittisombut, C.; Kat, A.; Adachi, I. Antioxidant, free radical-scavenging activity and cytotoxicity of different solvent extracts and their phenolic constituents from the fruit hull of mangosteen (*Garcinia mangostana*). *Pharm. Biol.* **2010**, *48*, 55–62. [CrossRef]

processes

MDPI

Article

Discrete and Continuous Glucose Monitoring Systems: The Point of View of a Patient Affected by Type-1 Diabetes

Paolo Trucillo

Department of Chemical, Material and Industrial Production Engineering, Piazzale Vincenzo Tecchio, 80-80125 Napoli, Italy; paolo.trucillo@unina.it

Abstract: This work represents the point of view of a diabetic patient with an indirect experience in this specific field of research. As a chemical engineer and researcher in drug carrier production, he has always approached type-1 diabetes (T1D) in a scientific manner. Therefore, this work represents a description of almost 20 years of this illness treatment using a multi-injection insulin system, compared with the experience acquired with a newly adopted micro-infusion system, allowing automatized insulin administration. The use of the continuous system reduced significantly the Hb1Ac average values, from 8.8% to 6.6%, in less than 2 years. Moreover, a full 24 h control guaranteed the almost total elimination of the hypoglycemia risk, thanks to the automated control system, that can stop insulin administration in order to prevent critical situations. It is also important to note that the point of view underlined in this work does not presume to be that of a doctor or of a researcher who works closely in the field of medicine or diabetology. However, the author wants to highlight that doctors could try to educate patients to a scientific approach to treat illnesses correctly. The author experienced the very common difficulties related to the use of insulin with multi-injection administration for many years; then, he was proposed to start treatment with the automated pump mechanism. In this work, the author provides comments on the physical and psychological advantages and disadvantages of both insulin release systems, in order to define their impact on a patient's daily life. This work may also represent a vademecum for patients during the beginning of diabetes treatment, helped by the constant support and advice of a medical doctor.

Keywords: glucose levels; insulin administration; glycosylated hemoglobin; carbohydrates; continuous glucose monitoring; delivery systems

Citation: Trucillo, P. Discrete and Continuous Glucose Monitoring Systems: The Point of View of a Patient Affected by Type-1 Diabetes. *Processes* **2022**, *10*, 2706. https://doi.org/10.3390/pr10122706

Academic Editor: Yi Lu

Received: 27 October 2022
Accepted: 13 December 2022
Published: 15 December 2022

1. Introduction

As a researcher in the fields of chemical, process and material engineering, the author has always had a scientific approach to drug release systems, with the aim of producing optimized drug carriers for the efficient treatment of illnesses. Type-1 diabetes (T1D) is defined as an autoimmune disorder, characterized by the own immune cells attacking pancreatic beta cells, that are responsible for insulin production. This results in a partial and, then, total stop of insulin secretion and circulation [1–6].

The author was diagnosed with type 1 diabetes on 9 June 2003, at the age of 15. The main side effects of untreated diabetes (in the first period) included, in his specific case, the triad of polyuria, polydipsia and weight loss [7]; therefore, he experienced a significant weight loss from 65 kg to 55 kg in a few months, from winter to spring [8]. In details, the specific author's case was characterized by the need to urinate that occurred almost 5 times per hour, even at night; this also led to a not negligible urinary tract infection, strongly indicated as a principal symptom for this diagnosis [9–12]. As millions of people affected by non-diagnosed diabetes can confirm, the sense of thirst was always unsatisfied; and, for this reason, also the need to urinate was never satisfied. This period lasted approximately 2 months in springtime, during which the author was occasionally involved in some sessions of physical activity. Then, following a capillary blood glucose check, a value of

507 mg/dL was obtained, much higher than the optimal glucose value for non-diabetic patients (generally, between 70 and 110 mg/dL). These conditions were almost close to those that usually require hospitalization for young people affected by type-1 diabetes. Fortunately, soon after this diagnosis, the author urgently started the common therapy, consisting in the administration of rapid insulin before each meal and basal long-lasting (about 24 h) insulin at night, before the usual sleeping hours.

The weight loss of about 10 kg, from the initial 65 kg, was still a macroscopic sign of physical not recovered conditions, also reported in the literature as a side effect of untreated illness [13]. Furthermore, the first blood tests reported that all the values were perfectly in line with a male patient of 15 years of age. This consideration confirmed that diabetes had not caused any irreversible damage or side effects yet. Of course, the only value significantly above the required limits was that of glycosylated hemoglobin (HbA1c). The first value of HbA1c measured using High-Performance Liquid Chromatography was 12% (June 2003). According to current scales, this value corresponded to about 108 mmol/mol. This high value, after the following summer and several episodes of hypoglycemia, fell back to a value of 4.9%.

A year later, in June 2004, even though the level of glycosylated hemoglobin stabilized in the range between 6% and 7% (considered preferable for type-1 diabetic patients), there was no significant recovery of the body weight. Following further analysis, carried out between May and June 2004, the author was also diagnosed with celiac disease. To that date, there were various theories and schools of thought on the possible links between celiac disease and diabetes [14–16]. In particular, celiac disease could be due to viral or bacterial infections, genetics, autoimmunity, or even the simple introduction of gluten in the daily patient diet [17]. Indeed, a link between the two diseases is considered highly probable, but it is not yet clear, in this specific case, which of the two diseases could have triggered the other. However, a recent study found that almost 6% of patients affected by celiac disease are also affected by type-1 diabetes [18].

The transition to a gluten-free diet required a remodeling of the insulin doses administered, also because gluten-free foods, on average, have a higher fat and carbohydrate content than traditional foods [19]. After the first 2 months of the gluten-free diet, there was a significant recovery of weight, which reached about 67 kg and, after 1 year, a plateau value of almost 70 kg. Today, at the age of 34, he has reached an ideal weight in the range between 72 and 74 kg, having a height of about 1.76 cm, which corresponds to a Body Mass Index between 23.2 and 23.9.

In this work, the intent of the author was to describe the advantages and disadvantages of the drug delivery systems commonly employed for the treatment of type-1 diabetes. The comments and opinions here reported should be considered as the point of view of a patient with a scientific background, describing the advantages and disadvantages of using the traditional administration process with syringes and the novel automatized pumping system. To support these comments, the author used data obtained from his first treatment period of type-1 diabetes, which ran from 2003 (the year of his diagnosis) to March 2021, the year in which he started the use of the automatized pump, coupled with a glucose sensor monitoring system. Both physical and psychological features deriving from the use of insulin with multi-injection administration were analyzed. This modality was then compared with the automated administration modality, implemented by the patient starting from March 2021.

2. Description of the HbA1c and Cholesterol Values Monitored over Time

In order to support the opinions proposed in this work, data related to the patient's glycosylated hemoglobin are provided, from June 2003 to December 2022 (see Figure 1). The value of 12%, which was measured in a public hospital when the disease was first diagnosed, is omitted in Figure 1 in order to avoid the presentation of an out-of-scale diagram. Therefore, all the data shown in Figure 1 concern the measurement of HbA1c during the whole period of treatment with insulin therapy. In all other cases, the measurements were carried out in private centers with proper instrumentation.

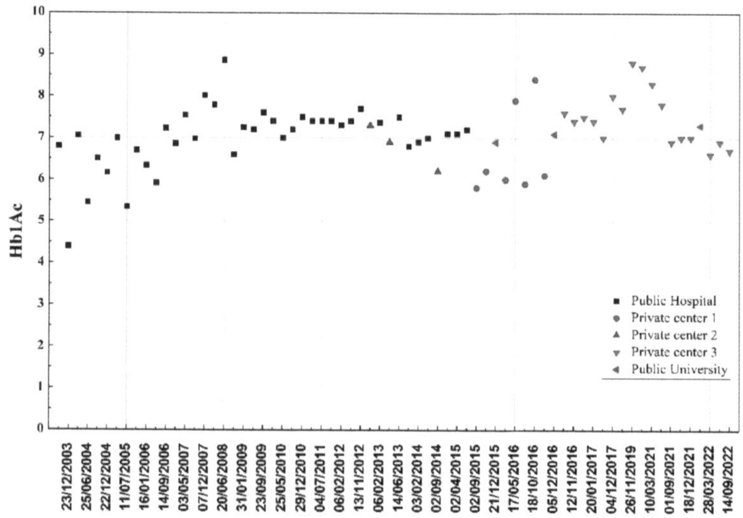

Figure 1. Evolution of the Hb1Ac values during the patient's 21 years of insulin therapy.

As it is possible to see in Figure 1, there is a huge variability of HbA1c values between the year 2003 and the year 2006. Therefore, in the following Table 1, the author excluded the HbA1c values measured before 2006; this is explained by the fact that the first years of insulin-therapy require to find a new metabolic balance for the patient. This causes a significant variation of HbA1c during the first years of treatment.. After finding a metabolic equilibrium, starting from the year 2007, the values measured in different centers were compared. In particular, the values of mean size and standard deviations of HbA1c are reported, corresponding to the different instruments used for their determination or the centers where they were obtained. Assuming that the average HbA1c level became stable with an average global value of 7.08 ± 0.78 %, it appears evident that in public institutes, more accurate measurements are made than in private centers. However, a general trend can be easily detected.

Table 1. Comparison of the mean values and standard deviations of HbA1c measured in different institutions.

Center	Mean Value, %	Standard Deviation, %
Public hospital	7.35	0.42
Private center 1	6.61	0.99
Private center 2	6.80	0.45
Private center 3	7.49	0.65
Public university	7.10	0.16
Total average	7.08	0.78

Figure 1 reports collection of data from one single patient. Therefore, this diagram does not represent a general trend for type-1 diabetic patients, but a report of a case study, useful to support the author's opinion and comments about the traditional and novel insulin delivery systems. However, this data may represent a guideline to achieve a good treatment of type-1 diabetes; moreover, diagrams and trends may be shared with other patients in order to discuss and compare the physical implications of different treatments and measurement devices.

Medical doctors could share or criticize these opinions, finding possibly some inspiration in order to show patients what they should do and what they should avoid during the therapy; this work could be also a support to teach patients how to approach this illness scientifically and without panic.

The first 3 years of diabetes treatment (between 2003 and 2006) were characterized by a not negligible variability, with an average HbA1c value of 6.29 ± 0.79 %. On the one hand, this fell perfectly within the suggested range of HbA1c for diabetic patients; on the other hand, it denoted a clear settling phase, characterized by both several hyperglycemia and hypoglycemia episodes, probably due to a still unconscious management of the disease, but also to a difficult control of the glucose level using the traditional syringe delivery system.

A value of HbA1c of about 9% was also reported in a trimester of the year 2008, when a study period was spent in another Italian city, lasting 1 year. This resulted in a poor glycemic control, essentially due to a type of unregulated and different diet compared to that of the previous years. After this year, diabetes control returned to quite good levels. In the subsequent period, from the end of 2008 to 2015, the average value of glycosylated hemoglobin was approximately 7.08 ± 0.43%, corresponding to an average value higher than that measured in the previous period, but with a lower variability.

On average, with reference to the period of high school (2003–2007) and the period of study at the University (2008–2015), glycosylated hemoglobin was kept relatively under control, due to a moderate lifestyle and very few diet excesses. Then, starting from 2015, the beginning of the working life led to an increase in stress due to new responsibilities, which inevitably had an impact on the average HbA1c values. Furthermore, the lack of time due to the reduced spare time and to the subsequent work commitments, led the patient to neglect the daily reflections on the amount of insulin to administer, meal by meal. Often, the described situation led to the choice to perform HbA1c levels checks in private laboratories, where sometimes the accuracy seemed to be quite lower than that achieved in public institutes (see Table 1). In these last cases, the measurements were not always carried out using High-Performance Liquid Chromatography (HPLC), which is the most appropriate instrument and method (see Private center Nr. 1 in Figure 1). However, this first period of stress increase was overcome, with HbA1c mean values of 6.94 ± 0.79%.

Starting from July 2017, monitoring with a glucose sensor [20] began, allowing the patient to move from the classic fingertip sampling for measuring capillary glucose to the continuous (24 h) measurement of interstitial glucose. On the one hand, a more accurate and increased number of data could be acquired, with the possibility to observe the precise value of the maximum glucose peak, reached after meals; however, on the other hand, the use of sensors triggered a psychological fear of reaching a low level of glucose concentration, feeling the classic symptoms of hypoglycemia [21]. This resulted in an average reduction in the insulin volumes administered before meals and in a consequent considerable glycemic decompensation. Starting from the second half of 2017, and until April 2020, HbA1c was measured in a different private laboratory, which ensured a proper measurement using the High-Performance Liquid Chromatography (HPCL) method (see Private center Nr. 3 in Figure 1). Therefore, in the period from 2017 to 2020, a significant worse glycemic control was recorded, which resulted in an average value of 8.30 ± 0.41%, with dangerous peaks of 8.8% and 8.7%, measured, respectively, in November 2019 and April 2020.

Aware that the problem was essentially psychological [22] and having to continue to carry out the same working activity, with all the associated stress, the author decided to contact the diabetes center of the Secondo Policlinico di Napoli (University of Naples Federico II, Naples, Italy), specialized in innovative techniques for diabetes care and insulin delivery. In this department, the author was proposed to use a new continuous glucose-monitoring device (Dexcom G6, Dex Com Inc., San Diego, CA, USA), coupled with an insulin pump (Tandem Diabetes Care, San Diego, CA, USA). In that case, the sensor would not simply record and display data continuously, it would also communicate constantly with the pump, taking constant therapeutic decisions in order to improve the management of type-1 diabetes. With this system, the only patient's decision is related to the insulin units of volume to be administered. This delivery system is constantly controlled by an algorithm that also allows the patient to automatically correct high glucose concentrations in the blood by calculating a correction bolus, taking into account the active insulin already administered to the body and also considering the glucose target, set (generally) at 110 mg/dL. When using the continuous system, another element could cause fear of hypoglycemia. This kind of fear could be due to the possibility of the system failure, causing the administration of hundreds of insulin units stocked in the pump's pot. Fortunately, this automated pump has a safety system. Therefore, the infusion set is separated from the insulin pot by a camera of micrometric volume; in case of failure, this camera only administers a negligible insulin volume, eliminating the tubing communication with the main pot. This avoids the risk of hypoglycemia due to a failure of the infusion system.

Moreover, the pumping delivery system simulates the normal body release rate, administering micro-volumes of insulin per unit time and avoiding the on-shot administration of insulin permeating the adipose tissues, in order to be absorbed by the body. This guarantees the elimination of the traditional administration of basal insulin, which is generally characterized by a huge volume of insulin that lasts around 24 h.

Following the use of the insulin pump, the glycosylated hemoglobin value slowly decreased from May 2021 (7.8%) to March 2022 (6.6%). Considering the peak reached in November 2019 (8.8%), the benefits of the new delivery systems were particularly evident (see Figure 2). In the period of treatment using the insulin pump (from March 2021 to the present), an average value of 7.1 ± 0.37% was recorded, indicating a clear improvement of the glycosylated hemoglobin control, with respect not only to its decreased average values, but also to its reduced variability.

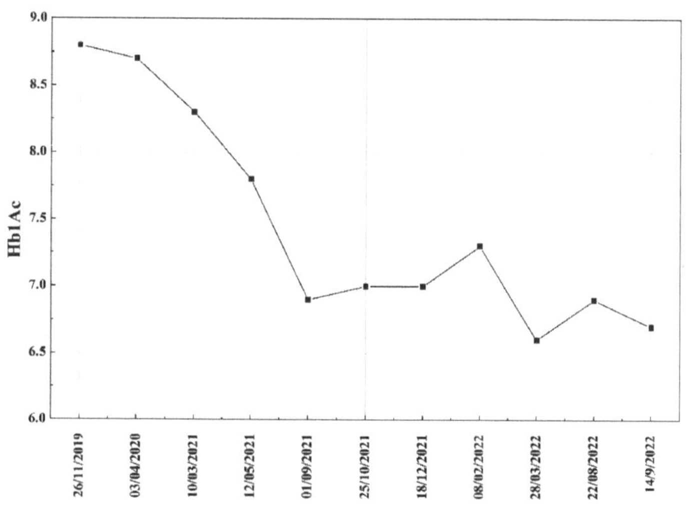

Figure 2. Evolution of HbA1c levels from November 2019 to the present day.

For the completeness of the data, Figure 3 shows the values of total cholesterol, high-density lipoprotein (HDL) and low-density lipoprotein (LDL) measured in almost 20 years of diabetes treatment.

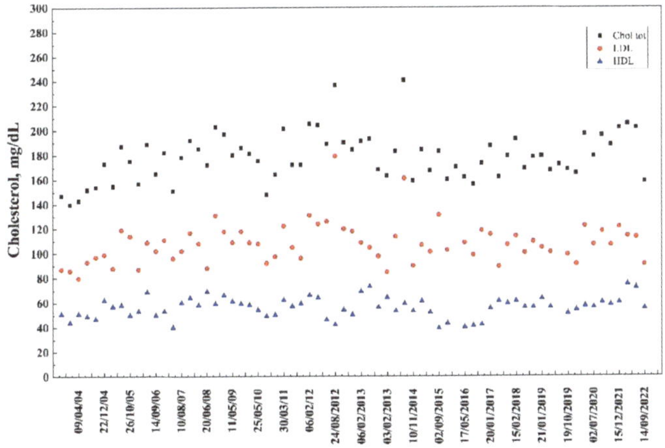

Figure 3. Evolution of total cholesterol, HDL and LDL values during the patient's 21 years of insulin therapy.

Looking at Figure 3, over the observed years, total cholesterol showed an average value of 178 ± 19 mg/dL, while HDL cholesterol had an average value of 56 ± 8 mg/dL, and LDL cholesterol of 108 ± 19 mg/dL. During the treatment period of type-1 diabetes, cholesterol monitoring recorded only two out-of-average episodes. In these two cases, the total value was close to 240 mg/dL, while the corresponding LDL value was about 180 mg/dL, probably because of incorrect compliance with the hours of fasting before carrying out these analyses. According to the most diffused guidelines, the total cholesterol level should be under 200 mg/dL; the LDL level should be under 110 mg/dL (diabetic people are suggested to keep it under 100 mg/dL); the HDL level should be less than 40 mg/dL for men and less than 50 mg/dL for women.

After listening to the point of view of the consulted medical doctors, it is also worth noting that, for the containment of cholesterol, this patient has not yet started any pharmacological cure. However, according to the guidelines for type-1 diabetes, his average levels of cholesterol are slightly above the recommended limit values. Indeed, in order to avoid serious cardiological risks in the future, the LDL value must be kept constantly under 100 mg/L.

3. Discussion

After about 20 years of experience in the treatment of type-1 diabetes, what appears evident is the possibility of acquiring more or less information on blood sugar or glucose values, depending on the monitoring system employed. In particular, with the classic method of finger sticking to obtain capillary blood, it is necessary to reduce the blood glucose measurements to no more than 6 measurements per day, i.e., before and after each meal (breakfast, lunch and dinner). In theory, the decision to have another meal would imply two additional fingertip samplings (before and after). The reduced number of measurements is also related to the fact that each one involves the use of a single stripe (device for capillary glucose measurement), which cannot be reused and is necessarily disposed of. Moreover, increasing the number of fingertip measurements increases the cost of the treatment and its specific environmental impact, as well as the patient sufferance (see Table 2).

Table 2. Advantages and disadvantages of continuous vs. discrete glucose monitoring systems.

Type of Glucose Monitoring	Advantages	Disadvantages
Discrete	6 g of waste per day A 24 h control of the glucose levels	4+ insulin administrations per day Up to 6 fingertip samplings per day More pain for the patient High probability of hypoglycemia
Continuous	1 infusion substitution every 3 days No fingertip sampling Reduced pain for the patient Reduced risk of hypoglycemia Automated reduced administration in exercise mode	32 g of waste every 3 days A spot-like glucose control (up to 6 points per day)

In addition to this, fingertip sampling only estimates the efficacy of the administered insulin bolus only 2 or 3 h after a meal. However, according to the author's personal opinion, by using the classical capillary glucose measurement, it appears more difficult to predict information on the maximum peak reached at the apex of meal digestion. Therefore, a good glucose level in the blood before and after a meal could hide a peak of 250 mg/dL or even, in some cases, 300 mg/dL or more. Furthermore, these spot measurements performed on capillary blood cannot give precise information on the efficacy of insulin during the night hours of sleep.

In Figure 4, the red dots represent the spot analysis of capillary blood glucose concentration (only one point), and the black line represent the data obtained using the Continuous Glucose Monitoring (CGM) system, which was programmed to collect a new glucose value every 5 min. In this last case, the data are much more interesting and useful, because they allow the patients and the medical doctors to understand how long the body had to manage hyperglycemia (high level of glucose concentration), over the daily course of 24 h.

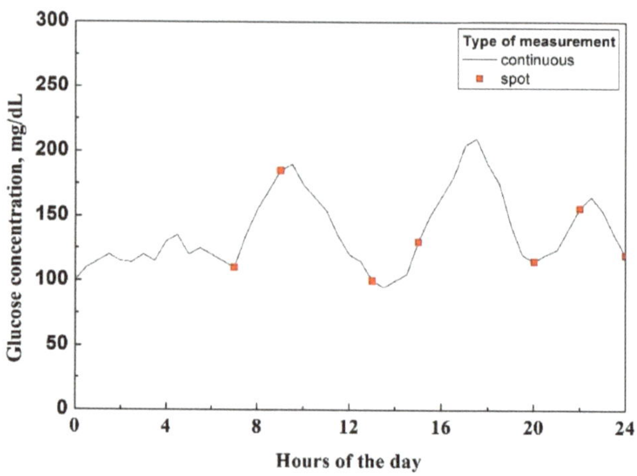

Figure 4. Comparison of type 1-diabetes continuous monitoring and spot-like monitoring.

Moreover, hypoglycemia (low level of glucose concentration) can be easily avoided by the use of sensors coupled with an insulin pumping system. If the glucose level tends to the minimum accepted value (generally below 60–70 mg/dL), the sensor sends warnings (also acoustic) to the patient and suggests assuming sugars, while communicating to the pump to stop insulin administration. The stop of insulin administration helps the body to increase the glucose concentration up to an average 80 mg/dL, establishing a new balance of normal values. In details, Figure 4 shows that, working with the spot metering system, all information on blood glucose fluctuations in the early morning would have been lost. Furthermore, from the finger stick sampling before lunch, a good glycemia level would have been recorded, followed by another good measurement (2 h later). Indeed, the blood glucose spiked at around 220 mg/dL, returning to a normal average level at 8:00 pm before dinner.

The advantages and the disadvantages deriving from the use of discrete or continuous glucose monitoring systems are listed in Table 2:

As it is possible to see, the continuous monitoring system introduces a significant innovation in the treatment of diabetes, since it provides a 24 h continuous control of the glucose level, with the possibility to obtain information on the glucose increase or decrease speed.

Moreover, the patient suffers less, since he/she needs to provide only one substitution of the infusion set every 3 days, on average. This means that the patient does not have to perform needle administration up to 4 times per day (breakfast, lunch, dinner and basal insulin), without taking into account any eventual corrections.

The use of conventional needles causes suffering. Often, it also causes fear in children affected by diabetes. In some cases, research in microneedles administration could make a difference in terms of pain reduction [23].

One specific disadvantage may be the overall environmental impact of the continuous monitoring system, since it creates about 32 g of waste materials every 3 days, compared with 18 g of waste (6 g per day, on average) for the conventional monitoring system. These materials are not reusable and not recyclable, nowadays; therefore, this waste weight reducing issue should be taken into account in future research.

Regarding the performance of physical activity by diabetic patients, the automated system gives the possibility to activate the "exercise mode" using the pump, inducing the system to reduce the amount of continuously administered insulin while physical activity is performed, as reported for fully automated systems [24]; this is of course not possible with the conventional spot-like glucose measurements.

4. Conclusions

To the eyes of a patient with a scientifically assisted approach, the spot monitoring system appears obsolete with respect to the loss of information on diabetes management and the effectiveness of administered insulin boluses.

Continuous monitoring, on the other hand, provides full awareness of the variations and trends of blood sugar, making the patient also much more aware of diabetes management and therapeutic choices. Of course, continuous monitoring has its beneficial effects if coupled with an insulin micro-infusion system, which regulates the delivery of the bolus according to the variation in blood sugar over time.

By eliminating the administration of basal insulin, the pump delivers only rapid insulin, emitting micro-boluses that continuously balance the glucose value around the set point (generally 110 mg/dL). Ultimately, a modern monitoring system is certainly the one resulting from the use of a sensor coupled with an insulin pump. Thanks to this system, the author has almost forgotten the problem of diabetes; this new treatment method guarantees the possibility for a diabetic patient to live a quiet life, almost totally free from the psychological implications of carrying out continuous checks and corrections of the blood glucose concentration with the use of syringes.

Indeed, the insulin pump almost perfectly replaces the human pancreas. However, there is still a lot to improve with this kind of insulin delivery; process engineers together with biotechnologists and researchers in medical fields can further optimize these systems. A synergistic approach would generate a system available to predict the amount of insulin to administer just before meals. In this last case, the patients will leave to the pump not only the possibility to monitor glucose level, releasing insulin constantly and correcting high glucose levels, but also the therapeutic choice of managing meals from their beginning. So far, the physical and psychological benefits of the actual system seem to be clearly significant and give us the hope of creating a fully automated pancreas-like working device, in the future.

Funding: This research received no external funding.

Institutional Review Board Statement: Not applicable.

Informed Consent Statement: Not applicable.

Data Availability Statement: The data presented in this study are available on request from the corresponding author. The data are not publicly available due to patient's own privacy restrictions.

Conflicts of Interest: The author declares no conflict of interest.

Abbreviations

T1D: Type-1 Diabetes; CGM: Continuous Glucose Monitoring; HbA1c: glycosylated hemoglobin; HPLC: High-Performance Liquid Chromatography; HDL: High-Density Lipoprotein; LDL: Low-Density Lipoprotein.

References

1. Swenne, I. Pancreatic beta-cell growth and diabetes mellitus. *Diabetologia* **1992**, *35*, 193–201. [CrossRef]
2. Oram, R.A.; Sims, E.K.; Evans-Molina, C. Beta cells in type 1 diabetes: Mass and function; sleeping or dead? *Diabetologia* **2019**, *62*, 567–577. [CrossRef]
3. Johannesson, B.; Sui, L.; Freytes, D.O.; Creusot, R.J.; Egli, D. Toward beta cell replacement for diabetes. *EMBO J.* **2015**, *34*, 841–855. [CrossRef]
4. Ahmad, K. Insulin sources and types: A review of insulin in terms of its mode on diabetes mellitus. *J. Tradit. Chin. Med.* **2014**, *34*, 234–237. [CrossRef]
5. Stolpe, S.; Kroes, M.A.; Webb, N.; Wisniewski, T. A systematic review of insulin adherence measures in patients with diabetes. *J. Manag. Care Spec. Pharm.* **2016**, *22*, 1224–1246. [CrossRef]
6. Quianzon, C.C.; Cheikh, I. History of insulin. *J. Community Hosp. Intern. Med. Perspect.* **2012**, *2*, 18701. [CrossRef]
7. Roche, E.F.; Menon, A.; Gill, D.; Hoey, H. Clinical presentation of type 1 diabetes. *Pediatr. Diabetes* **2005**, *6*, 75–78. [CrossRef]
8. Deshmukh, C.D.; Jain, A.; Nahata, B. Diabetes mellitus: A review. *Int. J. Pure App. Biosci.* **2015**, *3*, 224–230.
9. Geerlings, S.E.; Stolk, R.P.; Camps, M.J.; Netten, P.M.; Collet, T.J.; Hoepelman, A.I. Diabetes Women Asymptomatic Bacteriuria Utrecht Study Group. Risk factors for symptomatic urinary tract infection in women with diabetes. *Diabetes Care* **2000**, *23*, 1737–1741. [CrossRef]
10. Stapleton, A. Urinar y tract infections in patients with diabetes. *Am. J. Med.* **2002**, *113*, 80–84. [CrossRef]
11. Al-Rubeaan, K.A.; Moharram, O.; Al-Naqeb, D.; Hassan, A.; Rafiullah, M.R.M. Prevalence of urinary tract infection and risk factors among Saudi patients with diabetes. *World J. Urol.* **2013**, *31*, 573–578. [CrossRef]
12. Neu, A.; Willasch, A.; Ehehalt, S.; Hub, R.; Ranke, M.B. Ketoacidosis at onset of type 1 diabetes mellitus in children–frequency and clinical presentation. *Pediatr. Diabetes* **2003**, *4*, 77–81. [CrossRef]
13. Haller, M.J.; Atkinson, M.A.; Schatz, D. Type 1 diabetes mellitus: Etiology, presentation, and management. *Pediatr. Clin.* **2005**, *52*, 1553–1578. [CrossRef]
14. Holmes, G.K.T. Coeliac disease and type 1 diabetes mellitus–the case for screening. *Diabet. Med.* **2001**, *18*, 169–177. [CrossRef]
15. Holmes, G.K.T. Screening for coeliac disease in type 1 diabetes. *Arch. Dis. Child.* **2002**, *87*, 495–498. [CrossRef]
16. Volta, U.; Tovoli, F.; Caio, G. Clinical and immunological features of celiac disease in patients with type 1 diabetes mellitus. *Expert Rev. Gastroenterol. Hepatol.* **2011**, *5*, 479–487. [CrossRef]
17. Schuppan, D.; Hahn, E.G. Celiac disease and its link to type 1 diabetes mellitus. *J. Pediatr. Endocrinol. Metab.* **2001**, *14*, 597–606. [CrossRef]
18. Elfström, P.; Sundström, J.; Ludvigsson, J.F. Systematic review with meta-analysis: Associations between coeliac disease and type 1 diabetes. *Aliment. Pharmacol. Ther.* **2014**, *40*, 1123–1132. [CrossRef]
19. Cornicelli, M.; Saba, M.; Machello, N.; Silano, M.; Neuhold, S. Nutritional composition of gluten-free food versus regular food sold in the Italian market. *Dig. Liver Dis.* **2018**, *50*, 1305–1308. [CrossRef]
20. Massa, G.G.; Gys, I.; Op't Eyndt, A.; Bevilacqua, E.; Wijnands, A.; Declercq, P.; Zeevaert, R. Evaluation of the FreeStyle® Libre flash glucose monitoring system in children and adolescents with type 1 diabetes. *Horm. Res. Paediatr.* **2018**, *89*, 189–199. [CrossRef]
21. Gonder-Frederick, L.A.; Schmidt, K.M.; Vajda, K.A.; Greear, M.L.; Singh, H.; Shepard, J.A.; Cox, D.J. Psychometric properties of the hypoglycemia fear survey-ii for adults with type 1 diabetes. *Diabetes Care* **2011**, *34*, 801–806. [CrossRef]
22. Johnson, B.; Elliott, J.; Scott, A.; Heller, S.; Eiser, C. Medical and psychological outcomes for young adults with type 1 diabetes: No improvement despite recent advances in diabetes care. *Diabet. Med.* **2014**, *31*, 227–231. [CrossRef]

23. Sabbagh, F.; Kim, B.S. Microneedles for transdermal drug delivery using clay-based composites. *Expert Opin. Drug Deliv.* **2022**, *19*, 1099–1113. [CrossRef]
24. Turksoy, K.; Hajizadeh, I.; Hobbs, N.; Kilkus, J.; Littlejohn, E.; Samadi, S.; Feng, J.; Sevil, M.; Lazaro, C.; Ritthaler, J.; et al. Multivariable artificial pancreas for various exercise types and intensities. *Diabetes Technol. Ther.* **2018**, *20*, 662–671. [CrossRef]

processes

MDPI

Review

Innovative and Patented Liposome-Based Drug Carriers

Snežana Ilić-Stojanović [1,*], Suzana Cakić [1], Nada Nikolić [1] and Slobodan Petrović [2]

[1] Faculty of Technology, University of Niš, Bulevar oslobodjenja 124, 16000 Leskovac, Serbia; cakics@tf.ni.ac.rs (S.C.); nikolicnada@tf.ni.ac.rs (N.N.)
[2] Faculty of Technology and Metallurgy, University of Belgrade, Karnegijeva 4, 11000 Belgrade, Serbia; sloba@tmf.bg.ac.rs
* Correspondence: snezanai@tf.ni.ac.rs

Abstract: Liposome-based drug carriers are multipurpose colloidal drug delivery systems developed mainly for targeted therapy. Researchers have expanded their research on liposomes due to their unique characteristics (e.g., non-toxicity, biodegradability, biocompatibility, and non-immunogenicity). This review summarizes historical advances, from the first scientific papers and patents to the latest inventive solutions, in the field of liposome-based drug carriers and their production processes. Various bibliometric studies on the use of liposomes as drug carriers have been published; nevertheless, they focus on published scientific works rather than patent documents. Patent information is important for the pharmaceutical, nutraceutical, and cosmetic industries because technical knowledge in patent documentation is often not published in any other document. The research in this review was conducted using the Espacenet—European Patent Office database, with keywords and classification codes defined by the International Patent Classification. Innovative formulations, including the usage and administration route, are broadly researched to produce effective and safe drug delivery systems with negligible side effects. Global patenting trends in liposome drug carriers' production process were also discussed, and this evaluation unifies up-to-date development in this field. Patent database reviews and analyses could help as inspiration for future investigations as well as for problem-solving resources.

Keywords: drug delivery systems; liposomes; nanoparticles

Citation: Ilić-Stojanović, S.; Cakić, S.; Nikolić, N.; Petrović, S. Innovative and Patented Liposome-Based Drug Carriers. *Processes* **2024**, *12*, 1970. https://doi.org/10.3390/pr12091970

Academic Editor: Paolo Trucillo

Received: 1 August 2024
Revised: 9 September 2024
Accepted: 11 September 2024
Published: 13 September 2024

1. Introduction

The goal of pharmaceutical and technological progress is to discover novel processes for converting an active ingredient (e.g., drug) into a dosage form acceptable for application, which enables the concentration of the active ingredient in the site of action to quickly reach a therapeutic level and maintain an approximately constant level over time, according to the prescribed therapeutic needs [1]. The processes applied to reach this aim are very complex and challenging and demand sophisticated and functional drug delivery systems (DDS). Progress from the existing drug molecule, through the conventional drug dosage form to a new and innovative drug delivery system, could significantly improve features of active ingredients, considering efficiency, safety, and compliance (acceptability for patients). The drug targeting or targeted therapy concept implies an adaptable carrier (polymeric or colloidal) at which the active ingredient is bound (by sorption or chemical bonding). Colloidal drug delivery systems are mainly developed as target systems, i.e., carriers for achieving the target result [2,3]. Drug carriers should transport and/or direct the active ingredient precisely at a controlled rate to a specific group of cells or organs in which it should be released and act [4,5]. The key benefit of carriers is their capability of incorporating drugs, enhancing their bioavailability and selectivity; thus, reducing side effects to humans. The main task in formulating a drug delivery system is that the selected carrier should protect the incorporated/adsorbed drug from the negative effects of the organism on the way from the application site to the target site (diseased organ or cells) and release the active substance to the intended target site [6]. Nanoparticles used as DDS exhibit slow kinetic transportation in diseased cells and tissues due to vascular barriers [7]. In order to increase the possibility of crossing the vascular wall, a relatively long circulation time of nanomaterials is needed. The prototypical nanoparticles are blocked in the bloodstream due to rapid clearance by the reticuloendothelial system (RES).

Liposomes are among the most studied advanced pharmaceutical drug delivery systems. They represent microparticulate lipid vesicles, i.e., spherical self-enclosed structures, composed of curved lipid bilayers in which a part of the solvent (water) is enclosed, in which these structures freely float [3,4]. Liposomes were discovered in 1965 by the scientist Bangham, as a system similar to cell membranes [8]. Some years later, scientists Gregoriadis, Leathwood, and Ryman found novel pharmaceutical applications for the enzymes fructofuranosidase and amyloglucosidase, as well as ^{131}I-labeled albumin and human serum albumin delivery, using liposomes in genetic disorder treatments [9–11]. In the following decades, numerous studies were developed in the area of novel and improved liposome production processes and their innovative possibility of applications in humans [12]. Conventional liposomes are usually made only of phospholipids (neutral or negatively charged) and/or cholesterol and are characterized by a relatively short blood circulation time [13,14]. Drug delivery systems found solutions to improve colloidal stability in aqueous media, e.g., charge-repulsion-based stabilization and steric stabilization using hydrophilic, neutral, and flexible polymers, e.g., polyethylene glycol (PEG), as well as many others [7]. The applications of liposomes as carriers to protect insulin against proteolysis in the gastrointestinal tract, from the first study in 1976 to today, is at the preclinical level because of its limited stability in gastrointestinal conditions [15,16]. In contrast, in this period, many liposomal technologies were significantly improved and reached large-scale production thanks to plentiful investigations about possibilities in drug delivery, their stability in physiological fluids, many clinical trials, and approved products, and many are in various stages of testing, which confirms the remarkable progress [17–20]. In numerous cases, lipid complexes and liposomal drug forms provide less toxicity and better efficacy than the drug itself or the conventional pharmaceutical form of the same drug [4,6]. Due to their relatively high biocompatibility, liposomes are acceptable for therapeutic purposes and in vivo diagnostics. Liposomes are used to improve the "transfer/delivery" as carriers of various drugs, such as immunomodulators, chemotherapeutic agents, diagnostic agents, antigenic, genetic material, etc. (e.g., doxorubicin, epirubicin, amphotericin B, calcitonin, interferon, or technetium-99 as imaging agents) [6]. Encapsulation of drugs in liposomes has become a notable strategy for the invention of pharmaceutical preparations applied through different administration routes, such as ocular, oral, vaginal, urinary, or rectal routes [5,21]. Nowadays, there are many pharmaceutical preparations available on the market, developed on liposomes as carriers, mainly for parenteral administration (for intravenous, intramuscular, subcutaneous, epidural, intraperitoneal, and intrathecal applications) [22–24], and dermatological products for topical usage [25–27]. Encapsulation of active therapeutic ingredients in liposomes could enhance their penetration into the skin's deeper layers with the accumulation and drug "depot" formation, which provides a sustained release and reduces side effects and systemic absorption [1,28]. The stability of the active substance also increases with encapsulation in liposomes, which delays their shelf life [29].

Liposomes are generally formed of phospholipids (neutral or negatively charged) and/or cholesterol and characterized by a relatively short circulation time in the blood [8,21]. Phospholipids, as basic ingredients of plant, animal, and bacterial membranes, make liposomes non-toxic, biodegradable, and non-immunogenic carriers for bioactive ingredients. The phosphate part, called the polar "head", is the hydrophilic part, while the long chain of saturated and/or unsaturated fatty acids is the hydrophobic/non-polar part of the phospholipid molecule [1]. Phospholipids are solvable in organic solvents and their mixtures. As a result of their amphiphilic nature, they form aggregates in an aqueous medium, directly above the critical micellar concentration (CMC, about 10^{-8} M), the structure of which depends on the chemical structure of the dimension and saturation degree of the fatty acids chain and the polar "head", and the pH and ionic strength of the aqueous medium. Stealth (long-circulating)/sterically stabilized liposomes were developed to overcome the short retention time in the circulation of conventional liposomes. They have a coating/polymer protective layer, formed from polyethylene glycol (polyoxyethylene derivatives) on the surfaces of liposomes. It avoids rapid clearance by the immune system and prevents liposomes from leaving the circulation [5,6]. The prolonged presence in the blood allows them to pass to places with permeable vascularization, often in tumors. They have a potential application in the therapy of malignant lung cancer cells.

Target liposomes (immuno-liposomes) were additionally improved by incorporating monoclonal antibodies or antibody fragments into the membranes of conventional or sterically stabilized liposomes (by covalent or other bonds) for precise and specific targeting of tumors cells and to deliver drugs [6]. Cationic liposomes were developed to improve the transfer of genetic material, as the cationic lipid components react with deoxyribonucleic acid (DNA) and neutralize their negative charge. The lipid–DNA complex provides protection and helps direct the condensed plasmid to the target cell's surface and release the genetic material into the cytoplasm or nucleus. The new generation of stimulus-responsive liposomes was invented to overcome the disadvantages of conventional liposomes. Smart liposomes could be sensitive to the change in thermo, light, pH, ion, enzyme, and magnet stimulus, or the dual- or multi-sensitive liposomes. They can encapsulate different active ingredients, with low toxicity, high levels of viability for healthy cells, and increased bioavailability in damaged tissues. Strategies of surface modification of liposome-based carriers, their classification, and the possibility of active ingredients' encapsulation (drugs, proteins, small molecules, and targeting moieties, e.g., antibodies, peptides, aptamers, etc.), conjugated on the carrier's surfaces by various linkers, covalent or non-covalent bonds, and electrostatic interactions, are schematically presented in Figure 1 [19]. Many products based on liposomes as carriers have been approved and are on the drug market, and numerous others are in various testing stages [21].

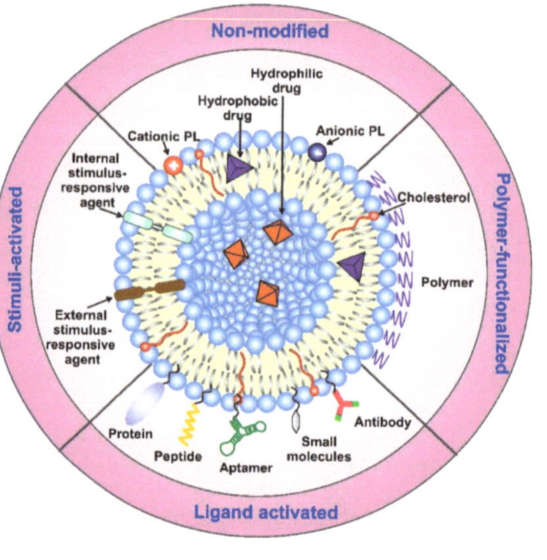

Figure 1. Surface modification strategies of liposomes, together with their classification. The modified carriers can contain active components, such as drugs, small molecules, proteins, and/or targeting moieties, such as antibodies, peptides, aptamers, etc., conjugated on the surface of the vehicles through different linkers, non-covalent or covalent bonds, and electrostatic interactions. Abbreviation: PL—phospholipid. Reprinted from [19] under an open-access creative common CC-BY license.

A few scientific review papers provide short reports on patent documents for liposome-based carriers, without a whole, up-to-date patent analysis [30,31]. This overview aims to provide a patent analysis of liposome-based carriers, from the first patent to the newest inventive patented solutions, focused on the aspects of procedures for obtaining them and pharmaceutical applications. Patent activity and filing trends of liposome-based drug carriers can be used as inspiration for future research and data sources for problem-solving in pharmaceutical or cosmetic applications.

2. Patent Searching for Liposome-Based Drug Carriers

The search methodology for relevant patent information was conducted on the reference database Espacenet, with free access to over 150 million patent documents from over 100 countries, provided by the European Patent Office (EPO) [32]. This search was carried out by including keywords and classification symbols (using both the International Patent Classification (IPC) and Cooperative Patent Classification (CPC)). The IPC was provided by the World Intellectual Property Organization (WIPO) [33], used by above 100 National Patent Offices and frequently used for searching. It is a hierarchical system for the classification of patents and utility models according to the different areas of technology to which they pertain, independent from language symbols [34]. The CPC is a classification system developed by the EPO and the United States Patent and Trademark Office (USPTO), based on the previous European classification system (ECLA), a more detailed version of the ICP [35]. Both classifications, the IPC and CPC, are hierarchical classification systems that separate all technical knowledge using the descending order of hierarchical levels: section, class, subclass, group, and subgroup [36]. The hierarchy among subgroups is determined solely by the number of dots (•) before their titles, not by the numbering of the subgroups. To avoid repetition, dots before a group title are used instead of the titles of its hierarchically upper-level groups. Liposomes and their derivatives are classified in the A61K9/00 group and A61K9/127 subgroup, while the classification symbol for processes for preparing is A61K9/1277. Medicinal preparations containing active ingredients are classified into group A61K31/00, while those characterized by the non-active ingredients used, e.g., carriers or inert additives, targeting or modifying agents chemically bound to the active ingredient, are classified into group A61K47/00. The main groups and particularly relevant subgroups of IPC/CPC codes, selected and applied for this search, are listed in Table 1.

Table 1. List of IPC/CPC codes used for the liposome drug delivery systems patent search [34,35].

IPC Code	IPC Description, as Indicated by the Relevant CPC Symbol
A61K9/00	Medicinal preparations characterized by special physical form (...)
A61K9/10	• Dispersions; Emulsions (...)
A61K9/127	•• Liposomes
A61K9/1271	••• Non-conventional liposomes, e.g., PEGylated liposomes, liposomes coated with polymers (...)
A61K9/1272	•••• With substantial amounts of non-phosphatidyl, i.e., non-acylglycerophosphate, surfactants as bilayer-forming substances, e.g., cationic lipids (...)
A61K9/1273	•••• Polymersomes; Liposomes with polymerizable or polymerized bilayer-forming substances (...)
A61K9/1274	••• Non-vesicle bilayer structures, e.g., liquid crystals, tubules, cubic phases, cochleates; Sponge phases
A61K9/1275	•••• Lipoproteins; Chylomicrons; Artificial HDL, LDL, VLDL, protein-free species thereof; Precursors thereof
A61K9/1276	•••• Globules of milk or constituents thereof
A61K9/1277	••• Processes for preparing; Proliposomes
A61K9/1278	•••• Post-loading, e.g., by ion or pH gradient
A61K31/00	Medicinal preparations containing organic active ingredients
A61P35/00	Antineoplastic agents

Table 1. *Cont.*

IPC Code	IPC Description, as Indicated by the Relevant CPC Symbol
A61K38/00	Medicinal preparations containing peptides (enzymes, hormones, porphyrin hemoglobin)
A61K43/00	Drugs for specific purposes, not provided for in groups A61P1/00–A61P41/00
A61K45/00	Medicinal preparations containing active ingredients not provided for in groups A61K31/00–A61K41/00
A61K45/06	• Mixtures of active ingredients without chemical characterization, e.g., antiphlogistic and cardiac
A61K47/00	Medicinal preparations characterized by the non-active ingredients used, e.g., carriers or inert additives; Targeting or modifying agents chemically bound to the active ingredient
A61K47/06	• Organic compounds, e.g., natural or synthetic hydrocarbons, polyolefins, mineral oil, petrolatum, or ozokerite
A61K47/08	•• Containing oxygen, e.g., ethers, acetals, ketones, quinones, aldehydes, peroxides
A61K47/10	••• Alcohols; Phenols; Salts thereof, e.g., glycerol; Polyethylene glycols (PEG); Poloxamers; PEG/POE alkyl ethers
A61K47/16	•• Containing nitrogen, e.g., nitro-, nitroso-, azo-compounds, nitriles, cyanates
A61K47/18	••• Amines; Amides; Urea; Quaternary ammonium compounds; Amino acids; Oligopeptides having up to five amino acids
A61K47/24	•• Containing atoms other than carbon, hydrogen, oxygen, halogen, nitrogen, or sulfur, e.g., cyclomethicone or phospholipids
A61K47/28	•• Steroids, e.g., cholesterol, bile acids, or glycyrrhetinic acid
A61K47/34	•• Macromolecular compounds obtained other than by reactions only involving carbon-to-carbon unsaturated bonds, e.g., polyesters, polyamino acids, polysiloxanes, polyphosphazines, copolymers of polyalkylene glycol, or poloxamers (...)
A61K47/48	• The non-active ingredient being chemically bound to the active ingredient, e.g., polymer drug conjugates (5)
A61K48/00	Medicinal preparations containing genetic material, which is inserted into cells of the living body to treat genetic diseases; Gene therapy

The final patent search results were obtained using the keywords ("liposome" and "drug delivery system" or "drug carrier" and "process" or "procedure" or "obtaining" and "method") in the search fields: full text, title, abstract, and claims.

Using a combination of the keywords with the classification codes, according to the abovementioned search criteria, 6296 patent documents were found from 1971 to July 2024, graphically presented in Figure 2 (with data for the first 10 countries, applicants, and IPC subgroups).

The first patent application, DE 2249552, named: "Encapsulation of chemical substances—in liposomes for medicinal use", claimed by a German company, Bayer AG (DE) (after the transmission of property from the Swiss company Inchema S.A., Wadenswill), was filed on 12 October 1971, published on 30 May 1973, and granted on 9 September 1976.

The patent application distribution shows a peak in the period 2016–2021, with more than 300 yearly, and 2021 is the year with the highest number of published patent applications, with 447 documents. Today, it seems to be decreasing, with "only" 19 records in the first 7 months of 2024. The United States is the country with the greatest number of priority patent applications filed in this technical field (3298), followed by WO patent applications (3150) using the international patent application system (PCT) administered by the WIPO [37], the Republic of China (2907), European Patent (EP; 2259), Japan (1636), Canada (1477), Australia (1343), the Republic of Korea (1008), Spain (570), and Brazil (436), as the first 10 countries. The largest number of patent applications come from universities (The University of Texas System (US), The Regents of the University of California (US), Shenyang Pharmaceutical University (CN), and Sichuan University (CN)) and companies (The Liposome Company, Inc. (US), Translate Bio Inc. (US), Alza Corp. (US), Yissum Research Development Company of the Hebrew University of Jerusalem Ltd. (IL), and L.E.A.F. Holdings Group LLC (US)), as the first ten applicants.

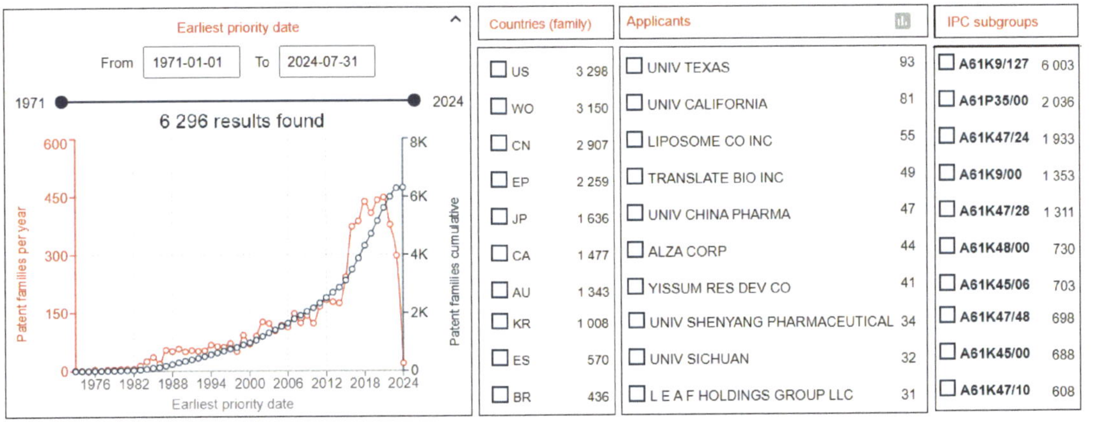

Figure 2. Liposome-based drug delivery systems: cumulative and yearly number of patent applications depending on earliest priority date, for the first ten countries, applicants, and IPC subgroups. Two-letter codes: US, WO, CN, EP, JP, CA, AU, KR, ES, and BR are abbreviations for the following countries: United States of America, the international publication of patent application using the Patent Cooperation Treaty (PCT) of the World Intellectual Property Organization (WIPO), Republic of China, European Patent, Japan, Canada, Australia, Republic of Korea, Spain, and Brazil, respectively. Data were obtained using the Espacenet database [32].

Searching criteria were narrowed to include only granted patents. During the time from the first application, a total of 2057 patents were granted in the period 1976 to July 2024, focusing on liposome drug delivery systems (Figure 2). Regarding countries, the Republic of China has the largest number of granted patents in this search criteria (2231), followed by Australia (1184), Japan, (1092), the Republic of Korea (793), and Taiwan Province of China (145), as the first five. The greatest number of granted patents also comes from universities (The University of Texas System (US), Shenyang Pharmaceutical University (CN), China Pharmaceutical University (CN), Zhejiang University (CN), The Regents of the University of California (US), and Shanghai Jiao Tong University (CN)) and companies (Translate Bio Inc. (US), The Liposome Company, Inc. (US), Shire Human Genetic Therapies, Inc. (US), Alza Corp. (US), and Insmed Inc. (US)), as the first 11 patent owners. The distribution shows a peak in the period 2016–2022 with more than 100 granted patents yearly, and the highest number of patents (186) was granted in 2018 (Figures 3 and 4), with a decreasing trend this year.

During the last ten years (from 2014 to 2024), 1306 patents were granted. The most numerous results and activities come from companies and universities in the Republic of China (Figure 4). This result suggests that the liposome-based drug delivery systems and correlated technologies are near the peak of research. By analyzing the obtained data, it was observed that many of the applications were based on liposomes and their application for medicinal preparations containing active ingredients. The most common were anti-cancer drugs, with a large and diverse class of medications. Data analysis indicated that the medicinal preparations containing genetic material (inserted into the cells of the living body) for genetic disease treatment, gene therapy, and mixtures of active ingredients were objects of patent protection. Also, objects of protection were drugs for specific purposes. Additional narrowed search criteria were included only for granted patents in specific fields, e.g., for different obtaining methods (classical or supercritical) or their pharmaceutical applications.

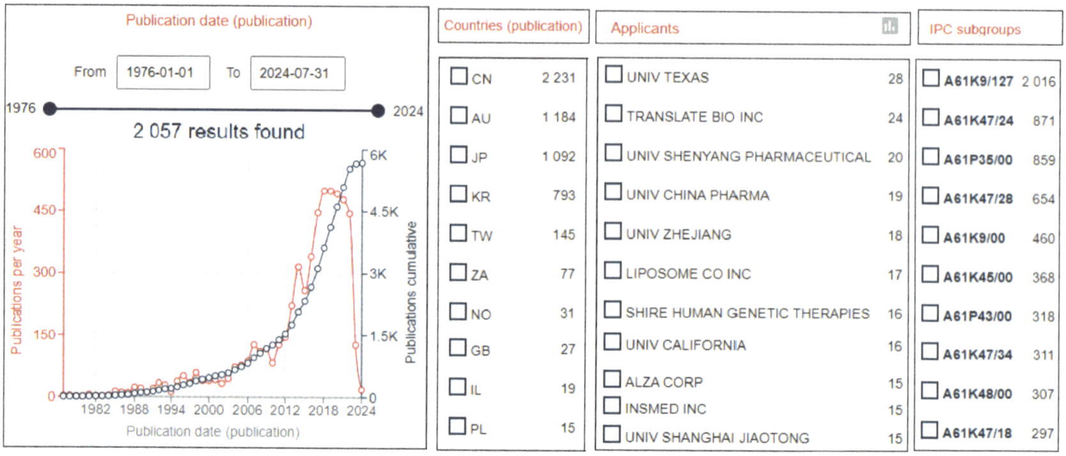

Figure 3. Liposome drug delivery systems: cumulative and yearly numbers of granted patents, depending on publication date, showing the first ten applicant origin countries, applicants, and IPC subgroups. Two-letter codes: CN, AU, JP, KR, TW, ZA, NO, GB, IL, and PL are abbreviations for the following countries: Republic of China, Australia, Japan, Republic of Korea, Taiwan Province of China, South Africa, Norway, United Kingdom, Israel, and Poland, respectively. Data were obtained using the Espacenet database [32].

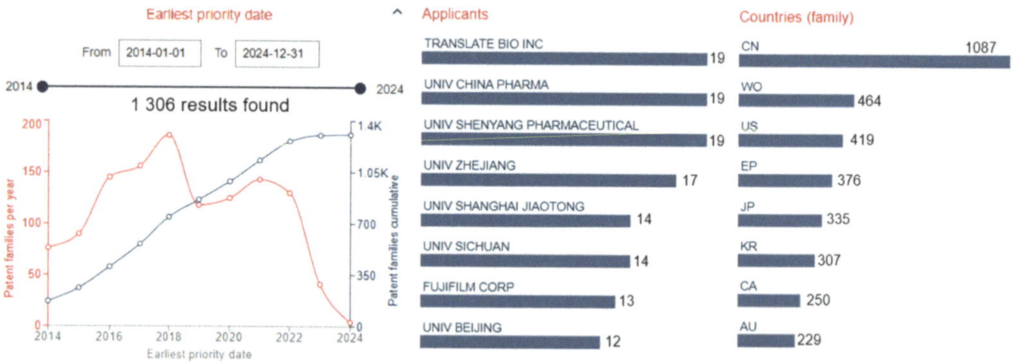

Figure 4. Liposome drug delivery systems: cumulative and yearly numbers of granted patents during the last ten years (from 2014 to 2024), depending on the earliest priority date, showing the first eight applicant origin countries and applicants. Two-letter codes: CN, WO, US, EP, JP, KR, CA, and AU are abbreviations for the countries: Republic of China, the international publication of patent application using the Patent Cooperation Treaty (PCT), United States of America, European Patent, Japan, Republic of Korea, Canada, and Australia, respectively. Data were obtained using the Espacenet database [32].

3. Processes for the Liposomes' Production

Engineering approaches for liposome formation were developed during the past five decades and include both "bottom-up" and "top-down" techniques (Figure 5 [19]).

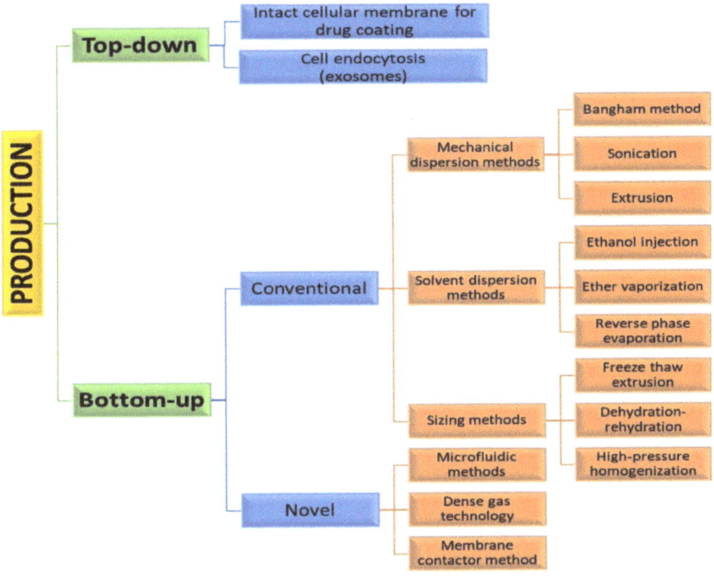

Figure 5. Diagram of various manufacturing engineering approaches and processes used for the synthesis of liposome vesicles. Reprinted from [19] under an open-access creative common CC-BY license.

The "bottom-up" strategy for liposome fabrication processes includes conventional methods (mechanical dispersion, solvent dispersion, or size-adjusting) or novel methods (microfluidic systems [38], dense gas technology, or membrane contactor methods). The conventional methods of mechanical dispersion include the well-known Bangham method (thin-layer dehydration rehydration), sonication, as well as the extrusion technique. With solvent dispersion methods, it is possible to apply ethanol injection, ether vaporization, or reverse-phase evaporation [39]. The size-adjusting conventional methods comprise improvement by applying freeze–thaw extrusion, dehydration–rehydration, and high-pressure homogenization techniques to obtain a narrower particle size distribution of liposomes. In general, all traditional procedures, which work at ambient temperature and pressure, lead to improvements with expensive post-processing stages, to obtain the nanometric dimensions of liposomes. However, these techniques have various disadvantages, e.g., low average encapsulation efficiencies of active ingredients (about 20–40%), loss of drugs, high residue of the solvent, problematic control of particle size distribution, large micrometric dimensions, expensive operating costs, and difficult repeatability in batch mode [40].

To overcome the problems related to conventional liposome production techniques, novel processes were proposed (microfluidic systems, dense gas technology, or membrane contactor methods). High-pressure systems, e.g., supercritical anti-solvent [41], reverse-phase supercritical evaporation [42], or depressurization of the expanded solution in an aqueous medium [43], were developed in semi-continuous formations to solve the mentioned problems. The crucial difference is the inversion of the traditional production phases so that the droplets of water were obtained first, and then they were rapidly encircled by phospholipids in supercritical conditions using high-pressure carbon (IV) oxide (in the range of 100–200 bar at 35–45 °C). By applying supercritical conditions, the encapsulation efficiency was increased by about 50–60%, with better particle size distribution control. A novel supercritical-assisted process was invented as a one-shot production method, with improved liposome stability over six months, excellent encapsulation efficacy of hydrophilic or lipophilic compounds (up to 99%), and lessened residual solvent [44]. Inverted, solvent-free liposomes were formed, with a second layer of phospholipids around the first one.

The "top-down" strategy comprises a cellular membrane/biomembrane for the drug-coating technique or cell endocytosis (exosome membranes) [19].

Even though novel, efficient production processes of liposomes were developed, batch-mode techniques with mostly low repeatability and expensive raw materials were industrialized [45].

3.1. Overview of Patented Liposome Production Processes

The patent search in the Espacenet database was further narrowed by adding field IPC or CPC A61K9/127/low and A61K9/1277 keywords: "liposom", "process", "procedure", "obtaining" or "method", and technique", in the claim search field, according to the abovementioned production techniques, "bottom-up" or "top-down". Selected granted patents for the liposome production processes, both conventional and novel, summarized in Tables 2–4, respectively, are briefly analyzed.

3.1.1. Conventional Liposome Production Processes

The advanced search in the Espacenet database was narrowed by adding keywords that describe conventional liposome production processes: "mechanical dispersion", "solvent dispersion", "size-adjusting", "thin-layer hydration", "sonication", "extrusion", "freeze–thaw extrusion", "dehydration–rehydration", and "high-pressure homogenization". Also, keywords for novel processes: "microfluidic systems", "dense gas technology", "membrane contactor", cellular membrane", "biomembrane", "cell endocytosis", and "exosome membrane", were excluded from the search criteria. As a search result, conventional liposome production processes were the subject of 518 patent documents, which included patent applications ("A" documents) and granted patents ("B" documents) from 1974 to July 2024 in the Espacenet database according to title, abstract, and claims, from which a total of 186 patents were granted (including filter "B" for publication number). After narrowing this analysis by excluding abstracts and titles in the query, 170 patents were found. During the last decade, 22 granted patents presented expansion with narrowing obtained results (publication date from 2014 to 2024). Finally, the search was narrowed to involve only granted patents with publication dates after 2020, and the documents were sorted by relevance. Several granted patents for conventional liposome production processes, selected by relevance and patenting levels, are summarized in Table 2 and briefly analyzed.

Table 2. Selected relevant patents for conventional liposome production processes.

Title	Publication No. [Reference]	Applicants	Earlies Priority
Encapsulation of chemical substances—in liposomes for medicinal use	FR2221122B1 [46]	Bayer AG	12 October 1971
Fusogenic liposomes and methods of making and using same	EP0758883B1 [47]	The Liposome Company, Inc.	12 April 1994
Liposome sensitive to pH or reductive condition and processes for the preparation thereof	KR100853172B1 [48]	Postech Acad Ind Found	4 April 2007
Improved lipid formulation	US8158601B2 [49]	Chen J.; Ansell S.; Akinc A; Dorkin J.R.; Qin X.; Cantley W.; Manoharan M.; Rajeev K.G.; Narayanannair J.K.; Jayaraman M.; Alnylam Pharmaceuticals, Inc.	10 June 2009

Table 2. *Cont.*

Title	Publication No. [Reference]	Applicants	Earliest Priority
Improved lipid formulation	US8802644B2 [50]	Chen J.; Ansell S.; Akinc A; Dorkin J.R.; Qin X.; Cantley W.; Manoharan M.; Rajeev K.G.; Narayanannair J.K.; Jayaraman M.; Tekmira Pharmaceuticals Corp	10 June 2009
Liposome formulation and manufacture	EP3135274B1 [51]	Biorest Ltd.	14 March 2013
Preparing method of proliposomes	KR101682821B1 [52]	Seoul National University R&DB Foundation	21 July 2015
Platinum-modified liposome as well as preparation method and application thereof	CN107550865B [53]	Guangzhou Heji Biological Tech Co., Ltd.	21 June 2017
Compositions and systems comprising transfection-competent vesicles free of organic solvents and detergents and methods related thereto	EP3864163B1 [54]	The University of British Columbia	9 October 2018
Alkannin active drug-loading liposome as well as preparation and application thereof	CN115105473B [55]	Suzhou Yutai Pharmaceutical Tech Co., Ltd.	19 Marc 2021
Hydrogel–liposome combined drug delivery system as well as preparation method and application thereof	CN115192581B [56]	The First Affiliated Hospital of Shandong First Medical Univ. Shandong Qianfoshan Hospital	21 July 2022
Preparation method and application of acid-response-type drug delivery platform based on liposome nano-vesicles	CN115337267B [57]	University of Science and Technology China	15 August 2022
Probucol liposome for hypercholesterolemia, preparation and preparation method of probucol liposome	CN115364056B [58]	Pingdingshan No. 2 People's Hospital, Shangdong New Time Pharmaceutical Co., Ltd.	26 August 2022
Thermo-sensitive nano-liposome capable of realizing stepped release of active matters and application of thermo-sensitive nano-liposome	CN115590818B [59]	University of Jiangnan	9 September 2022
Alendronate sodium liposome and preparation thereof	CN116650418B [60]	Zhengzhou Central Hospital	5 June 2023
A stabilized nano-liposome delivery carrier comprising nucleic acid and a method for preparing the same	KR102678046B1 [61]	Binotec Co., Ltd.	22 October 2023
Anti-alopecia hair-strengthening nano-liposome and preparation method thereof	CN118001207B [62]	Wangshuhe Biomedical Wuhan Co., Ltd.	2 February 2024

The first patent, FR2221122B1, relates to the encapsulation of chemical substances in liposomes for medicinal use and liposomes' preparation with a maximum diameter of 1000 Å, preferably from 200 to 500 Å [46]. The liposome walls consist of mono-, bi-, or multi-molecular layers, preferably bimolecular with a thickness of 30–100 Å. Two simple methods were used for the liposome preparation. In the first method, a lipid was added to the aqueous phase, the mixture was slightly heated, then energetically shaken, followed by sonication until liposomes were formed, manifested by partial illumination and a slight bluish opalescence appearance. In the second method, the dispersion system, as a linear colloid, was obtained most simply by transforming the dispersed phase into thin films, evaporating the solvent, and bringing these films into contact with a continuous aqueous or non-aqueous phase, with vigorous mixing and treating with ultrasound for a long period.

Fusogenic liposomes' composition, production methods, and applications were disclosed in patent EP0758883B1 [47]. They were obtained according to known methods by drying chloroform solutions of lipids under nitrogen, then removal of the residual solvent under a high vacuum. Formed lipid films were hydrated by vortex-mixing with buffers to produce multilamellar liposomes (MLVs), and five freeze–thaw cycles were used to achieve homogeneous mixtures. The MLVs were extruded ten times to make large unilamellar liposomes. Composition was comprised of a liposome with (i) an outermost lipid bilayer comprising a fusion-promoting ionizable lipid with a cationic protonatable headgroup and unsaturated acyl chain, and (ii) a section contiguous to the terminal lipid bilayer, which comprised an aqueous solute with the first pH. The first pH was lower than the pKa of the ionizable lipid in an outermost lipid bilayer. Outside of the liposome, in the structure there was an aqueous solution with the second pH (bigger than the pKa of the ionizable lipid in the outermost lipid bilayer), such that there was a pH gradient across the outermost lipid bilayer. An ionizable lipid was accumulated in the inner monolayer of the outermost lipid bilayer in response to the gradient.

The proliposome preparation method by patent KR101682821B1 comprised of the following steps: (1) the first solution was prepared by dissolving a sugar-based carrier in water, (2) the second solution was prepared by dissolving a bioactive ingredient, lipids, and a stabilizer in an organic solvent, (3) the first and the second solutions were mixed and condensation-stirred to a uniform consistency, and (4) the condensation-stirred mixed solution was freeze-dried [52]. The condensation-stirring method could significantly reduce the excipients amount, compared to conventional proliposomes. The obtained proliposome can enhance the bioavailability of the encapsulated insoluble drug by promoting intestinal elution and absorption.

Korean patent KR100853172B1 disclosed pH-sensitive liposomes, used as a desirable drug delivery system based on a self-assembling cucurbituril derivative, stable in blood, and processes for their preparation, wherein the dispersing step is performed by ultrasonic wave using a sonication method [48]. The obtained pH-sensitive liposomes disintegrated after absorption in the cell, allowing the drug to act on a tissue of a desired target portion, while preventing side effects of a systemic drug by encapsulating.

The granted US patents US8158601B2 and US8802644B2 relate to the cationic lipids, and lipid particles comprising these lipid formulations are prepared by an extrusion method [49,50]. Moreover, these patents provide production methods of these compositions, and methods of introducing nucleic acids into cells through obtained compositions, e.g., for treatments of various disease conditions. The invention presented in the EP3135274B1 patent relates to a novel process for obtaining liposomal formulations of uniform size with desirable, independently controllable features [51]. This formulation for use in the prevention of restenosis was manufactured via a method of extruding the vesicles, consisting essentially of repeatedly extruding through a single filter 10–18 times before ultrafiltration of the vesicles. Compositions and systems comprising transfection-competent vesicles free of organic solvents and detergents, and production methods via extrusion, were disclosed in the granted patent EP3864163B1 [54]. An alkannin/shikonin active drug-loading liposome (CN115105473B) was prepared by dissolving phospholipids, cholesterol, and PEGylated phospholipids in a volatile organic solvent, evaporating the organic solvent under reduced pressure to form a dry lipid film, adding copper ion salt solution, and performing hydration treatment and extrusion to reduce the particle size of the liposomes to form nano-sized liposomes [55]. After that, agarose gel chromatography columns were used to exchange the extracellular aqueous phase of the liposomes, the blank liposomes were incubated with the dimethyl sulfoxide solution of shikonin at 55–65 °C for 30 min, and then they were cooled for 5 min. The shikonin active drug-carrying liposome can be applied in the preparation of antitumor drugs. CN115337267B protects the pH-responsive lipid nanovesicles that are prepared by using acidic phospholipids, acid-sensitive silyl ether prodrug Chol-R848, and 1,2-distearoyl-rac-glycero-3-PE-N-polyethyleneglycol-2000 (DSPE-mPEG2000) as raw materials through film hydration and extrusion [57].

The specific preparation method to obtain a hydrogel–liposome combined drug delivery system includes dissolving hydrogenated soybean phosphatidylcholine, cholesterol, cyclic arginine-glycine-aspartic acid-modified DSPE-PEG, and elastane in an organic solvent, obtaining crude liposomes after sonication, and mixing the crude liposomes with an aqueous solution of temozolomide (TMZ) to form an emulsion. After removing the organic solvent, the liposome was processed by ultrasonic wave and extrusion (CN115192581B) [56]. Probucol liposome for hypercholesterolemia (CN115364056B) [58] was prepared via several steps, which included (1) preparation of the initial mixture of phospholipids, cholesterol, polysorbate, and probucol, (2) the sonication step and formation of liposomes of 30–80 nm by the French extrusion method, and (3) freeze-drying to obtain the final product.

The CN107550865B patent describes a high-pressure homogenization method at a frequency of 500–1500 bar and 20–60 Hz for preparing a liposome, wherein the phospholipid and the platinum-containing aqueous solvent are used in a weight ratio of 1:30 to 80 [53]. The liposome obtained by the invention has excellent stability and can be stored for a long time, while still maintaining a small particle size. Granted patent KR102678046B1 [61] described a stabilized nano-liposome as a carrier for encapsulated nucleic acid delivery obtained by the high-pressure homogenization method. A thermo-sensitive nano-liposome capable of realizing stepped release of active matters (CN115590818B) [59], anti-alopecia hair-strengthening nano-liposome (CN118001207B) [62], and alendronate sodium liposome (CN116650418B) [60] were also prepared using a high-pressure homogenization method.

Analysis of the granted patents showed that sonication, extrusion, dehydration–rehydration, and high-pressure homogenization methods were most usually applied for novel pharmaceutical applications.

3.1.2. Novel Liposome Production Processes

New technologies have been developed to support less scalable liposome production technologies, including self-assembling liposome systems and microfluidic production.

Microfluidic methods allow precise control of many factors during preparation, e.g., liposome size distribution and fluids in a constrained volume (by laminar flow and lipid concentrations in microfluidic channels), pH, temperature, vesicle size, salinity, and osmolarity [63]. This is an effective drug encapsulation method to achieve self-assembled liposomes. Besides the advantages of this simple and low-cost method, the main disadvantages are unsuitability for bulk production and difficulty in removal of organic solvents. In the membrane contactor method, a dissolved lipid phase in alcohol was pushed throughout a porous membrane in an aqueous phase flow, from where lipid molecules were self-assembled in homogenous-sized liposomes [64]. Dense gas technology is one more novel method for liposome preparation, which utilizes supercritical fluids (e.g., supercritical carbon dioxide) as excellent solvents for many lipids. After mixing with the water phase, liposomes with a narrow size distribution are synthesized [65].

Search criteria in the Espacenet database were carried out using keywords that describe novel processes: "microfluidic systems", "dense gas technology", "membrane contactor", "cellular membrane", "biomembrane", "cell endocytosis", and "exosome membrane". All previously mentioned keywords for conventional liposome production processes (Section 3.1.1) were excluded from the search criteria. Novel liposome production processes were the subject of 661 patent documents (patent applications and granted patents) from 1973 to July 2024 according to the title, abstract, and claims, and 195 patents were granted therefrom. Throughout the last 10 years, 431 patents were found, and after narrowing this result by claims only in the query, 82 granted patents were found. A narrowed search included only granted patents chosen by relevance. Selected granted patents for novel liposome production processes are summarized in Tables 3 and 4, respectively, and briefly described.

Table 3. Selected relevant patents for novel liposome production processes.

Title	Publication No. [Reference]	Applicants	Earliest Priority
Novel functionalized liposomes and a process for production thereof	EP0247497B1 [66]	Wako Pure Chemical Industries, Ltd.	20 May 1986
Synergistic liposomal formulation for the treatment of cancer	EP3046542B1 [67]	Council of Scientific and Industrial Research	18 September 2014
Liposome compositions and methods of use thereof	EP2603201B1 [68]	Rhode Island Board of Governors for Higher Education; Yale University, Inc.	12 August 2011
Targeted hybrid exosome loaded with curcumin and miR140 as well as preparation method and application of targeted hybrid exosome	CN117965429B [69]	Sichuan University	29 March 2024
Hybridosomes, compositions comprising the same, processes for their production, and uses thereof	US10561610B2 [70]	Anjarium Biosciences AG	20 January 2015
Hybridosomes, compositions comprising the same, processes for their production, and uses thereof	US11484500B2 [71]	Anjarium Biosciences AG	2 January 2020
Process for the production of hybridosomes	EP3096741B1 [72]	Anjarium Biosciences AG	21 January 2014
Preparation method of exosome bionic preparation for synergistically promoting wound healing and preparation thereof	CN113577272B [73]	China Pharmaceutical University	28 July 2021
Co-loaded liposome and preparation method thereof	CN116270473B [74]	Chengdu Jinrui Found Biotechnology Co., Ltd.	25 May 2023
A method for preparing a functional synthetic cell in form of a giant unilamellar vesicle	EP3630068B1 [75]	Max-Planck-Gesellschaft zur Förderung der Wissenschaften e.V.	16 June 2017
PEG2, N-lipid derivative-modified nano-carrier, preparation method and application	CN113350512B [76]	Shenyang Pharmaceutical University	7 June 2021
Compound responding to endocytosis and release and application thereof	CN113683769B [77]	Shanghai Jiao Tong University	19 May 2020
All-trans retinoic acid liposome preparation and preparation and application thereof	EP3501500B1 [78]	Shanghai Jiao Tong University	18 August 2016
Granular-type adjuvant as well as preparation method and application thereof	CN108324938B [79]	Institute of Process Engineering, Chinese Academy of Sciences	1 March 2018
mRNA–liposome complex and application thereof	CN112107680B [80]	Zhejiang Zhida Pharmaceutical Co., Ltd.	21 June 2019
Drug delivery vector and pharmaceutical formulation	CN114007653B [81]	Fudan University; JSR Corporation	5 May 2019

Novel functionalized liposomes containing a high-molecular-weight amphiphilic compound, as one of the matrix materials, were obtained by the reverse-phase evaporation method, described in patent EP0247497B1 [66]. They have a very high encapsulation efficiency and readily undergo lysis. An immunological substance (antigens, antibodies, etc.) or a physiologically active substance could be immobilized on the liposomes efficiently, with a sufficient binding rate, using the amphiphilic compound as a spacer.

The synergistic liposomal formulation for use in the treatment of cancer is prepared by reverse-phase evaporation vesicles or multilamellar vesicles (EP3046542B1) [67]. The reverse-phase evaporation technique encapsulates up to 50% of solute (the first to use 'water-in-oil' emulsions). The preparation method involves a rapid injection of aqueous solution into an organic solvent containing dissolved lipids. After water droplet formation ('water-in-oil' emulsion) by bath sonication of the two-phase mixture, the emulsion is dried down to a semi-solid gel in a rotary evaporator. The following step is vigorous mechanical shaking of the gel to induce a phase change from a "water-in-oil" emulsion to a vesicle suspension. Some water droplets collapse, attach to adjacent, intact vesicles, and form the outer leaflet of the bilayer of a large unilamellar liposome (0.1–1 μm diameter).

The invention is based on the discovery that pH (Low) Insertion Peptide (pHLIP) liposomes target acidic tissue, and release liposome content, i.e., cargo, into a cell (EP2603201B1) [68]. In the method of delivering cargo into a target cell (tumor cell, ischemic cell, inflamed cell, bacterially infected cell, fungus-infected cell, or virally infected cell), pHLIP+ liposome both fuses with a cell membrane of an endosomal compartment of a target cell and is taken up by cell endocytosis.

The granted Korean patent CN117965429B disclosed a targeted hybrid exosome loaded with microRNA 140 (miR140) and curcumin [69]. The obtaining method comprises the following phases: (1) obtaining a cell strain capable of stably expressing chondrocyte-targeted peptide, (2) addition of miR-140, carrying out co-culture, separation, and extraction to obtain an exosome loaded with miR-140-5p, (3) addition of the curcumin into the cholesterol and soybean lecithin solutions, addition of buffer solution, and ultrasonication after rotary evaporation to form a curcumin-loaded liposome buffer solution, adding sucrose, and (4) the exosome obtained in phase 2 and the solution obtained in phase 3 were subjected to a freezing circulation method, and the targeted hybrid exosome was obtained.

Granted patents US10561610B2, US11484500B2, and EP3096741B1 describe hybrid biocompatible carriers (hybridosomes). The process for manufacturing hybridosomes, hybrid biocompatible carriers, is comprised of contact and uniting of a first vesicle and a second vesicle [70–72]. The first vesicle, produced in vitro, comprises a membrane, a therapeutic agent, and a fusogenic, ionizable, cationic lipid at a molar concentration of at least 30% of the total lipid of the first vesicle. The second vesicle, produced in vivo, comprises a lipid bilayer and is released into the extracellular environment. These inventions further provided pharmaceutical compositions comprising pharmaceutical applications and appropriated pharmaceutical methods.

The other example is an exosome bionic preparation, which synergistically promotes wound healing, prepared by the method described in Chinese patent CN113577272B [73]. This procedure comprises several steps: preparing a catalase-photosensitizer micelle, preparing an reactive oxygen species (ROS) response liposome carrying the catalase-photosensitizer micelle, extracting, separating, and purifying the exosome, mixing the exosome with the liposome, mediating membrane fusion between the liposome and exosome using the extrusion method, mixing the membrane fusion carrier and the gel, and homogenously stirring the mixture to prepare the final product.

The invention CN116270473B relates to adopting a film dispersion method of a co-loaded liposome preparation, which comprises 5–35% magnolol and the sum of honokiol (as active pharmaceutical ingredients), 50–80% phospholipid, 8–15% cholesterol, 3–10% polyethylene glycol or PEGylated phospholipid, 1–5% cholesterol, and 1–5% polyethylene glycol [74]. The raw materials and the auxiliary materials are mixed and then dissolved with an organic solvent, rotary evaporation is performed to form a film, and after that, hydration and homogenization are performed to obtain the co-loaded liposome. The encapsulation efficacy reaches 90% or above, the particle size distribution is uniform, and the stability of the preparation is remarkably improved.

European patent EP3630068B1 outlines a preparation method for a protocell in the form of a giant unilamellar vesicle by a microfluidic device or bulk technique by adding destabilizing molecules. It comprises the following stages: (1) forming a water-based droplet (0.5–1000 μm) encapsulated by a surface polymer shell, which borders the inner space of the droplet, containing at least one lipid, (2) transforming the droplet lipid content into a lipid bilayer, which covers the inner surface of the polymer shell and oil phase to form a polymer-shell-stabilized giant unilamellar vesicle, (3) alternatively, incorporating one or more nuclei and/or proteins into the polymer-shell-stabilized giant unilamellar vesicle, and (4) alternatively, removing the polymer shell and oil phase from the polymer-shell-stabilized giant unilamellar vesicle and transferring it from the oil to the water phase [75].

The PEG2, N-lipid derivative-modified nano-carrier, was prepared using a method described in patent CN113350512B, capable of eliminating an ABC phenomenon caused by a PEGylation nano-carrier [76]. PEG2 forms a compact hydration layer on the carrier surface, with improved physical and biological stabilities. In addition, this novel nano-carrier overcame insufficient circulation time defects of many PEG substitute materials, and a firmer foundation was laid for its clinical transformation.

A compound having a response to the endocytic release of cells, and its usage, were described in the patent CN113683769B [77]. This compound was composed of a lipophilic head, a hydrophilic chain, and a group having a response to the pH. The modified complex is stable in body fluid circulation and has a high transfection efficacy.

The invention EP3501500B1 relates to an all-trans retinoic acid liposome and a liposome vector, its preparation, and its application [78]. The all-trans retinoic acid liposome preparation obtained through this active drug-loading method significantly improves the plasma drug concentration in all-trans retinoic acid liposome and prolongs the half-life. The granular-type adjuvant was prepared from a biocompatible wall material, biocompatible grease, and all-trans retinoic acid according to patent CN108324938B [79]. The biocompatible wall material includes a biocompatible polymer poly(lactic-co-glycolic acid (PLGA) and liposomes, and the mass percentage of the cationic liposome in the biocompatible wall material is less than or equal to 80%. An inner core formed by the biocompatible grease was covered in a shell, formed by the biocompatible wall material. All-trans retinoic acid was coated in the inner core. Systemic and mucosal immunizations were activated, and a dual-activation effect was realized, according to this invention.

According to patent CN112107680B [80], the method of obtaining the messenger ribonucleic acid (mRNA) and liposome complex comprised diluting the mRNA, adding the liposome, or preparing it in a continuous online mixing mode, or using a microfluidic mixing system.

The Chinese granted patent CN114007653B described a drug delivery vector containing a membrane-penetrating peptide-modified cationic liposome with 1,2-dioleoyl-3-trimethylammonium-propane (cationic lipid), 1,2-di-(9Z-octadecanoyl)-sn-glyceryl-3-phosphoethanolamine (non-cationic lipid), cholesterol, and polyethylene glycol phospholipids conjugated to membrane-penetrating peptides (PEGylated phospholipids) [81].

Table 4 highlights a few granted patents for novel supercritical-assisted processes.

Table 4. Selected patents for supercritical-assisted liposome production processes.

Patent Title	Publication No. [Reference]	Applicants	Earliest Priority
A delivery device	EP3334413B1 [82]	PTT Holding Aps	11 August 2015
Pharmaceutical composition, preparation, and uses thereof	EP3236934B1 [83]	Curadigm SAS	25 November 2014
Method for preparing medicinal extract containing intermediate peashrub seeds	CN104382984B [84]	Northwest Institute of Plateau Biology Chinese Academy of Sciences	1 December 2014

Table 4. *Cont.*

Patent Title	Publication No. [Reference]	Applicants	Earliest Priority
Preparation method of temperature and fluorescence probe of liposome loaded with gold nanocluster and anti-cancer drug	CN103599070B [85]	Shanghai Jiao Tong University	26 November 2013
Antitumor liposome formulation and method for producing the same	JP4715133B2 [86]	Konica Minolta Med and Graphic	26 August 2004
Method for producing liposome-containing formulation and liposome-containing formulation	JP4599849B2 [87]	Konica Minolta Med and Graphic	18 February 2004
Method for producing liposome-containing formulation and liposome-containing formulation	JP4649841B2 [88]	Konica Minolta Med and Graphic	18 February 2004

The invention EP3334413B1 relates to a delivery device suitable for delivering a chemical substance, e.g., a medical device, such as a catheter, a microcapsule, an implantable capsule, or a P-ring [84]. This device is formed by loading a drug into the matrix using a solvent in its supercritical, near supercritical, or dense gas states. A preferred drug carrier includes CO_2 in a liquid and/or supercritical state. The delivery device for the chemical substance comprises a drug (for treating symptoms of Alzheimer's, e.g., dual agonist of the peroxisome proliferator activated nuclear receptor delta/gamma, aka PPARδ/γ (T3D-959), donepezil, galantamine, memantine, rivastigmine, or any combinations thereof) trapped in liposomes. Together with the metal particles, the liposomes are dispersed in a polar liquid, e.g., an aqueous liquid.

The invention EP3236934 (B1) relates to a pharmaceutical composition comprising the combination of at least one (i) biocompatible nanoparticle and (ii) carrier comprising at least one pharmaceutical compound, a polymer selected from dextran, polysialic acid, hyaluronic acid, chitosan, heparin, polyvinyl pyrrolidone, polyvinyl alcohol, polyacrylamide, or poly(ethylene glycol), and a PEG-based copolymer [85]. The carrier can respond to an intracellular or extracellular stimulus when it is exposed to electromagnetic radiations, ultrasounds, and a magnetic field.

The drug-containing liposome nanoparticles and medicinal extract of *Caragana korshinskii* seeds were prepared by the patent CN104382984B [84]. The *Caragana korshinskii* seed extract, with membrane material, cholesterol, lecithin, and deionized water, at a temperature of 35~40 °C and a pressure of 25~30 MPa, was dispersed and mixed in supercritical CO_2 medium for 35–45 min. The mass ratio of lecithin and cholesterol was 2:1. The mass ratio of the film material and the seed medicinal extract was 3:1.

The object of patent CN103599070B is the preparation method of fluorescence and temperature probes of liposome membrane, loaded with the water-soluble anti-cancer drug and gold nanocluster dispersion [85]. They are incubated by a supercritical CO_2 method at a defined temperature and pressure.

A highly safe antitumor liposome formulation with a stabilized structure, affording retention stability of an anti-cancerous compound and achieving efficient delivery of an anti-cancer agent, was described in Japanese patent JP4715133B2 [88]. The liposome with an average particle size diameter of 0.05 to 0.7 μm was produced by mixing the lipid membrane components, constituting the phospholipid membrane, with carbon dioxide in a supercritical or subcritical state, in the presence of at least one compound having a polyethylene glycol (PEG) group.

Patent JP4599849B2 provides a production method for liposome-containing formulations, with improved content of an enclosed substance in liposomes under supercritical conditions [89]. A mixed solution of supercritical carbon dioxide, an aqueous solution containing an encapsulating substance and liposome suspension, was subjected to a heat and pressure treatment (115–135 °C and 0.17–0.31 MPa). The method includes the encapsulated anti-cancer substance or a contrast-enhancing compound.

The JP4649841B2 patent relates to a method for producing a liposome-containing preparation consisting of a micronized liposome with a high encapsulation rate of an encapsulated substance [90]. In the described effective method for their production, even though the main solvent is supercritical carbon dioxide, organic solvent (as a solvent aid) addition is unavoidable, and the organic solvent tends to lower the film strength. The obtained preparation can be applied to internal medicines, external preparations for skin, cosmetics, and contrast agents, and similar applications.

Inventive designs of novel liposomal production processes contribute to drug activity optimization in vitro and, consequently, in vivo. Published patents showed new possibilities to develop stable liposomes, which exhibited excellent abilities as drug carriers. Novel liposome preparation processes developed in the laboratory are not easily translated to the semi-industrial and industrial levels. This may slow down the further development and application of new liposome systems.

4. Liposome-Based Drug Delivery Systems

Any drug delivery system aims to beneficially modulate the pharmacokinetics parameters of the drug and/or distribution on the tissue. Nanoparticles as drug carriers show kinetically slow transportation in diseased tissues due to vascular barriers. A relatively long circulation time is needed to extend the time of crossing the vascular wall [7]. Liposome-based drug carriers have become increasingly important due to their ability to protect drugs and improve delivery across biological barriers. A key aspect is their unique structure that allows them to encapsulate both hydrophilic and hydrophobic drugs. Liposomes pass through biological barriers, of which the blood–brain barrier is one of the most challenging because it protects the brain from potentially harmful substances, but also limits the entry of many drugs. Mechanisms for overcoming biological barriers include surface functionalization with ligands or antibodies that bind to specific receptors on target cells. Stimuli-responsive liposomes overcame biological barriers by releasing the drug in response to specific stimuli in the body (temperature, pH, light, concentration, and pressure). Stealth liposomes have a central role in improving pharmacokinetics (biodistribution, blood circulation, and tissue targeting) [7]. The stealth effect shows the ability of nanoparticles to be invisible to the immune system, especially the reticuloendothelial system (RES), as a result leading to lessening clearance and increased retention time in the bloodstream. The most important effect of liposome PEGylation is shifting blood pharmacokinetics from non-linear saturable to linear non-saturable kinetics [7,89,90]. For non-linear kinetics, the saturation can happen during any kinetic clearance processes; however, it is generally in correlation with saturating RES clearance, thus bringing a net effect of the RES-blockade [91]. Due to the RES-blockade, non-PEGylated liposomes exhibit biphasic clearance (short α-phase, followed by a relatively long β-phase) until the PEGylated liposomes tend to exhibit a long half-life monophasic clearance. Furthermore, monophasic clearance of PEGylated liposomes exhibits a constant clearance rate, dose-independent, with above-wide dosage ranges. The analysis demonstrated that 85% of the reported stealth nanomaterials/drug delivery systems encounter a sharp α-phase clearance in blood circulation (a rapid drop in blood concentration to half of the administered dose within 1 h post-administration), while a relatively long β-phase could occur. The term pseudo-stealth effect describes this common pharmacokinetics behavior. The stealth effect and pseudo-stealth effect improve pharmacokinetics [7]. It is important to keep a good balance between the stealth effect and the interaction with diseased tissue. Dynamic modulation of the stealth effect throughout a stimuli-responsive strategy can further advance the functionality and increase the therapeutic efficacy [91].

Analysis of the obtained results from the Espacenet database search showed numerous applications based on liposomes with encapsulated medicinal active ingredients. The patent search for liposome-based carriers was further narrowed by adding keywords for specific fields, e.g., "anti-cancer", "pulmonary", "ocular", "RNA", "neural", "orthopedic", "internal", "anti-inflammation", and "dermal". Selected patents for liposome-based carriers are summarized in Tables 5–10 and subsequently analyzed.

4.1. Liposome-Based Drug Carriers for Cancer Treatments

Keywords "anti-cancer", "cancer", "tumor", and "tumour" were added to the search criteria. A total of 1822 patent documents (applications and granted patents) from 1983 to July 2024 according to the title, abstract, and claims, and 637 granted patents, were found. Obtained data are graphically presented in Figure 6 as cumulative and yearly numbers of patent applications and granted patents, the first seven applicant origins and countries, and the IPC/CPC subgroups for granted patents. During the last 10 years, 476 patents were granted. Table 5 highlights the selected granted patents for liposome-based drug carriers used in drug delivery for cancer treatments, chosen by relevance.

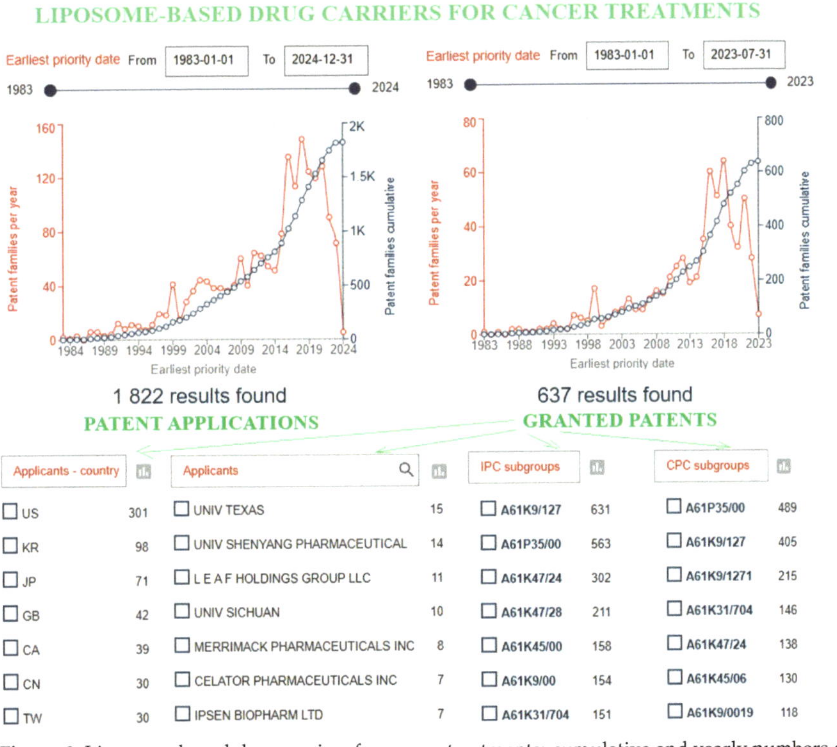

Figure 6. Liposome-based drug carriers for cancer treatments: cumulative and yearly numbers of patent applications and granted patents depending on the earliest priority date (from 1983 to 2024), the first seven applicant origins and countries, and the IPC and CPC subgroups for granted patents. Two-letter codes: US, KR, JP, GB, CA, CN, and TW, are abbreviations for the following: United States of America, Republic of Korea, Japan, Canada, United Kingdom, Republic of China, and Taiwan Province of China, respectively. Data were obtained using the Espacenet database [32].

Table 5. Selected granted patents for liposome-based drug carriers for cancer treatments.

Patent Title	Publication No. [Reference]	Applicants	Earliest Priority
Controlled drug release liposome composition	EP2968146B1 [92]	Taiwan Liposome Company Ltd.; TLC	15 March 2013
Method of reconstituting liposomal annamycin	US11980634B2 [93]	Board of Regents, The University of Texas System	28 June 2019
Protamine short-peptide-modified paclitaxel liposome and preparation method thereof	NL2033676B1 [94]	Institute of Biological Resources, Jiangxi Academy of Sciences; Anhui Zishanyuan Ecological Agricultural Technology Co., Ltd.	6 December 2022
Paclitaxel liposome composition for treatment of cancer and preparation thereof	EP1332755B1 [95]	Nanjing Zhenzhong Bioengineering Comp. Ltd.	19 October 2000
Liposome of irinotecan or its hydrochloride and preparation method thereof	EP2508170B1 [96]	Jiangsu Hengrui Medicine Co., Ltd.; Shanghai Hengrui Pharmaceuticals Co., Ltd.	3 December 2009
Irinotecan liposome preparation and preparation and application thereof	CN109260155B [97]	HighField BioPharmaceutical Corporation	5 November 2018
Pegylated lipid, liposome modified by pegylated lipid, pharmaceutical composition containing liposome, and preparation and application of pharmaceutical composition	CN115197078B [98]	Xiamen Sinopeg Biotech Co., Ltd.	8 April 2021
Liposomes with ginsenoside as membrane material and preparations and use thereof	EP3337456B1 [99]	Shanghai Ginposome Pharmatech Co., Ltd.	19 August 2015
Dianhydrogalactitol, diacetyldianhydrogalactitol or dibromodulcitol to treat non-small-cell carcinoma of the lung and ovarian cancer	EP3125920B1 [100]	Del Mar Pharmaceuticals Inc.	4 April 2014
Multicellular targeting liposome	JP7470418B2 [101]	Shanghai Jiao Tong University	14 September 2017
Tumor-targeted therapeutic drug carrier as well as preparation method and application thereof	CN102133404B [102]	Chengdu Nuoen Biological Technology Co., Ltd.	23 March 2011
Application of redox-responsive chitosan-liposome	CN107982547B [103]	University Dalian Minzu	27 July 2017
Pharmaceutical composition comprising jasmonates	KR101751918B1 [104]	Fehr Pereira Lopes, Jose E.	15 July 2008
Liposomal delivery of vitamin-E-based compounds	AU2002361812B2 [105]	Research Development Foundation	19 December 2001
Liposomes useful for drug delivery	US8147867B2 [106]	Hermes Biosciences Inc.	3 May 2004
Liposomes useful for drug delivery	US8703181B2 [107]	Merrimack Pharmaceuticals	3 May 2004
Liposomes useful for drug delivery	US9782349B2 [108]	Ipsen Biopharma Ltd.	11 December 2015
Preparation method of temperature and fluorescence probe of liposome loaded with gold nanocluster and anti-cancer drug	CN103599070B [85]	Shanghai Jiao Tong University	26 November 2013

Table 5. *Cont.*

Patent Title	Publication No. [Reference]	Applicants	Earliest Priority
Granular-type adjuvant as well as preparation method and application thereof	CN108324938B [79]	Institute of Process Engineering, Chinese Academy of Sciences	1 March 2018
Interleukin-2 variant proteins fused to human IgG4 Fc and uses thereof	US11896648B2 [109]	Gilead Sciences, Inc.	22 October 2020

The pharmaceutical composition for inhibiting the cancer cell growth, according to patent EP2968146B1, comprised: (1) at least one liposome having a particle-forming component, selected from (i) phospholipid and (ii) a mixture of at least one phospholipid and cholesterol, (2) dextran sulfate or a pharmaceutically acceptable salt, (3) ammonium sulfate, and (4) a vinca alkaloid [92]. The method of reconstituting pre-liposomal annamycin, a cancer chemotherapeutic agent, lyophilized to form a liposomal annamycin suspension, was presented in patent US11980634B2 [93].

The disclosure in patent NL2033676B1 provides protamine, short-peptide-modified paclitaxel liposome, pharmaceutical formulations, and their preparation method [94]. The method includes mixing paclitaxel, lecithin, and cholesterol, dissolving the mixture in a solvent, and forming a film by removing the solvent after sterilization, then washing the film with an aqueous glucose solution. Paclitaxel liposome suspension was obtained by homogenization and lyophilized as a powder by freeze-drying. Protamine peptide combined with liposome increases the water solubility of paclitaxel, increases paclitaxel content in cells, and significantly reduces the vascular density in tumors, which has antitumor effects.

The other patent, EP1332755B1 [95], disclosed a paclitaxel-based liposome composition for the treatment of cancer, which consists of paclitaxel (2–5 parts), phosphatide (20–200 parts), cholesterol (2–30 parts), amino acids (0.3–4 parts), and lyophilized excipient (10–75 parts). According to this invention, the products do not contain polyoxyethylated castor oil, the toxic and expensive adjuvant, which was substituted with nontoxic media and an easily obtained adjuvant. This invention has low toxicity, better stability, good water solubility, and good patient tolerance, and can be realized in the industry.

The object of patent EP2508170B1 was the liposome comprising irinotecan or its hydrochloride (widely used in the treatment of malignant tumor), neutral phospholipid, and cholesterol, wherein the weight ratio of cholesterol to neutral phospholipid was 1:3–5. Said liposome was prepared using the ion gradient method [96].

The irinotecan liposome preparation provided by the invention CN109260155B includes monovalent sulfonate and disulfonate as internal water phases, and stably encapsulates irinotecan in the form of insoluble sulfonate and disulfonate in the aqueous phase of the lipid body and obtains a good sustained-release effect [97]. The irinotecan liposome preparation has the advantages of a high drug loading amount, easy preparation, good biocompatibility and targeting modification, and is favorable for improving drug efficacy. The irinotecan liposome of the invention has a long half-life in vivo and high bioavailability.

Registered patent CN115197078B [98] protects cationic-modified liposomes, which contain cationic lipids, used as a drug delivery system to cells, specifically, antitumor drugs and nucleic acids.

A blank liposome with ginsenoside as a membrane material as a carrier for active substances (e.g., drugs, cosmetically active substances, and substances with healthcare function), their preparation, and usage, especially against human gastric cancer, are outlined by patent EP3337456B1 [99]. Their preparations can comprise a thermo-sensitive or pH-sensitive excipient, a surfactant, or an ion additive.

Granted patent EP3125920B1 [100] provides a novel therapeutic modality for non-small-cell lung carcinoma and ovarian cancer treatment, based on dianhydrogalactitol, an alkylating agent on DNA that produces N7 methylation. Dianhydrogalactitol is active against tumors that are refractory to temozolomide, cisplatin, and thymidine kinase inhibitors, is also effective in suppressing cancer stem cell growth, and acts independently of the repair mechanism. Liposomes and other delivery vehicles or carriers for hydrophobic drugs were used to improve the therapeutic potential. A neutral liposome is incorporated into a pegylated substituted hexitol derivative, wherein the polyethylene glycol strand is conjugated to at least one mobile peptide or targeting agent.

JP7470418B2 [101] describes the multicellular-targeting liposomes used in antitumor drug preparation. One liposome with multiple antibodies against different cells, with a stable structure and long circulation characteristics in vivo, was formed by molecular self-assembly. It can simultaneously or sequentially bind to multiple different cells using the action of antibodies to promote the cytotoxicity, communication between, the identification, and the clearance or pro-apoptosis of diverse cells.

A tumor-targeted therapeutic drug carrier consisting of liposome, lecithin, 1,2-bis (oleoyloxy)-3-(trimethylammonium)propane (DOTAP), or methylsulfated DOTAP, its preparation method, and its application were described in the granted Chinese patent CN102133404B [102].

The chitosan-liposome, redox-responsive to a disulfide bond and encapsulation of superparamagnetic ferro-ferric oxide nanoparticles, used as a drug carrier for anti-human non-small lung cancer cell A549, was described in patent CN107982547B [103]. Dithiobissuccinimidyl-substituted ester was used for the synthesis of the double fatty chain substituent phosphatidylethanolamine-s-s-chitosan. The chitosan-liposomes have strong cell adhesion characteristics, high drug delivery efficiency, and antiserum abilities, and they are biocompatible and suitable for intravenous injections.

A pharmaceutical formulation that contains methyl jasmonate or jasmonic acid as an active ingredient is described in patent KR101751918B1 [104]. Jasmonates are contained within, associated with, or linked to nano-carriers or microcarriers, such as cyclodextrins, pegylated liposomes, or ceramide liposomes, liposomes selected from nano-emulsions resembling low-density lipoprotein (LDE), poly(ethylene glycol)-600-hydroxystearate, which minimize side effects. The pharmaceutical formulation could be used in chemotherapeutic treatment for cancer in humans to decrease the side effects, as a single treatment or in association with mixed therapies.

A composition based on vitamin E and an anti-cancer compound contained within a delivery vesicle was presented in Australian patent AU2002361812B2 and provides a method for treating a cell proliferative disease [105]. The vitamin-E-based compounds are 2,5,7,8-tetramethyl-(2R-(4R,8R,12-trimethyltridecyl) chroman-6-yloxy) acetic acid, α-tocotrienol, β-tocotrienols, γ-tocotrienol, and δ-tocotrienol, and the lipid is 1,2-dilauroyl-sn-glycero-3-phosphocholine. Anti-cancer drugs are 9-nitrocamptothecin, cisplatin, paclitaxel, doxorubicin, or celecoxib. This composition comprises a liposome and is to be administered by aerosol delivery.

The objects of patents US8147867B2, US8703181B2, and US9782349B2 are compositions comprising liposomes with an interior space separated from the medium by membranes, containing one or more lipids, in which the inner space of the liposome contains substituted ammonium [106–108]. Liposomal compositions comprise one or more lipids and encapsulated anti-cancer therapeutic agents (e.g., camptothecin, topoisomerase inhibitor, and irinotecan) and sucrose hexasulfate. The toxicity of the liposomal composition administered to the mammal is at least 2–4-fold less than the substance administered in free non-liposomal form. There is a half-release time from the liposome of at least 10–24 h.

Patent CN103599070B described the method of fluorescence and temperature probes of the liposome membrane, loaded with the water-soluble anti-cancer drug (paclitaxel, doxorubicin, berberine, or cisplatin) and gold nanocluster dispersion [85].

Hydrophobic immunoregulation substances, all-trans retinoic acid and antigens, are loaded, delivered, and released at the same time and act as a therapeutic vaccine against mucosa-associated tumors, according to patent CN108324938B [79].

Interleukin-2 variant proteins fused to human immunoglobulin heavy constant gamma 4 (IgG4 Fc) (US11896648B2) [109] are used for preventing, reducing, and/or inhibiting the recurrence, growth, proliferation, migration, and/or metastasis of a cancer cell or population of cancer cells in a subject in need. An effective amount of the heterodimer is administered depending on the cancer cell type or population of cancer cells, selected from melanoma and colon cancer.

4.2. Liposome-Based Drug Carriers for Nucleic Acid Delivery

Search criteria were conducted using the keywords "nucleic acid", "RNA", and "DNA". From 1975 to July 2024, were found 1580 patent documents (applications and granted patents) according to the title, abstract, and claims, and 637 granted patents. Cumulative and yearly patent applications and granted patents according to these results, the first seven applicant origins and countries, and the IPC and CPC subgroups only for granted patents are graphically presented in Figure 7. During the last 10 years, 476 patents were granted. Selected patents for liposome-based drug carriers for RNA delivery are presented in Table 6 and analyzed afterward.

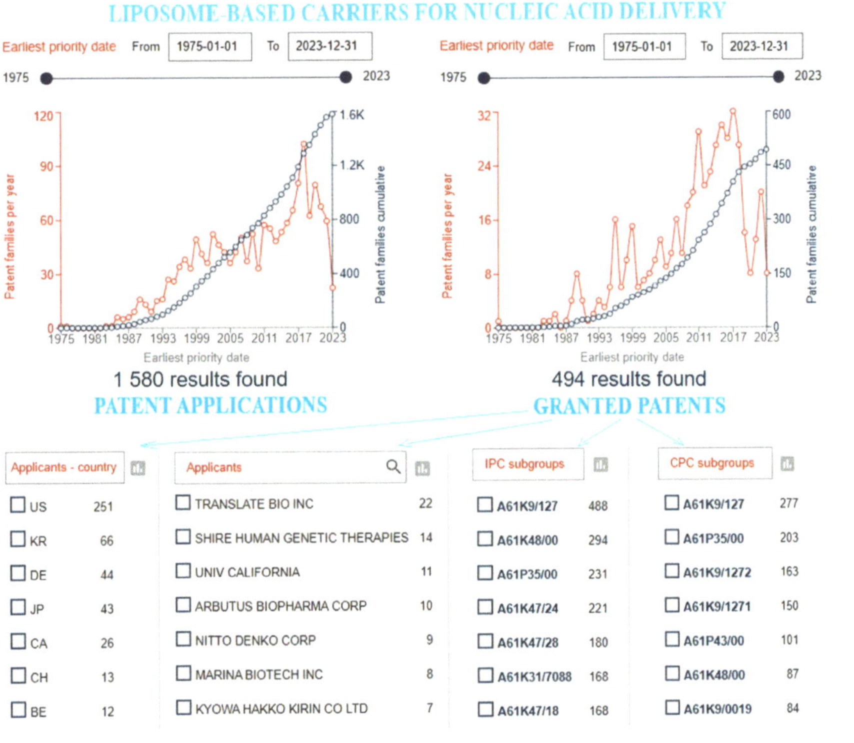

Figure 7. Liposome-based drug carriers for nucleic acid delivery: cumulative and yearly numbers of patent applications and granted patents depending on the earliest priority date (from 1975 to 2024), the first seven applicant origins and countries, and the IPC and CPC subgroups for granted patents. Two-letter codes: US, KR, DE, JP, CA, CH, and BE, are abbreviations for the following: United States of America, Republic of Korea, Germany, Japan, Canada, Switzerland, and Belgium, respectively. Data were obtained using the Espacenet database [32].

Table 6. Selected patents of liposome-based carriers for nucleic acid delivery.

Patent Title	Publication No. [Reference]	Applicants	Earliest Priority
Messenger RNA vaccines and uses thereof	CN112384205B [110]	Translate Bio, Inc.	30 May 2018
Methods and compositions for delivering mRNA-coded antibodies	EP2970456B1 [111]	Translate Bio, Inc.	14 March 2013
Preparation and storage of liposomal RNA formulations suitable for therapy	US11395799B2 [112]	BioNTech SE	20 October 2017
Compositions and systems comprising transfection-competent vesicles free of organic solvents and detergents and methods related thereto	EP3864163B1 [54]	The University of British Columbia	9 October 2018
Cationic lipid, liposome containing cationic lipid, and nucleic acid, pharmaceutical composition containing liposome, and preparation and application thereof	CN115197078B [100]	Xiamen Sinopeg Biotech Co., Ltd.	8 April 2021
Reduced and oxidized polysaccharides and methods of use thereof	CN108430458B [113]	Harvard College	26 October 2015

The patent CN112384205B outlined compositions comprising a messenger RNA (mRNA) encoding an antigen encapsulated in a PEG lipid nanoparticle and the method of preparation of a vaccine for inducing an immune response in vivo, wherein the lipid nanoparticle comprises one or more cationic lipids [110]. Obtained vaccines could be administered subcutaneously, intradermally, intramuscularly, or intravenously.

The other patent, EP2970456B [111], demonstrates compositions for delivering mRNA-coded antibodies comprised of the first mRNA encoding the antibody heavy chain (of approximately 50–70 kD), and the second mRNA encoding the antibody light chain (of approximately 25 kD), for use in therapy in a patient administered intravenously. The first and the second mRNA (i) each comprised a 5′ cap structure, a 3′ poly-A tail, a 5′ untranslated region, and a 3′ untranslated region, and (ii) encapsulated within liposomes comprising a cationic lipid, a neutral lipid, a cholesterol-based lipid, and a PEG-modified lipid (up to 150 nm). The first mRNA encoding the heavy chain and the second mRNA encoding the light chain are encapsulated in the same liposome, and the antibody is a tetramer composed of two identical pairs of polypeptide chains.

Preparation and storage of liposomal RNA formulations for delivery of RNA to target tissues after parenteral or intravenous administration were described in the granted patent US11395799B2 [112].

Methods were outlined for preparing RNA lipoplex particles in an industrial GMP-compliant manner. Additionally, they protect methods and compositions for storing RNA lipoplex particles without significant loss of product quality and RNA activity. Systems comprising transfection-competent vesicles free of organic solvents and detergents configured to efficiently and safely deliver nucleic acid, e.g., DNA, RNA, ribonucleoprotein (RNP), and protein cargoes in target cells, were disclosed in the granted patent EP3864163B1 [54].

Pharmaceutical compositions based on cationic liposomes and nucleic acid drugs (i.e., DNA, plasmid, aptamer, interfering nucleic acid, antisense nucleic acid, mRNA, small interfering RNA (siRNA), microRNA (miRNA), antagomir, and ribozyme) have good biocompatibility and higher gene transfection efficiency (patent CN115197078B) [98].

According to Chinese patent CN108430458B, the therapeutic or diagnostic agents were encapsulated in lipid-based nanoparticles (liposomes or virosomes), highly oxidized polysaccharides—alginates, human mesenchymal stem cells, small molecules, and biological agents (peptides, antibodies, or fragments thereof, vaccines, DNA, RNA, or peptide nucleic acid (PNA) molecules) [113]—and formulated as the implantable or injectable device, or the drug delivery composition.

4.3. Liposome-Based Drug Carriers for Pulmonary and Ocular Treatment

Keywords "pulmonary", "lung", "inhal", and "respiratory" were added to the search criteria. A total of 3187 patent documents (applications and granted patents) from 1975 to July 2024 according to the title, abstract, and claims, and 928 granted patents, were found (Figure 8). Using the keywords "ocular", "eye", "sclera", and "retina", a total of 1752 patent documents (applications and granted patents) were found from 1975 to July 2024 according to the title, abstract, and claims, and 542 granted patents (Figure 9). During the last 10 years, 476 patents were granted. Table 7 presents certain patents for liposome-based drug carriers for pulmonary and ocular treatment that are subsequently analyzed.

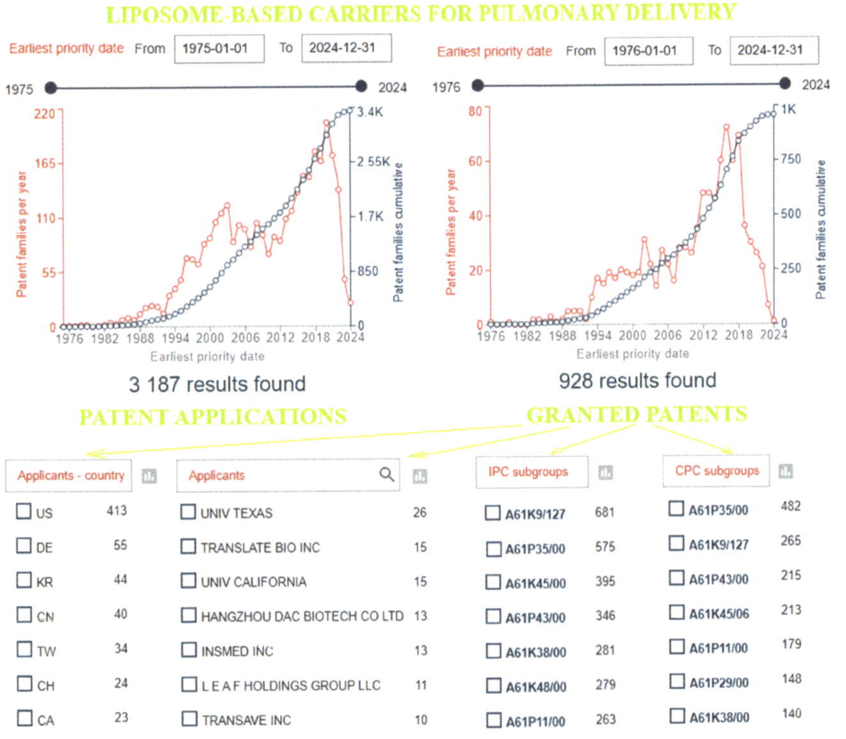

Figure 8. Liposome-based drug carriers for pulmonary treatments: cumulative and yearly numbers of patent applications and granted patents depending on the earliest priority date (from 1983 to 2024), the first seven applicant origins and countries, and the IPC and CPC subgroups for granted patents. Two-letter codes: US, DE, KR, CN, TW, CH, and CA, are abbreviations for the following: United States of America, Germany, Republic of Korea, Republic of China, Taiwan Province of China, Switzerland, and Canada, respectively. Data were obtained using the Espacenet database [32].

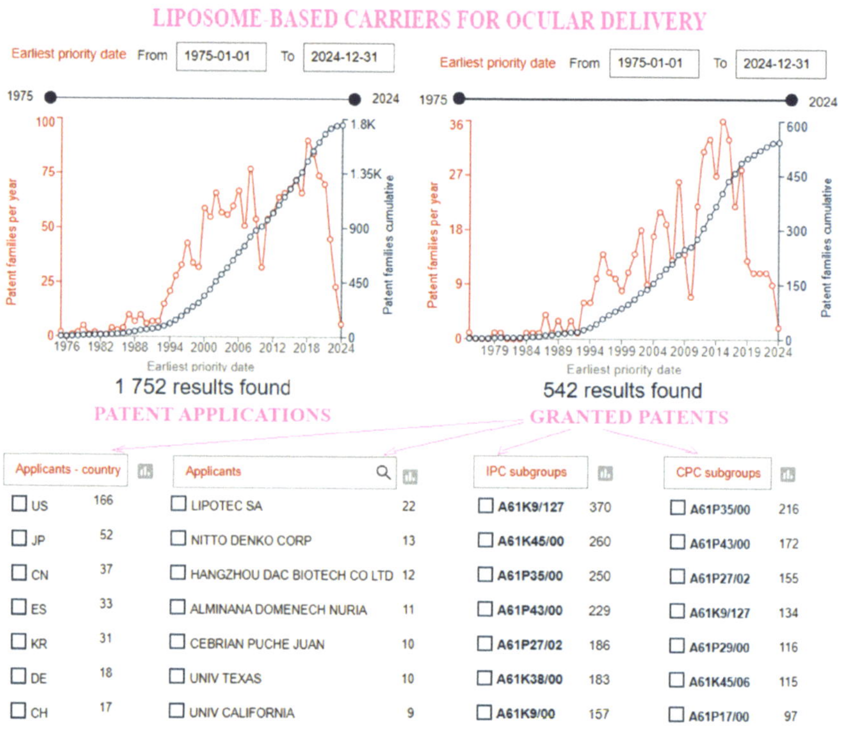

Figure 9. Liposome-based drug carriers for ocular treatments: cumulative and yearly numbers of patent applications and granted patents depending on the earliest priority date (from 1983 to 2024), the first seven applicant origins and countries, and the IPC and CPC subgroups for granted patents. Two-letter codes: US, JP, CN, ES, KR, DE, and CH, are abbreviations for the following: United States of America, Japan, Republic of China, Spain, Republic of Korea, Germany, and Switzerland, respectively. Data were obtained using the Espacenet database [32].

Table 7. Selected patents of liposome-based drug carriers for pulmonary and ocular treatment.

Patent Title	Publication No. [Reference]	Applicants	Earliest Priority
Inhalable compositions for use in the treatment of pulmonary diseases	EP3890767B1 [114]	Breath Therapeutics GmbH	4 December 2018
Slow-release salbutamol sulfate inhalation preparation and production process thereof	CN111700883B [115]	Shenzhen Daphne Medicine Co., Ltd.	23 July 2020
Tobramycin liposome used for aerosol inhalation and production method thereof	CN111228243B [116]	Zhuhai Essex Bio-Pharmaceutical Co., Ltd.	24 December 2019
Liposome atomization treatment preparation containing acetaldehyde dehydrogenase as well as preparation method and application thereof	CN108434446B [117]	Hangzhou Hibio Technology Co., Ltd.	21 March 2018
Liposomes containing steroid esters	EP0170642B1 [118]	Draco Ab	30 July 1984

Table 7. *Cont.*

Patent Title	Publication No. [Reference]	Applicants	Earliest Priority
Liposomal eye drops solution and uses thereof for the treatment of dry eye syndrome	EP3673896B1 [119]	Dr. Rolf Lambert Pharma-Consulting GmbH	28 December 2018
Sustained-release ophthalmic pharmaceutical compositions and uses thereof	TWI786328B [120]	Taiwan Liposome Company, Ltd.; TLC Biopharmaceuticals, Inc.	10 September 2018

The aerosol inhalation technology has the advantages that a drug can be directly delivered to a respiratory tract, quickly absorbed, takes effect quickly, and the bioavailability and the stability are high. A pharmaceutical composition comprising cyclosporine A, as an inhalable immunosuppressive macrocyclic active ingredient for the prevention or treatment of pulmonary disease or condition, administered by inhalation in the form of an aerosol, generated by nebulization of the pharmaceutical composition, were disclosed in patent EP3890767B1 [114].

A slow-release salbutamol sulfate inhalation preparation (capable of relieving acute attack of asthma) and its production process are objects of patent CN111700883B [115]. The invention disclosed in patent CN111228243B is a tobramycin liposome used for aerosol inhalation, composed of 0.1–15.0% tobramycin, 0.5–36.0% of a phospholipid, 0.05–20.0% of a stabilizer, 0.01–10.0% of a charge modifier, 0.01–5.0% of an antioxidant, 5.0–50.0% of an organic-phase medium, and the balance of an aqueous-phase medium [116]. The production process is simple, easy, and rational for industrial production.

The invention CN108434446B provides a therapeutic preparation process for a liposomal aerosol [117]. By using the pulmonary method of administration, it is possible to avoid enzyme destruction in the digestive tract of oral preparations, as well as the insufficient activity and safety problems of injectable preparations. The proportion of inhalable particles after administration via the aerosol inhalation route satisfies the pharmacopoeia. This preparation has good application forecasts for treating drunkenness, alcoholism, and other acetate dehydrogenase deficiency.

A pharmaceutical composition for the local administration primarily to the respiratory tract when treating and controlling anti-inflammatory and anti-allergic conditions comprising lyophilized liposomes in the presence of lactose (range of 0 to 95% of the final composition), in combination with active steroid esters and budesonide-21-palmitate, was described in patent EP0170642B1 [118]. The lecithin (derived from egg, soybean, or synthetic) has different lengths of fatty acid chains and, therefore, has different phase-transition temperatures. The administration routes involve powder aerosols, nebulization, instillation, and aerosols.

Ophthalmic formulations for the treatment of dry eye syndrome, based on an eye drop solution composed of liposomes, were protected by patent EP3673896B1 [119]. Eye drop solutions are comprised of liposomes built with non-hydrogenated phospholipids (containing linseed oil, Vitamin E tocopherol polyethylene glycol succinate (TPGS), and Vitamin A Palmitate) and water phase (with Vitamin B12 and Pycnogenol). Vitamin E TPGS inside the liposomes and Vitamin B12 and Pycnogenol outside the liposomes have a protective effect against UVA/UVB rays. A liposomal eye drop solution containing a specific, peculiar system composed of 2-amino-2(hydroxymethyl) propane-1,3-diol, which makes an agent for Pycnogenol and borate buffer to improve the filterability, has a satisfying filtration procedure to sterilize liposomal eyes drops (up to 0.2 μm), avoiding the steam sterilization (which can destroy components and liposome's structure).

Sustained-release ophthalmic pharmaceutical compositions based on liposomes (comprising a bilayer membrane) and a therapeutic agent for treating an eye disease with a high drug-to-lipid ratio and encapsulation efficiency were described in the granted patent

TWI786328B [120]. It provides a method for treating age-related macular degeneration or diabetic eye disease using this ophthalmic composition, which can be administered by injection (intravitreal or suprachoroidal) or topical (by eye drop or ointment).

4.4. Liposome-Based Drug Carriers for Some Neural, Orthopedic, and Internal Diseases

Using the keywords "neural", "brain", "cerebral", "orthopedic", "bone", "cartilage", "osteoarthritis", "arthritis", "internal diseases", "gastric", "intestinal", and "liver", a total of 1809 patent documents (applications and granted patents) from 1975 to July 2024 according to the title, abstract, and claims, and 549 granted patents, were found (Figure 10). In the last 10 years, 476 patents were granted. A few examples of patents for drug carriers based on liposomes for neural, orthopedic, and some internal diseases are analyzed and summarized in Table 8.

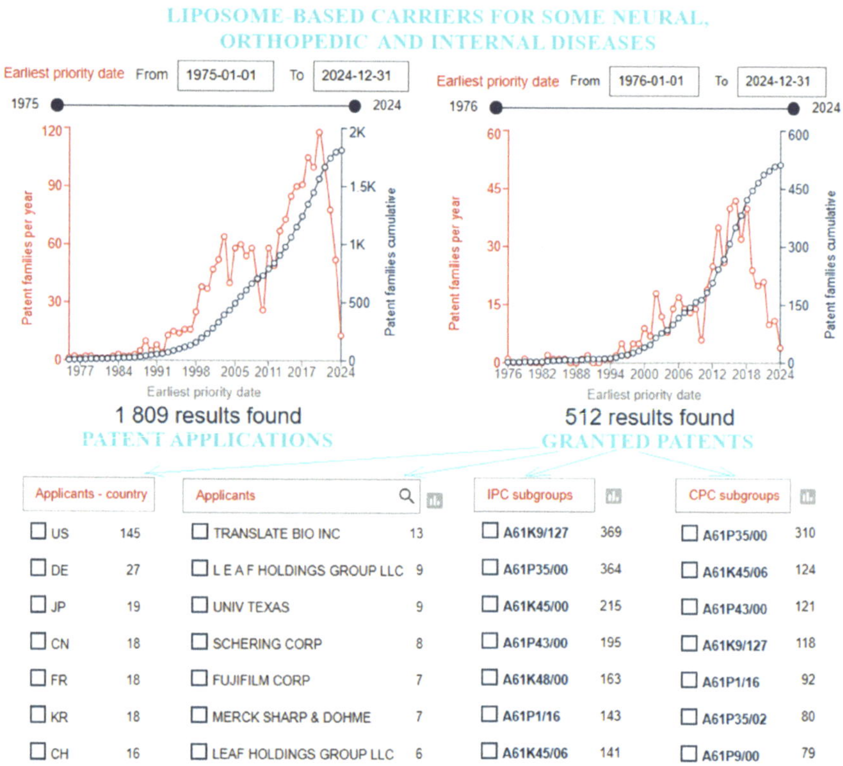

Figure 10. Liposome-based drug carriers for some neural, orthopedic, and internal diseases: cumulative and yearly numbers of patent applications and granted patents depending on the earliest priority date (from 1975 to 2024), the first seven applicant origins and countries, and the IPC and CPC subgroups for granted patents. Two-letter codes: US, DE, JP, CN, FR, KR, and CH, are abbreviations for the following: United States of America, Germany, Japan, Republic of China, France, Republic of Korea, and Switzerland, respectively. Data were obtained using the Espacenet database [32].

Table 8. Selected patents of liposome-based drug carriers for neural, orthopedic, and internal diseases.

Patent Title	Publication No. [Reference]	Applicants	Earliest Priority
A formulation useful for delivery of neuro-protecting agent	EP3200770B1 [121]	Council of Scientific and Industrial Research	29 September 2014
Sustained-release pharmaceutical compositions comprising a therapeutic agent for treating diseases due to reduced bone density or cartilage loss and uses thereof	JP7431419B2 [122]	Taiwan Liposome Company Ltd.; TLC Biopharmaceuticals, Inc.	14 November 2018
Methods of treating arthritis	EP2854769B1 [123]	Taiwan Liposome Company Ltd.; TLC Biopharmaceuticals, Inc.	5 July 2012
Manufacturing of bupivacaine multivesicular liposomes	GB2603047B [124]	Pacira Pharmaceuticals Inc.	22 November 2021
Formulations of volatile anesthetics and methods of use for reducing inflammation	US9744142B2 [125]	Spakevicius, D.; Ozsoy H.; Board of Regents, The University of Texas Systems	5 May 2009
Method for selecting cationic or anionic liposomes for treatment of a mucosa membrane, and kit comprising the same	EP1694298B1 [126]	Yissum Research Development Company	3 November 2003
Pharmaceutical composition for treating alcoholic fatty liver and preparation method thereof	CN117860678B [127]	China Agricultural University	13 March 2024
Cilostazol liposome solid agent	CN112006992B [128]	Pu-Pharma. Co., Ltd.	16 September 2020
Targeted hybrid exosome loaded with curcumin and miR140 as well as preparation method and application of targeted hybrid exosome	CN117965429B [69]	Sichuan University	29 March 2024

The EP3200770B1 patent [121] discloses a novel delivery system based on liposomes, with the potent and prudent therapeutic potential of a new formulation using the standardized extract fraction of a new NMITLI118RT+ chemotype of *Withania sonmifera* roots. It is useful for brain function restoration and neuroprotection against cerebral stroke. A new chemotype of *Withania somnifera* roots, a variety of Ashwagandha, is phytochemically characterized by a particular abundance of two withanolides, i.e., withanolide and/or withanone, for neuroprotection against cerebral stroke, prescribed for administering the formulation in a predetermined dose and period.

Pharmaceutical compositions with sustained release of a therapeutic agent for treating diseases due to reduced bone density or cartilage loss, and uses thereof, were described in patent JP7431419B2 [122]. Sustained-release compositions comprised (a) at least one liposome with a lipid bilayer membrane, (b) a scavenger (triethylammonium sucrose octasulphate and/or ammonium sulphate), and (c) a cathepsin K inhibitor. The injection with a therapeutic agent for treating reduced bone density or cartilage loss could be administered via an intraarticular, subcutaneous, subdermal, intradermal, or intramuscular route. The other sustained-release composition for use in the treatment of arthritis, comprising liposomes, cholesterol, and one or more therapeutic agents, e.g., the water-soluble steroid dexamethasone sodium phosphate, and the nonsteroidal anti-inflammatory drug indomethacin, a disease-modifying anti-rheumatic drug, were objects of the granted patent EP2854769B1 [123].

Composition of multivesicular liposomes (MVLs) with encapsulated bupivacaine, residing within a plurality of internal aqueous chambers of the MVLs, separated by lipid membranes, was disclosed in patent GB2603047B [124]. The internal aqueous chambers may comprise lysine and/or dextrose, or/and cholesterol and tricaprylin. These compositions could be useful in treating or ameliorating pain, particularly via the interscalene brachial plexus nerve block, via local infiltration to a surgical site, or via femoral nerve block to provide regional analgesia. Another volatile anesthetic for treating or reducing inflammation or a wound in need of wound treatment or inflammation treatment by delivering a volatile anesthetic to the wound or inflammation site was formulated for topical, mucosal, rectal, vaginal, or buccal administration.

The invention EP1694298B1 [126] concerns a method for selecting cationic or anionic liposomes, a medicament for the treatment/prevention of a gastrointestinal mucosa disorder, and a kit for its use. The application of charged lipid assemblies is based on the principles of the differential adhesion of negatively vs. positively charged lipid assemblies to diseased and healthy mucosa, respectively.

The invention CN117860678B [127] discloses a pharmaceutical composition for treating alcoholic fatty liver and an appropriate obtaining method. Ethyl *p*-methoxycinnamate and nano-enzymes are enclosed in liposomes to form a composite material. According to this invention and using a new strategy, liposomes are delivered, in a targeted manner, to the alcoholic fatty liver for precise treatment with improved bioavailability.

The invention CN112006992B discloses a cilostazol liposome solid agent, prepared by a subsequent method: cilostazol (improves blood flow) was dissolved in acetic acid, and mixed with dioleoyl phosphatidylcholine, dimyristoyl phosphatidylcholine, poloxamer P188, and cholesterol succinic acid monoester, to form a lipid membrane solution [128]. After that, an acetic acid sodium acetate buffer solution was added to disperse and emulsify to obtain a uniform liposome solution. Then, lactose, sodium carboxymethyl starch, microcrystalline cellulose, and magnesium stearate were added, mixed, refined and dispersed. Obtained a uniform mixed suspoemulsion was freeze-dryed. The cilostazol loading capacity and liposome encapsulation efficiency are improved, the technological process is shortened, and the raw materials are safe and non-toxic.

The hybridized exosome, described in Korean patent CN117965429B, showed a synergistic effect of loaded curcumin and miR140, inhibited activation of an NF-kappa B pathway, enhanced the anti-inflammatory effect, and reduced the cytotoxicity of the curcumin. It is used in the preparation of a drug for treating osteoarthritis [69].

4.5. Liposome-Based Drug Carriers with Anti-Inflammation and Antibiofilm Agents

Search criteria were narrowed using the keywords "anti-inflammation", "inflammation", "antibiofilm", "microbial", and "biofilm". From 1975 to July 2024, 1231 patent documents (applications and granted patents) were found according to the title, abstract, and claims, as well as 368 granted patents. The cumulative and yearly numbers of patent applications and granted patents, according to these search results, with the first seven applicant origins and countries and the IPC and CPC subgroups (only for granted patents), are graphically presented in Figure 11. A few examples of granted patents for drug carriers based on liposomes with anti-inflammation and antibiofilm agents (presented in Table 9) are briefly described.

Table 9. Selected granted patents for liposome-based drug carriers for anti-inflammation agents.

Patent Title	Publication No. [Reference]	Applicants	Earliest Priority
Sterile pharmaceutical composition and process for a solution of propofol emulsion having microbial growth retardation	US7468394B1 [129]	Amphastar Pharmaceuticals Inc.	11 March 2003
Liposomes for inhibiting biofilm formation	EP3589278A1 [130]	Combioxin SA	2 March 2017

Table 9. *Cont.*

Patent Title	Publication No. [Reference]	Applicants	Earliest Priority
Methods for the manufacture of liposomal drug formulations	AU2019262117B2 [131]	Insmed Inc. Worsham Robert	2 May 2018
Interdigitation-fusion liposomes containing arachidonic acid metabolites	EP0729352B1 [132]	Lipsome Co., Inc.	16 November 1993
Pharmaceutical composition comprising jasmonates	KR101751918B1 [104]	Fehr Pereira Lopes, Jose.E.	15 July 2008

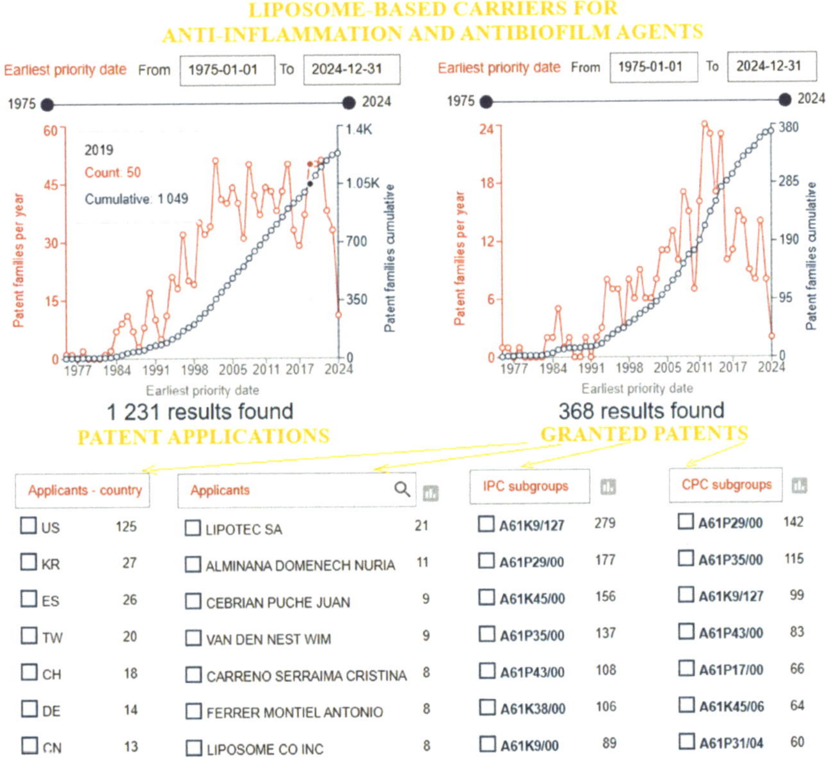

Figure 11. Liposome-based drug carriers for nucleic acid delivery: cumulative and yearly numbers of patent applications and granted patents depending on the earliest priority date (from 1975 to 2024), the first seven applicant origins and countries, and the IPC and CPC subgroups for granted patents. Two-letter codes: US, KR, ES, TW, CH, DE, and CN, are abbreviations for the following: United States of America, Republic of Korea, Spain, Taiwan Province of China, Switzerland, Germany, and Republic of China, respectively. Data were obtained using the Espacenet database [32].

A sterile pharmaceutical composition and process for an oil-in-water propofol emulsion having microbial growth retardation, without the side effects associated with other growth retardation additives, are disclosed in the US7468394B1 patent [129]. Several methods for the optimization of the innate microbial retardation capability of propofol are described. This composition contains liposome particles, emulsion particles, additives to act as a microbial growth retardation agent, and supporting aqueous phase. The oil-in-water propofol emulsion contains soybean oil and egg lecithin. The described process improves the propofol emulsion solution's safety by controlling microbial growth, without the side effects associated with growth retardation additives.

Liposomes for preventing or reducing biofilm formation, or for eradicating or reducing existing biofilm, are described in patent EP3589278A1 [130]. A method for a large-scale production of liposomal drug formulations, containing amikacin sulfate formulation as an aminoglycoside (antibiotics for serious bacterial infections, especially Gram-negative), with valuable lipid/drug characteristics, was presented in patent AU2019262117B2 [131]. The method employs a particular relative flow rate ratio of lipid to drug streams to obtain liposomes with a high aminoglycoside encapsulation efficiency. The invention EP0729352B1 offers an interdigitation-fusion liposome, including an arachidonic acid metabolite, a lipid bilayer containing a lipid, and an aqueous compartment comprising a release-inhibiting buffer [132]. The preferred arachidonic acid metabolites are the prostaglandins, particularly PGE1. The liposomal formulations can be used to treat animals, particularly humans, for diseases, disorders, or conditions that can be ameliorated by prostaglandins, e.g., cell activation/adhesion disorders and inflammatory disorders.

A pharmaceutical formulation for use in the treatment of bacterial or fungal diseases, containing methyl jasmonate or jasmonic acid as an active ingredient, is described in patent KR101751918B1 [104].

4.6. Liposome-Based Drug Carriers for Dermal Applications

Using the keywords "dermatological", "derma", "skin", "wound", "cosmetic", and "stratum coroneum", a total of 1809 patent documents (applications and granted patents) from 1975 to July 2024 according to the title, abstract, and claims, and 549 granted patents, were found (Figure 12). In the last 10 years, 476 patents were granted. Several examples of granted patents for drug carriers based on liposomes for dermal applications, shown in Table 10, are briefly described.

Table 10. Selected relevant patents of liposome-based carriers for dermal applications.

Patent Title	Publication No. [Reference]	Applicants	Earliest Priority
Method for preparing medicinal extract containing intermediate peashrub seeds	CN104382984B [84]	Northwest Institute of Plateau Biology of Chinese Academy of Sciences	1 December 2014
Composition for increasing expression of PGC-1 alpha	EP3403655B1 [133]	Benebiosis Co., Ltd.	13 January 2016
Peptides which inhibit activated receptors and their use in cosmetic or pharmaceutical compositions	EP2773366B1 [134]	Lipotec SA	4 November 2011
Cosmetological and pharmaceutical formulae for the rejuvenation and restoration of skin, including after surgical procedures	US8846064B2 [135]	Martynov, A.; Farber, B.S.; Farber, S.S.; Sitenko, A.	22 November 2010
Cosmetic or pharmaceutical compositions comprising metalloproteinase inhibitors	ES2330291B1 [136]	Lipotec SA	29 February 2008
Pharmaceutical composition, in particular dermatological or cosmetic, comprising hydrous lipidic lamellar phases or liposomes containing a retinoid or a structural analogue thereof, such as a carotenoid	FR2591105B1 [137]	Moet Hennessy Recherche	11 December 1985

Table 10. *Cont.*

Patent Title	Publication No. [Reference]	Applicants	Earliest Priority
Hyaluronic drug delivery system	EP0963196B1 [138]	Hyal Pharmaceutical Corporation; JagotecAg	29 September 1996
Preparation method of exosome bionic preparation for synergistically promoting wound healing and preparation thereof	CN113577272B [73]	China Pharmaceutical University	28 July 2021

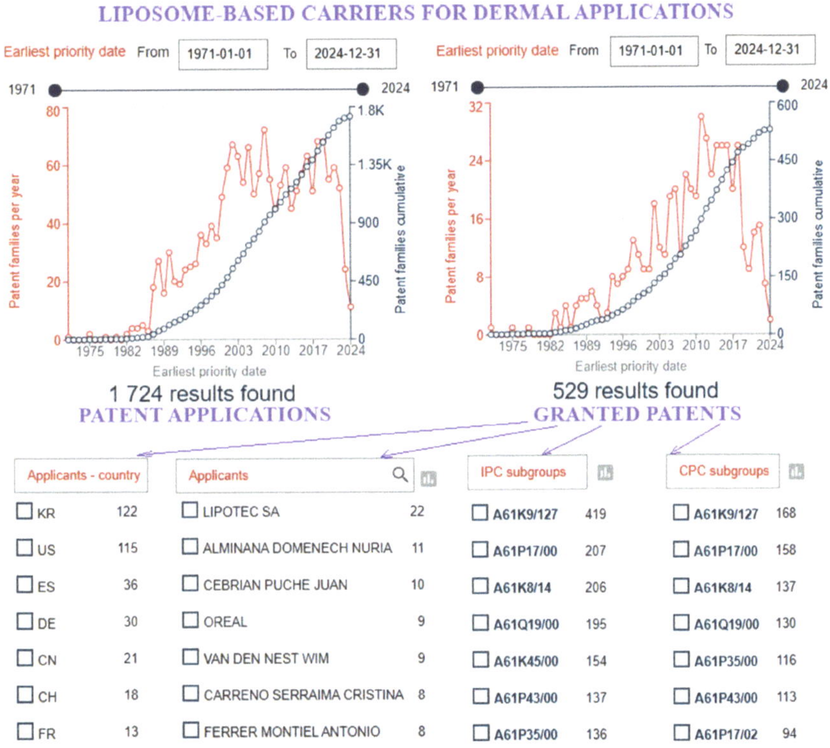

Figure 12. Liposome-based drug carriers for dermal applications: cumulative and yearly numbers of patent applications and granted patents depending on the earliest priority date (from 1971 to 2024), for granted patents the first 7 applicant names and countries, and the IPC and CPC subgroups. Two-letter codes: KR, US, ES, DE, CH, CN, and FR, are abbreviations for the following countries: Republic of Korea, United States of America, Spain, Germany, Republic of China, Switzerland, and France, respectively. Data were obtained using the Espacenet database [32].

Medicinal extract of *Caragana korshinskii* seeds for the drug-containing liposome nanoparticles prepared by the patent CN104382984B have the characteristics of targeted drug delivery and long-term release [84]. Due to their cell-like structure, they are cell-compatible and can release drugs through the cell membrane. Drug-containing liposome nanoparticles with *Caragana korshinskii* seed extract improve therapeutic index and are more effective in skin disease treatments, e.g., psoriasis, dermatitis, impetigo, and skin itching.

Registered patent EP3403655B1 relates to a composition for preventing or treating diseases or symptoms associated with a reduction in the expression of peroxisome proliferator-activated receptor coactivator 1-α (PGC-1α), consisting of an active ingredient, and its salt or solvate, encapsulated in liposomes [133]. The composition is used for reducing the subcutaneous adipose tissue volume (of the femoral region, a lower part of the neck, neckline, chest, buttocks, lips, face, cheeks, eyelids, and/or hands), or for the triglyceride content reduction in adipose tissue.

The object of patent EP2773366B1 [134] is a pharmaceutical or cosmetic composition with a sustained-release system, formed by liposomes, mixed liposomes, or some other carriers, and the peptide, which inhibits activated receptors.

Cosmetic and pharmaceutical compositions with liposomes and proteins were described in US patent US8846064B2 [135]. Used proteins were partially acylated (0.1–10% of their mass) and present in the supramolecular assembly. These partially acylated proteins were derived from collagenase, hyaluronidase, insulin, or their mixtures.

Spanish patent ES2330291B1 protects a pharmaceutical or cosmetic composition consisting of the peptide (its stereoisomers, mixtures, and cosmetically or pharmaceutically acceptable salts), incorporated in a vehicle or a system for sustained delivery, as well as an obtaining method [136]. Liposomes and mixed liposomes were chosen, among other cosmetic or pharmaceutically acceptable carriers. Obtained compositions were applied for the treatment and/or care of disorders, conditions, and/or skin pathologies, mucosae, and scalp that are a result of the matrix metalloproteinases (MMP) activity increase or an MMP overexpression.

Dermatological, pharmaceutical, or cosmetic compositions based on liposomes, or hydrated lipid lamellar phases, comprising retinoids (or analogues, e.g., carotenoids), are described in patent FR2591105B1 [137]. Defined compositions have additional effects against acne and are less irritating to the skin, and they have pharmaceutical and certain dermatological or cosmetic applications.

Pharmaceutical compositions based on multilamellar, negatively charged liposomes with encapsulated cyclosporin A and hyaluronic acid, described in patent EP0963196B1, were used for topical administrations for psoriasis treatment [138].

The obtained exosome bionic preparation synergistically helps wound healing according to Chinese patent CN113577272B [73].

The patent analysis of liposome-based drug carriers for cancer treatments, nucleic acid delivery, pulmonary and ocular treatment, neural, orthopedic, and internal diseases treatment, anti-inflammation and antibiofilm treatments, as well as dermal applications highlighted that published patents protect novel and alternative formulations, which are not found in scientific journals before their priority date because novelty in patent applications is the first demand for patentability and granted patent rights. Patent publications on liposomal formulations expanded new methods of drug delivery. Liposome formulations have been developed to establish novel approaches in drug delivery systems, e.g., to optimize the content of drug loading, stability, and release profiles, including long-circulating structures and further functionalization of liposomes for the co-delivery of drugs to the targeted site. New liposome-based compositions and strategies to optimize the stabilities and capabilities of drug vehiculation found more acceptable fabrication methods. Continued utilization of liposomes in disease treatments is based on defining stable, simpler, and cost-effective formulations, which could be applied in the pharmaceutical industry in the future to provide more beneficial therapeutic solutions for many diseases.

5. Conclusions

Many inventors have significantly contributed to improvements in liposome-based carriers by synthesizing more complex vesicles and developing new targeted and selective release mechanisms, without causing side effects. This has resulted in enhanced therapy efficiency and reduced treatment costs. New drug release methods have been identified to activate administration only at specific sites when needed. Drugs can be administered into the body via different methods, some of which are oral, topical, intranasal, sublingual, and intravenous. Future trends in liposome technology as drug carriers may include new patent applications aimed at improving the protection of encapsulated drugs and delivery efficiency by overcoming biological barriers. New inventive solutions may be directed toward development of stealth liposome technology, incorporating PEG or other polymers, so that liposomes avoid the immune system more effectively, achieving long circulation (sterically stabilized) in the bloodstream with optimized and increased stability. Also, further progress in the design of liposomes to overcome biological barriers, such as the blood–brain barrier, is important, as it represents a major challenge for the treatment of neurological disorders. For the treatment of pulmonary and gastrointestinal diseases, it is important to improve liposome-based carriers so that they improve and optimize the passages through the mucosal barriers for more efficient delivery of inhaled or orally administered drugs. "Smart" liposomes can be designed as bioreactive carriers that release contents in the body in a controlled manner in response to biological signals (such as pH, temperature, light, or enzymes) exclusively in targeted areas (such as infected tissues or tumors). Liposomal formulations are intensively developed for precision gene therapy to improve encapsulation of genetic materials (such as RNA and DNA), protection during delivery, and enhancement of cellular uptake. The trend of future developments will probably be directed toward new methods of obtaining nano-liposomes, at a nanoscale below 100 nm, which can penetrate deeper into tissues (e.g., solid tumors) to improve drug delivery and treatment outcomes. Significant progress is expected in the technology of lipid nanoparticles, closely related to liposomes, especially in the delivery of nucleotides (e.g., mRNA in vaccines) and their optimization. One of the key trends for future technologies may be targeted drug delivery by modifying and functionalizing the surface of liposomes with ligands, antibodies, or aptamers that can selectively bind to specific cell receptors. This is particularly important for the more precise delivery of drugs for the treatment of cancer and genetic disorders, improving therapeutic outcomes and reducing side effects. The future of liposome technology will focus on reproducible, scalable, and environmentally friendly manufacturing processes to meet the growing demand for liposomal drugs.

Patent data are authoritative for recognizing trends in technical development, assessing industrial technological attractiveness, monitoring the process of protecting competition's patent rights, and defining technological advantages and new strategies. The review of the patents could help in future investigations as a rich source of inspiration in several ways. Patent searching can help researchers avoid duplicating efforts and save time by analyzing and recognizing what has already been patented and, thus, focusing on novel aspects, strategies, or procedures. By informing with patents in force and existing patent rights, researchers can guarantee that their innovations do not infringe on the owners' patent rights (other companies). Patent applications often reveal trends (e.g., new processes, methods, or materials) and could motivate researchers to modify or combine current methods and investigate similar technologies. Through the patent analysis of liposome-based drug carriers in the period 1976–2024, it is interesting to point out that the activities were aimed at finding new, improved production processes, as well as formulations for the targeted treatment and prevention of diseases (mostly for pulmonary, dermal, ocular, neural, and gastric). Numerous scientific papers on various drug delivery systems have been published in scientific journals, but a smaller number requested patent applications and fewer have registered patents. It is important that companies and academic researchers first protect their inventions by patenting new products, production processes, and/or formulations, and then publish their research in peer-reviewed journals. Currently, the largest number of patents are from universities, and additional cooperation with companies is needed to improve the technological procedures in industry by applying efficient and modern technologies.

Author Contributions: Conceptualization, S.I.-S.; methodology, S.I.-S.; writing—original draft preparation, S.I.-S.; writing—review and editing, S.I.-S., S.C., N.N. and S.P. All authors have read and agreed to the published version of the manuscript.

Funding: This research was funded by the Ministry of Science and Technological Development of the Republic of Serbia, contract number 451-03-65/2024-03/200133.

Acknowledgments: The authors gratefully acknowledge the EPO for the Espacenet patent search database used in this study.

Conflicts of Interest: The authors declare no conflicts of interest.

References

1. Petrović, S.; Ilić-Stojanović, S.; Tačić, A.; Nikolić, L.; Nikolić, V. Chapter 5: Vesicular drug carriers as delivery system. In *Nanoconjugate Nanocarriers for Drug Delivery*; Keservani, R.K., Sharma, A.K., Eds.; Apple Academic Press Inc. Exclusive worldwide distribution by CRC Press, a Taylor & Francis Group: Waretown, NJ, USA, 2018; pp. 141–174. Available online: www.appleacademicpress.com/nanoconjugate-nanocarriers-for-drug-delivery-/9781771886772 (accessed on 31 August 2018).
2. Florence, T. New drug delivery systems. *Chem. Ind.* **1993**, *20*, 1000–1004.
3. Speiser, P.P. Poorly soluble drugs, a challenge in drug delivery. In *Emulsions and Nanosuspensions for the Formulation of Poorly Soluble Drugs*; Müller, R.H., Benita, S., Böhm, B.H., Eds.; CRC Press: Boca Raton, FL, USA; Medpharm Scientific Publisher: Stuttgart, Germany, 1998; pp. 15–28.
4. Rolland, A.; Dekker, M. (Eds.) *Pharmaceutical Particulate Carriers: Therapeutic Applications*; CRC Press: New York, NY, USA, 1993; pp. 1–367.
5. Nikolić, V.; Ilić-Stojanović, S.; Petrović, S.; Tačić, A.; Nikolić, L. Administration routes for nano-drugs and characterization of nano-drug loading. In *Characterization and Biology of Nanomaterials for Drug Delivery: Nanoscience and Nanotechnology in Drug Delivery*; Mohapatra, S., Ranjan, S., Dasgupta, N., Mishra, R.K., Thomas, S., Eds.; Elsevier Book: Amsterdam, The Netherlands, 2019; pp. 587–625. [CrossRef]
6. Milić, J.; Petrović, S.D. The characteristics of modern pharmaceutical forms. *Hem. Ind.* **2003**, *57*, 424–436. [CrossRef]
7. Wen, P.; Ke, W.; Dirisala, A.; Toh, K.; Tanaka, M.; Li, J. Stealth and pseudo-stealth nanocarriers. *Adv. Drug Deliv. Rev.* **2023**, *198*, 114895. [CrossRef]
8. Bangham, A.D.; Standish, M.M.; Weissmann, G. Diffusion of univalent ions across the lamellae of swollen phospholipids. *J. Mol. Biol.* **1965**, *13*, 238–252. [CrossRef] [PubMed]
9. Gregoriadis, G.; Leathwood, P.D.; Ryman, B.E. Enzyme entrapment in liposomes. *FEBS Lett.* **1971**, *14*, 95–99. [CrossRef]
10. Gregoriadis, G.; Ryman, B.E. Liposomal localisation of beta-fructofuranosidase-containing liposomes injected into rats. Some implications in the treatment of genetic disorders. *Biochem. J.* **1972**, *129*, 123–133. [CrossRef]
11. Gregoriadis, G.; Ryman, B.E. Fate of protein-containing liposomes injected into rats. An approach to the treatment of storage disease. *Eur. J. Biochem.* **1972**, *24*, 485–491. [CrossRef]
12. Gregoriadis, G. Liposomes in Drug Delivery: How it all happened. *Pharmaceutics* **2016**, *8*, 19. [CrossRef]
13. Perrie, Y. Gregory Gregoriadis: Introducing liposomes to drug delivery. *J. Drug Target.* **2008**, *16*, 518–519. [CrossRef]
14. Gregoriadis, G.; Florence, A.T. Liposomes in drug delivery. *Drugs* **1993**, *45*, 15–28. [CrossRef]
15. Djekic, L. Liposomes: Properties and therapeutic applications. In *Novel Approaches for Drug Delivery*; Keservani, R., Sharma, A., Kesharwani, R., Eds.; IGI Global: Hershey, PA, USA, 2017; pp. 27–51. [CrossRef]
16. Dapergolas, G.; Gregoriadis, G. Hypoglycaemic effect of liposome-entrapped insulin administered intragastrically into rats. *Lancet* **1976**, *308*, 824–827. [CrossRef] [PubMed]
17. Shah, S.; Dhawan, V.; Holm, R.; Nagarsenker, S.; Perrie, Y. Liposomes: Advancements and innovation in the manufacturing process. *Adv. Drug Deliv. Rev.* **2020**, *54*, 102–122. [CrossRef] [PubMed]
18. Pattni, B.S.; Chupin, V.V.; Torchilin, V.P. New developments in liposomal drug delivery. *Chem. Rev.* **2015**, *115*, 10938–10966. [CrossRef] [PubMed]
19. Nikolova, M.P.; Kumar, E.M.; Chavali, M.S. Updates on responsive drug delivery based on liposome vehicles for cancer treatment. *Pharmaceutics* **2022**, *14*, 2195. [CrossRef]
20. Liu, P.; Chen, G.; Zhang, J. A review of liposomes as a drug delivery system: Current status of approved products, regulatory environments, and future perspectives. *Molecules* **2022**, *27*, 1372. [CrossRef]
21. Storm, G.; Crommelin, J.A. Liposomes: Quo Vadis. *Pharm. Sci. Technol. Today* **1998**, *1*, 19–31. [CrossRef]
22. Wang, Y.; Grainger, D.W. Lyophilized liposome-based parenteral drug development: Reviewing complex product design strategies and current regulatory environments. *Adv. Drug Deliv. Rev.* **2019**, *151*, 56–71. [CrossRef]
23. Raviv, S.A.; Alyan, M.; Egorov, E.; Zano, A.; Harush, M.Y.; Pieters, C.; Korach-Rechtman, H.; Saadya, A.; Kaneti, G.; Nudelman, I.; et al. Lung targeted liposomes for treating ARDS. *J. Control. Release* **2022**, *346*, 421–433. [CrossRef]
24. Angst, M.S.; Drover, D.R. Pharmacology of drugs formulated with DepoFoam™: A sustained release drug delivery system for parenteral administration using multivesicular liposome technology. *Clin. Pharmacokinet.* **2006**, *45*, 1153–1176. [CrossRef]
25. Pierre, M.B.R.; dos Santos, M.C.I. Liposomal systems as drug delivery vehicles for dermal and transdermal applications. *Arch. Dermatol. Res.* **2011**, *303*, 607–621. [CrossRef]
26. Ahmed, K.S.; Hussein, S.A.; Ali, A.H.; Korma, S.A.; Lipeng, Q.; Jinghua, C. Liposome: Composition, characterisation, preparation, and recent innovation in clinical applications. *J. Drug Target.* **2019**, *27*, 742–761. [CrossRef] [PubMed]

27. Carita, A.C.; Eloy, J.O.; Chorilli, M.; Lee, R.J.; Leonardi, G.R. Recent advances and perspectives in liposomes for cutaneous drug delivery. *Curr. Med. Chem.* **2018**, *25*, 606–635. [CrossRef] [PubMed]
28. Kyong, K.Y.; Lee, C.K. Development and prospect of emulsion technology in cosmetics. *J. Soc. Cosmet. Sci. Korea* **2006**, *32*, 209–217.
29. Stuchlík, M.; Žák, S. Lipid-based vehicle for oral drug delivery. *Biomed. Papers* **2001**, *145*, 17–26. [CrossRef]
30. Andra, V.V.S.N.L.; Pammi, S.V.N.; Bhatraju, L.V.K.P.; Ruddaraju, L.K. A comprehensive review on novel liposomal methodologies, commercial formulations, clinical trials and patents. *Bionanoscience* **2022**, *12*, 274–291. [CrossRef]
31. El-Hammadi, M.M.; Arias, J.L. An update on liposomes in drug delivery: A patent review (2014–2018). *Expert Opin. Ther. Pat.* **2019**, *29*, 891–907. [CrossRef]
32. European Patent Office. Espacenet Patent Search. Available online: https://worldwide.espacenet.com (accessed on 1 July 2024).
33. World Intellectual Property Organization. Available online: https://www.wipo.int/patents/en/ (accessed on 1 July 2024).
34. International Patent Classification (IPC). Available online: https://www.wipo.int/classifications/ipc/en/ (accessed on 1 July 2024).
35. Cooperative Patent Classification. Available online: https://www.cooperativepatentclassification.org/home (accessed on 1 July 2024).
36. Guide to the International Patent Classification (2024). Available online: https://www.wipo.int/edocs/pubdocs/en/wipo-guide-ipc-2024-en-guide-to-the-international-patent-classification-2024.pdf (accessed on 1 July 2024).
37. PCT—The International Patent System. Available online: https://www.wipo.int/pct/en/ (accessed on 1 July 2024).
38. Pradhan, P.; Guan, J.; Lu, N.; Wang, P.G.; Lee, L.J.; Lee, R.J. A facile microfluidic method for production of liposomes. *Anticancer Res.* **2008**, *28*, 943–947.
39. Cortesi, R. Preparation of liposomes by reverse-phase evaporation using alternative organic solvents. *J. Microencapsul.* **1999**, *16*, 251–256. [CrossRef]
40. Coelho, J.; Trucillo, P.; Nobre, B.; Palavra, A.F.; Campardelli, R.; Reverchon, E. Extraction and bioprocessing with supercritical fluids. *Phys. Sci. Rev.* **2020**, *5*, 20180069. [CrossRef]
41. Xia, F.; Hu, D.; Jin, H.; Zhao, Y.; Liang, J. Preparation of lutein proliposomes by supercritical anti-solvent technique. *Food Hydrocoll.* **2012**, *26*, 456–463. [CrossRef]
42. Yamaguchi, S.; Tsuchiya, K.; Sakai, K.; Abe, M.; Sakai, H. Preparation of nonionic vesicles using the supercritical carbon dioxide reverse phase evaporation method and analysis of their solution properties. *J. Oleo Sci.* **2016**, *65*, 21–26. [CrossRef]
43. Meure, L.A.; Knott, R.; Foster, N.R.; Dehghani, F. The depressurization of an expanded solution into aqueous media for the bulk production of liposomes. *Langmuir* **2009**, *25*, 326–337. [CrossRef]
44. Trucillo, P.; Campardelli, R.; Reverchon, E. A versatile supercritical assisted process for the one-shot production of liposomes. *J. Supercrit. Fluids* **2019**, *146*, 136–143. [CrossRef]
45. Trucillo, P. Drug Carriers: Classification, Administration, Release Profiles, and Industrial Approach. *Processes* **2021**, *9*, 470. [CrossRef]
46. Encapsulation of Chemical Substances—In Liposomes for Medicinal Use. French Patent No. FR2221122B1, 9 April 1976.
47. Bailey, A.; Cullis, P.R. Fusogenic Liposomes and Methods of Making and Using Same. European Patent No. EP0758883B1, 19 March 2003.
48. Kim, K.M.; Park, K.M.; Lee, D.W.; Jung, M.S.; Jung, H.T.; Ko, Y.H. Liposome Sensitive to pH or Reductive Condition and Processes for the Preparation Thereof. Korean Patent No. KR100853172B1, 20 August 2008.
49. Chen, J.; Ansell, S.; Akinc, A.; Dorkin, J.R.; Qin, X.; Cantley, W.; Manoharan, M.R.; Kallanthottathil, G.; Narayanannair, J.K.; Jayaraman, M. Lipid Formulation. U.S. Patent No. US8158601B2, 17 April 2012.
50. Chen, J.; Ansell, S.; Akinc, A.; Dorkin, J.R.; Qin, X.; Cantley, W.; Manoharan, M.R.; Kallanthottathil, G.; Narayanannair, J.K.; Jayaraman, M. Lipid Formulation. U.S. Patent No. US8802644B2, 12 August 2014.
51. Richter, Y.; Zelig, Y.; Elmalak, O.; Eyal, D. Liposome Formulation and Manufacture. European Patent No. EP3135274B1, 14 March 2013.
52. Kim, D.D.; Jeon, D.E.; Kim, K.T.; Ubonvan, T.; Kang, M.J. Preparing Method of Proliposomes. Korean Patent No. KR101682821B1, 6 December 2016.
53. He, T.; Ai, Y.; Zhang, C.; Lu, Y. Platinum Modified Liposome as Well as Preparation Method and Application Thereof. Chinese Patent No. CN107550865B, 4 January 2019.
54. Leavitt, B.; Cullis, P.; Petkau, T.; Hill, A.; Wagner, P.; Kulkarni, J. Compositions and Systems Comprising Transfection-Competent Vesicles Free of Organic-Solvents and Detergents and Methods Related Thereto. European Patent No. EP3864163B1, 20 March 2024.
55. He, Z.; Wang, Y.; Qiu, S.; Ma, H.; Yang, J. Alkannin active Drug-Loading Liposome as Well as Preparation and Application Thereof. CN115105473B, 19 September 2023.
56. Liu, Q.; Xin, T.; Liu, Z.; Han, M.; Wang, Z.; Ji, X.; He, D.; Wang, X.; Wang, Z.; Wang, C. Hydrogel-Liposome Combined Drug Delivery System as Well as Preparation Method and Application Thereof. Chinese Patent No. CN115192581B, 28 April 2023.
57. Yuanzeng, M.; Ruijie, H. Preparation Method and Application of Acid Response Type Drug Delivery Platform Based on Liposome Nano-Vesicles. Chinese Patent No. CN115337267B, 20 October 2023.
58. Zhang, H. Probucol Liposome for Hypercholesterolemia, Preparation and Preparation Method of Probucol Liposome. Chinese Patent No. CN115364056B, 11 July 2023.
59. Zang, Y. Thermo-Sensitive Nano-Liposome Capable of Realizing Stepped Release of Active Matters and Application of Thermo-Sensitive Nano-Liposome. CN115590818B, 7 May 2024.

60. Li, Z. Alendronate Sodium Liposome and Preparation Thereof. Chinese Patent No. CN116650418B, 12 July 2024.
61. Yu, M.K.; Gi, H.J.; Ji, W.S.; Do, H.K. A Stabilized Nano Liposome Delivery Carrier Comprising Nucleic Acid and a Method for Preparing the Same. Korean Patent No. KR102678046B1, 10 October 2020.
62. Tao, C.; Ma, X.; Wang, F.; Chen, Y. Anti-Alopecia Hair-Strengthening Nano-Liposome and Preparation Method Thereof. Chinese Patent No. CN118001207B, 16 August 2024.
63. Jahn, A.; Vreeland, W.N.; Devoe, D.L.; Locascio, L.E.; Gaitan, M. Microfluidic directed formation of liposomes of controlled size. *Langmuir* **2007**, *23*, 6289–6293. [CrossRef] [PubMed]
64. Laouini, A.; Charcosset, C.; Fessi, H.; Holdich, R.; Vladisavljevic, G. Preparation of liposomes: A novel application of microengineered membranes-investigation of the process parameters and application to the encapsulation of vitamin E. *RSC Adv.* **2013**, *3*, 4985–4994. [CrossRef]
65. Meure, L.A.; Foster, N.R.; Dehghani, F. Conventional and dense gas techniques for the production of liposomes: A review. *AAPS PharmSciTech* **2008**, *9*, 798–809. [CrossRef]
66. Kida, M.; Kitabata, I.; Kubotsu, K.; Sakata, Y. Novel Functionalized Liposomes and a Process for Production Thereof. European Patent No. EP0247497B1, 4 March 1992.
67. Ali, N.; De, M.; Sen, T. Synergistic Liposomal Formulation for the Treatment of Cancer. EP3046542B1, 6 January 2021.
68. Reshetnyak, Y.K.; Andreev, O.A.; Engelman, D.M. Liposome Compositions and Methods of Use Thereof. European Patent No. EP2603201B1, 19 June 2013.
69. Chen, M.; Fan, Y.; Lu, Y.; Wang, Q. Targeted Hybrid Exosome Loaded with Curcumin and miR140 as Well as Preparation Method and Application of Targeted Hybrid Exosome. Chinese Patent No. CN117965429B, 14 June 2024.
70. De Beer, J. Hybridosomes, Compositions Comprising the Same, Processes for Their Production and Uses Thereof. U.S. Patent No. US10561610B2, 18 February 2020.
71. De Beer, J. Hybridosomes, Compositions Comprising the Same, Processes for Their Production and Uses Thereof. U.S. Patent No. US11484500B2, 1 November 2022.
72. De Beer, J. Process for the Production of Hybridosomes. European Patent No. EP3096741B1, 17 March 2021.
73. Dai, Y.; Li, S.; Li, Y. Preparation Method of Exosome Bionic Preparation for Synergistically Promoting Wound Healing and Preparation Thereof. Chinese Patent No. CN113577272B, 12 April 2022.
74. Qiu, N.; Kong, F.; Ye, H.; Li, Z.; Ding, L. Co-loaded liposome and preparation method thereof. Chinese Patent No. CN116270473B, 19 December 2023.
75. Spatz, J.P.; Benk, L.T.; Frohnmayer, J.P.; Haller, B.; Janiesch, J.-W.; Plazman, Y.; Weiss, M. A Method for Preparing a Functional Synthetic Cell in Form of a Giant Unilamellar Vesicle. European Patent No. EP3630068B1, 27 October 2021.
76. Deng, Y.; Liu, M.; Li, J.; Yan, N.; Zhao, D.; Liu, M.; Zhang, H.; Tang, X.; Song, Y.; Liu, X. PEG2, n-Lipid Derivative Modified Nano-Carrier, Preparation Method and Application. Chinese Patent No. CN113350512B, 27 October 2023.
77. Xu, Y.; Liang, Y. Compound Responding to Endocytosis and Release and Application Thereof. Chinese Patent No. CN113683769B, 11 November 2022.
78. Xu, Y.; Zheng, A.; Chen, X. All-Trans Retinoic Acid Liposome Preparation and Preparation and Application Thereof. European Patent No. EP3501500B1, 11 May 2022.
79. Ma, G.; Wu, J.; Xia, Y.; Miao, C.; Du, Y.; Zhou, W. Granular Type Adjuvant as Well as Preparation method and Application Thereof. Chinese Patent No. CN108324938B, 4 May 2021.
80. Smriti, A.; Gao, X.; Huang, J. mRNA-Liposome Complex and Application Thereof. Chinese Patent No. CN112107680B, 8 April 2022.
81. Wei, G.; Hu, Y.; Jiang, K.; Wang, Y.; Koshikari, Y. Drug Delivery Vector and Pharmaceutical Formulation. Chinese Patent No. CN114007653B, 12 November 2020.
82. Thomsen, P.T.; Alm, M. A Delivery Device. European Patent No. EP3334413B1, 18 May 2022.
83. Germain, M.; Meyre, M.-E.; Pottier, A.; Levy, L. Pharmaceutical Composition, Preparation and Uses Thereof. European Patent No. EP3236934B1, 22 May 2024.
84. Wang, X.; Wang, H.; Suo, Y. A kind of Preparation Method of the Medicinal Extract Containing Intermediate Peashrub Seed. Chinese Patent No. CN104382984B, 8 May 2018.
85. Wan, K.; Gui, R. Preparation Method of Temperature and Fluorescence Probe of Liposome Loaded with Gold Nanocluster and Anti-Cancer Drug. Chinese Patent No. CN103599070B, 20 May 2015.
86. Aoki, Y.; Ueda, E. Antitumor Liposome Formulation and Method for Producing the Same. Japan Patent No. JP4715133B2, 9 June 2006.
87. Nakajima, A.; Ueda, E.; Nagaike, C. Method for Producing Liposome-Containing Formulation and Liposome-Containing Formulation. Japan Patent No. JP4599849B2, 15 December 2010.
88. Nakajima, A.; Ueda, E.; Nagaike, C. Method for Producing Liposome-Containing Formulation and Liposome-Containing Formulation. Japan Patent No. JP4649841B2, 16 March 2011.
89. Allen, T.M.; Hansen, C.B.; de Menezes, D.E.L. Pharmacokinetics of longcirculating liposomes. *Adv. Drug Deliv. Rev.* **1995**, *16*, 267–284. [CrossRef]
90. Allen, T.M.; Hansen, C. Pharmacokinetics of stealth versus conventional liposomes: Effect of dose. *Biochim. Biophys. Acta* **1991**, *1068*, 133–141. [CrossRef]

91. Han, Y.; Wen, P.; Li, J.; Kataoka, K. Targeted nanomedicine in cisplatin-based cancer therapeutics. *J. Control. Release* **2022**, *345*, 709–720. [CrossRef]
92. Kan, P.; Tseng, Y.-L.; Ou, H.C. Controlled Drug Release Liposome Composition. European Patent No. EP2968146B1, 19 February 2020.
93. Donald, P.; Waldemar, P. Method of Reconstituting Liposomal Annamycin. U.S. Patent No. US11980634B2, 14 May 2024.
94. Ouyang, K.; Guo, X.; Cai, L.; Yue, T.; Liu, J.; Yang, C. Protamine Short Peptide Modified Paclitaxel Liposome and Preparation Method Thereof. Netherlands Patent No. NL2033676B1, 3 May 2024.
95. Weng, G.; Zhou, W.; Cheng, G.; Cheng, P. Paclitaxel Liposome Composition for Treatment of Cancer and Preparation Thereof. European Patent No. EP1332755B1, 6 August 2003.
96. Tong, X.; Lei, G.; Yu, C.; Chen, L. Liposome of Irinotecan or Its Hydrochloride and Preparation Method Thereof. European Patent No. EP2508170B1, 29 July 2015.
97. Wei, X.; Jiang, B. Irinotecan liposome preparation and preparation and application thereof. Chinese Patent No. CN109260155B, 4 April 2021.
98. Weng, W.; Liu, C.; Wang, A.; Lin, C.; Liu, Q.; Wang, L.; Lin, S.; Liu, Y. Pegylated Lipid, Liposome Modified by Pegylated Lipid, Pharmaceutical Composition Containing Liposome, and Preparation and Application of Pharmaceutical Composition. Chinese Patent No. CN115197078B, 4 June 2024.
99. Li, C.; Wang, Y.; Zhan, H. Liposomes with Ginsenoside as Membrane Material and Preparations and Use Thereof. European Patent No. EP3337456B1, 26 October 2022.
100. Bacha, J.A.; Brown, D.M.; Stein Oslash, A. Dianhydrogalactitol, Diacetyldianhydrogalactitol or Dibromodulcitol to Treat Non-Small-Cell Carcinoma of the Lung and Ovarian Cancer. European Patent No. EP3125920B1, 23 December 2020.
101. Xu, Y.; Xie, F. Multicellular Targeting Liposome. Japan Patent No. JP7470418B2, 21 March 2019.
102. Xiao, H. Tumor Targeted Therapeutic Drug Carrier as Well as Preparation Method and Application Thereof. Chinese Patent No. CN102133404B, 16 January 2013.
103. Zhang, S.; Chen, H.; Ma, Y.; Qin, X.; Lan, H.; Cui, S. Application of Redox-Responsive Chitosan-Liposome. Chinese Patent No. CN107982547B, 19 July 2019.
104. Fehr Pereira Lopes, J.E. Phamaceutical Composition Comprising Jasmonates. Korean Patent No. KR101751918B1, 30 June 2017.
105. Kline, K.; Lawson, K.A.; Menchaca, M.S.; Knight, J.V.; Wellen, C.W.; Sanders, B.G. Liposomal Delivery of Vitamin E Based Compounds. Australian Patent No. AU2002361812B2, 8 November 2007.
106. Hong, K.; Drummond, D.C.; Kirpotin, D.B. Liposome Useful for Drug Delivery. U.S. Patent No. US8147867B2, 3 April 2012.
107. Hong, K.; Drummond, D.C.; Kirpotin, D.B. Liposome Useful for Drug Delivery. U.S. Patent No. US8703181B2, 22 April 2014.
108. Hong, K.; Drummond, D.C.; Kirpotin, D.B. Liposome Useful for Drug Delivery. U.S. Patent No. US9782349B2, 10 October 2017.
109. Baca, M.; Gilmore, S.A.; Javanbakht, H.; Mukherjee, P.K. Interleukin-2 Variant Proteins Fused to Human IgG4 Fc and Uses Thereof. U.S. Patent No US11896648B2, 13 February 2024.
110. Karve, S.; Patel, Z.; Sarode, A.; Zhang, Y.; Derosa, F.; Heartlein, M. Messenger RNA Vaccines and Uses Thereof. CN112384205B, 3 May 2024.
111. Anusha, D.; Braydon, G.C.; Derosa, F.; Dias, A.; Derosa, F.; Guild, B.C.; Heartlein, M.; Heartlein, M. Methods and Compositions for Delivering mRNA Coded Antibodies. European Patent No. EP2970456B1, 19 May 2021.
112. Haas, H.; Hörner, S.; Esparza Borquez, I.H.; Hiller, T.M.; Bates, F. Preparation and Storage of Liposomal RNA Formulations Suitable for Therapy. U.S. Patent No. US11395799B2, 26 July 2022.
113. Mooney, D.J.; Stafford, A.; Desai, R.; Martinick, K. Reduced and Oxidized Polysaccharides and Methods of Use Thereof. Chinese Patent No. CN108430458B, 10 October 2020.
114. Denk, O. Inhalable Compositions for Use in the Treatment of Pulmonary Diseases. European Patent No. EP3890767B1, 12 October 2022.
115. Yang, R. Slow-Release Salbutamol Sulfate Inhalation Preparation and Production Process Thereof. Chinese Patent No. CN111700883B, 6 April 2021.
116. Zhang, S.; Yan, X.; Xue, Q.; Tian, X.; Niu, L.; Wang, Z. Tobramycin Liposome Used for Aerosol Inhalation and Production Method Thereof. Chinese Patent No. CN111228243B, 18 February 2022.
117. Yang, R. Liposome Atomization Treatment Preparation Containing Acetaldehyde Dehydrogenase as Well as Preparation Method and Application Thereof. Chinese Patent No. CN108434446B, 26 April 2022.
118. Axelsson, B.I.; Brattsand, R.L.; Dahlback, C.M.O.; Kallstrom, L.A.; Trofast, J.W. Liposomes Containing Steroid Esters. European Patent No. EP0170642B1, 27 February 1991.
119. Lambert, R.; Cavallo, G. Liposomial Eye Drops Solution and Uses Thereof for the Treatment of Dry Eye Syndrome. European Patent No. EP3673896B1, 12 January 2022.
120. Hong, K.; Kao, H.-W.; Lin, Y.-Y.; Fang, W. Sustained-Release Ophthalmic Pharmaceutical Compositions and Uses Thereof. Taiwan Province of China Patent No TWI786328B, 11 December 2022.
121. Dwivedi, A.K.; Ahmad, H.; Khandelwal, K.; Sangwan, R.S.; Sangwan, N.S.; Gayen, J.R.; Sarika, B.S.; Gaur, S.P.S.; Bhosale, V.V.; Rath, S.K.; et al. A formulation Useful for Delivery of Neuro Protecting Agent. European Patent No. EP3200770B1, 2 September 2020.

122. Hong, K.; Kao, H.-W.; Lin, Y.-Y.; Fang, J.; Fang, W. Sustained-Release Pharmaceutical Compositions Comprising a Therapeutic Agent for Treating Diseases Due to Reduced Bone Density or Cartilage Loss and Uses Thereof. JP7431419B2, 15 February 2024.
123. Hong, K.; Guo, L.S.S.; Tseng, Y.-L.; Shih, S.-F.; Chang, P.-C.; Tsai, C.-C.; Lin, H.-H. Methods of Treating Arthritis. European Patent No. EP2854769B1, 7 June 2023.
124. Hall, J.S.; Turnbull, D.J.; Grigsby, J.J.; Ardekani, S.M.; Davis, P.N.; Garcia, L.D.; Kurz, S.M.; Los, K.D.A. Manufacturing of Bupivacaine Multivesicular Liposomes. United Kingdom Patent No. GB2603047B, 10 January 2024.
125. Danguole, S.; Hatice, O. Formulations of Volatile Anesthetics and Methods of Use for Reducing Inflammation. U.S. Patent No. US9744142B2, 29 August 2017.
126. Barenholz, Y.; Rubinstein, A.; Jubeh, T. Method for Selecting Cationic or Anionic Liposomes for Treatment of a Mucosa Membrane, and Kit Comprising the Same. European Patent No. EP1694298B1, 9 June 2010.
127. Chen, J.; Fang, B.; Guo, H.; Jiang, W.; Kwak, H.; Li, Y.; Liu, R.; Liu, S.; Nie, C.; Wang, P.; et al. Pharmaceutical Composition for Treating Alcoholic Fatty Liver and Preparation Method Thereof. Chinese Patent No. CN117860678B, 28 May 2024.
128. Hao, J.; Zhou, G. Cilostazol Liposome Solid Agent. Chinese Patent No. CN112006992B, 25 March 2022.
129. Zhang, J.Y.; Luo, M.Z. Sterile Pharmaceutical Composition and Process for a Solution of Propofol Emulsion Having Microbial Growth Retardation. U.S. Patent No. US7468394B1, 23 December 2008.
130. Azeredo Da Silveira Lajaunias, S.; Lajaunias, F. Liposomes for Inhibiting Biofilm Formation. European Patent No. EP3589278A1, 8 January 2020.
131. Worsham, R. Methods for the Manufacture of Liposomal Drug Formulations. Australian Patent No. AU2019262117B2, 13 June 2024.
132. Janoff, A.S.; Minchey, S.R. Interdigitation-Fusion Liposomes Containing Arachidonic Acid Metabolites. European Patent No. EP0729352B1, 3 February 1999.
133. Kang, S.W. Composition for Increasing Expression of PGC-1 Alpha. European Patent No. EP3403655B1, 7 June 2023.
134. Ferrer, M.A.; Garcia, A.J.M.; Delgado, G.R. Peptides Which Inhibit Activated Receptors and Their Use in Cosmetic or Pharmaceutical Compositions. European Patent No. EP2773366B1, 1 March 2017.
135. Martynov, A.; Farber, B.S.; Farber, S.S.; Sitenko, A. Cosmetological and Pharmaceutical Formulae for the Rejuvenation and Restoration of Skin, Including After Surgical Procedures. U.S. Patent No. US8846064B2, 2 August 2012.
136. Carreno, S.C.; Van Den, N.W.; Cebrian, P.J.; Alminana, D.N.; Ferrer, M.A.; Garcia, S.N. Cosmetic or Pharmaceutical Compositions Comprising Metalloproteinase Inhibitors. Spanish Patent No. ES2330291B1, 18 October 2010.
137. Meybeck, A.; Michelon, P.; Montastier, C.; Redziniak, G. Composition Pharmaceutique, Notamment Dermatologique, ou cosmetique, a Base de Phases Lamellaires Lipidiques Hydratees ou de Liposomes Contenant un Retinoide ou un Analogue Structural Dudit Retinoide tel qu'un Carotenoide. French Patent No. FR2591105B1, 24 March 1989.
138. Marriott, C.; Martin, G.P.; Brown, M.B. Hyaluronic Drug Delivery System. European Patent No. EP0963196B1, 19 January 2005.

processes

Review

Transdermal Patches Containing Opioids in the Treatment of Patients with Chronic Pain

Iolanda De Marco [1,2]

[1] Department of Industrial Engineering, University of Salerno, Via Giovanni Paolo II, 132, 84084 Fisciano, SA, Italy; idemarco@unisa.it

[2] Research Centre for Biomaterials BIONAM, University of Salerno, Via Giovanni Paolo II, 132, 84084 Fisciano, SA, Italy

Abstract: Transdermal delivery is a non-invasive route, used as an alternative to the oral route, to administer drugs through the skin surface. One of the fields in which they are particularly used is that of pain therapy. In this treatment, transdermal patches, particularly those containing opioids, are used to complement or replace orally administered drugs. First-generation patches are constituted by reservoir systems, where the drug is dissolved in a solvent and gelled with a polymer. In contrast, the active principle is incorporated into the polymer adhesive in more recent matrix patches. In this review, the main papers related to the production and employment of transdermal patches containing the two most used opioids, i.e., fentanyl and buprenorphine, have been critically analyzed. From the analysis of the literature, it is possible to deduce that the type of drug and the amount of drug present in the patch must be chosen not according to the origin of the pain but to the age of the patient, the area where the patch is applied, and the frequency at which the patch is replaced.

Keywords: fentanyl; buprenorphine; transdermal patches; reservoir and matrix systems

Citation: De Marco, I. Transdermal Patches Containing Opioids in the Treatment of Patients with Chronic Pain. *Processes* **2023**, *11*, 2673. https://doi.org/10.3390/pr11092673

Academic Editors: Wei Wu and Yi Lu

Received: 4 July 2023
Revised: 17 August 2023
Accepted: 5 September 2023
Published: 6 September 2023

1. Introduction

The International Association for the Study of Pain (IASP), which is the central organization concerned with the study of pain and possible strategies to alleviate it, currently defines pain as "an unpleasant sensory and emotional experience associated with actual or potential tissue damage, or described in terms of such damage" [1]. Pain can be categorized regarding its time frame and pathophysiology [2]. Concerning the former classification, pain can be acute and chronic. It is defined as "acute" if it lasts for less than three months and chronic if it restricts daily activities for more than three months [3]. Sometimes an intermediate category, called subacute pain, is considered [4]. As for the pathophysiology, pain can be divided into nociceptive, neuropathic, and nociplastic. Nociceptive pain (somatic, visceral), which is the most common form of chronic pain, is associated with damage to organs and tissues with an intact nervous system. Neuropathic pain is associated with damage or disease of the nervous system, and nociplastic pain is due to abnormal processing of pain signals without clearly identifying tissue damage or diagnosing pathologies concerning the somatosensory system [5]. It can be of somatic origin (bone fractures, muscle spasms, osteoarthritis, and postoperative pain) or visceral (peptic ulcer, cancer, liver cirrhosis, and ischemia). Neuropathic pain can affect central nerves, such as from spinal injuries, Parkinson's disease, multiple sclerosis or transverse myelitis, or peripherally, such as with carpal tunnel syndrome, or local pain of a traumatic nature, nutritional deficiencies, and acute herpes zoster [6]. Nociplastic pain can have various causes, such as fibromyalgia, irritable bowel syndrome, or bladder pain [7]. Chronic pain, which lasts over time, is sometimes considered a real disease, with the need for pharmacological treatment (for example, to be operational although the pain) and psychological implications (for example, accepting pain) [7]. In some cases, a distinction is made according to whether the chronic pain originates from cancer or not; chronic cancer pain (CCP) and chronic non-cancer pain (CNCP) treatments are therefore considered. However, regardless of the origin, pain treatments are often common [8].

In 1986, the World Health Organization (WHO) proposed a three-step analgesic ladder (sketched in Figure 1) to provide adequate pain relief in the treatment of CCP [9], which, nowadays, is also used in the treatment of CNCP. The WHO analgesic ladder requires analgesic therapy to be established based on pain intensity:

- Mild pain can be treated with non-steroidal anti-inflammatory drugs (NSAIDs) or non-opioid analgesics such as acetaminophen or paracetamol and/or adjuvants such as antidepressants or anticonvulsants;
- Moderate pain requires the use of weak opioid analgesics, such as codeine, dihydrocodeine, tramadol, and/or adjuvants;
- Severe and persistent pain can be reduced by using strong opioids, such as morphine, methadone, diamorphine, oxycodone, fentanyl, and buprenorphine or adjuvants [10–12].

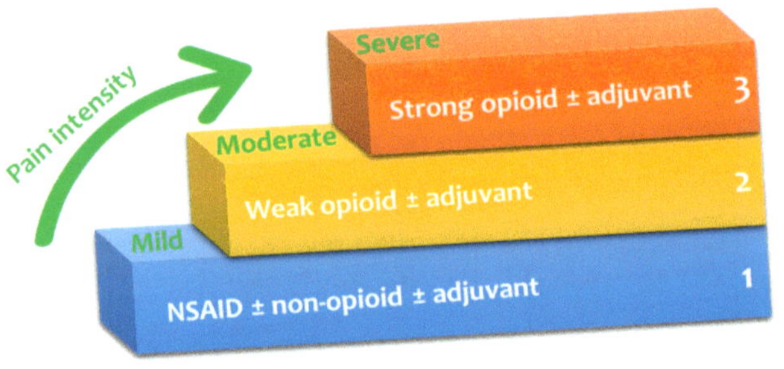

Figure 1. Adapted WHO three-step analgesic ladder.

It is, therefore, evident that opioids are the most potent analgesics and are well-established drugs for the treatment of severe acute [13], surgical [14], and chronic pain [15].

It is important to underline that in defining analgesic therapy, it is not only necessary to consider the choice between the different active ingredients [16] but also the different administration routes, such as oral, sublingual, inhaled, intravenous, intramuscular, rectal, transdermal and topical [17]. The oral route is preferred except when oral intake is not possible because of severe vomiting, problems with swallowing, constipation, and poor compliance. To overcome these drawbacks, transdermal drug delivery (TDD) is a good choice; in addition, opioid transdermal delivery results in a lower risk of addiction than oral and parenteral routes [2]. Different research articles and reviews have been published regarding treating chronic pain using opioids; they are generally focused on the mechanism of action of the different drugs in the case of CCP and CNCP. Less frequent is the publication of reviews focused on preparing and producing patches containing opioids [2,7,16,18–22].

Therefore, this review aims to underlie the aspects related to the production of opioid-based patches; in particular, this study is focused on two of the most frequently administered opioids via transdermal delivery: fentanyl and buprenorphine.

2. Transdermal Drug Delivery

TDD is a non-invasive and painless way to administer active ingredients at a predetermined rate across the dermis to achieve a local or systemic effect; it consists of the application of a pharmacological formulation (generally a transdermal patch containing the active ingredients) on intact skin [23–27]. On the other hand, the treatment of damaged skin is different; in this case, we speak of wound healing [28–31]. TDD has different advantages over both oral and parenteral administration. Compared to the first, there is no passage of the active principle through the gastrointestinal tract; therefore, there is no loss of efficacy due to first-pass hepatic metabolism, and drugs can be administered without pH changes, enzymes, and intestinal bacteria interference [32]. The dosing frequency is also reduced, and the adverse effects arising from peak plasma drug concentrations are minimized; in addition, this method improves patient compliance and is also advisable for patients who are unconscious or who have vomiting as a symptom of their pathology or as a side effect related to the medications taken. TDD is also suitable for patients who self-administer the drugs. Compared to the parenteral route, transdermal administration does not carry any risk of infection and overcomes problems related to poor acceptance of needles by the patients [33]. On the other hand, the drawbacks of TDD are linked to skin irritation, difficulties in delivering hydrophilic drugs, delayed absorption, and high costs [34].

Transdermal patches are constituted of different components [35]: (a) an impermeable backing layer (represented in light blue in Figure 2), constituted of elastomers, which has the aim of protecting the patch from the external environment and confers flexibility to the patch; (b) a membrane (represented in light pink in Figure 2), which controls the drug release; (c) an adhesive part (represented in light orange in Figure 2), which binds the components of the patch together and the patch to the skin; and (d) a liner (in light green in Figure 2), which serves to protect the patch during storage and is removed before application to the skin.

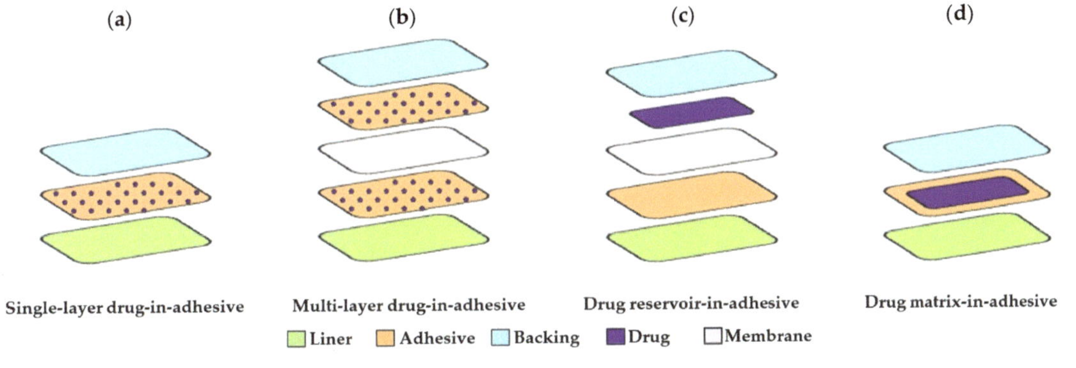

Figure 2. Different types of transdermal delivery patches. (**a**) single-layer drug-in-adhesive; (**b**) multi-layer drug-in-adhesive; (**c**) drug reservoir-in-adhesive; (**d**) drug matrix-in-adhesive. Adapted from [36].

Provided that the backing and the liner constitute the two external layers of the patch, different solutions are obtained depending on how the active ingredient is distributed into the adhesive layer, as represented in Figure 2 [36]. In particular, if the drug is confined directly into the adhesive layer, a drug-in-adhesive system is proposed; when it is dispersed in a unique layer, a "single-layer drug-in-adhesive" (Figure 2a) is prepared, whereas when the drug is dispersed into two adhesive layers parted by a membrane, a "multi-layer drug-in-adhesive" (Figure 2b) is arranged. The latter solution allows the administration of different drugs within the same patch or the possibility of having both an immediate drug release from one of the two layers and a controlled release from the other. In the "drug reservoir-in-adhesive" patch (Figure 2c), the drug is dissolved or dispersed in a liquid phase which, thanks to a semi-permeable membrane, is separated from the adhesive layer. Finally, the drug can be dissolved or dispersed in a hydrophilic or lipophilic polymer matrix, concentric with respect to the adhesive layer, which is positioned between the backing and the liner, forming a "drug matrix-in-adhesive" (Figure 2d).

Other types of transdermal patches based on the use of microneedles have also been proposed [37–39]. Such systems are used to administer drugs such as naltrexone [40], lidocaine [41], and insulin [42], but there are no papers involving the administration of opioids.

In the following subsections, the articles dedicated to producing devices containing the most transdermally administered opioids are critically and systematically analyzed. The classification has been made considering the different administered opioids.

2.1. Fentanyl

Fentanyl is an analgesic with a potency of at least 80 times higher than that of morphine, which is widely used as a surgical anesthetic and for the control of chronic pain. Depending on the application, it is administered by intravenous injection, transdermal patches, oral transmucosal tablets, and vestibular tabs. Transdermal administration of fentanyl is associated with ease of dosing and a significantly extended duration of analgesic effect compared to parenteral administration [43]. It is an excellent option for the long-term treatment of cancer pain and chronic non-cancer pain [44]. Moreover, fentanyl's characteristics, such as high potency, low molecular weight, lipid solubility, and excellent skin flux properties (approximately 1000 times greater than morphine), make it suitable for delivery via the transdermal therapeutic system [45,46]. The first generation of fentanyl-based patches consists of reservoir delivery systems that provide continuous, systemic delivery of fentanyl for 72 h in a constant amount per unit time, while second-generation patches have a matrix-type design. Reservoir systems are constituted of an occlusive backing (polyester/ethylene or polyester film), a drug reservoir (fentanyl base dissolved in ethyl alcohol and gelled with hydroxyethyl cellulose), a rate-controlling membrane (generally constituted of ethylene-vinyl acetate copolymer film) and a silicone adhesive layer. In matrix-designed systems, fentanyl is incorporated directly into the adhesive, which can be constituted by polyacrylate or acrylate vinyl acetate; sometimes, dypropylene glycol droplets are dispersed in a silicone matrix.

In Table 1, a summary of fentanyl-based patches obtained with the different technologies is reported.

Table 1. Transdermal patches containing fentanyl.

Type of Patch	Amount of Drug [μg/h]	Main Results	Reference
Reservoir	25–100	Comparison with fentanyl infusion; the drug half-life after patch removal was significantly longer than the half-life after infusion	[47]
	25–50	Use of Durotep® commercial patches on 60 Japanese cancer patients that previously used oral morphine or oxycodone; evaluation of the influence of two biomarkers on fentanyl pharmacokinetics	[48]
	25	Use of Duragesic® commercial patches to evaluate stability and skin permeation profiles as a function of the age of the patch	[49]
Matrix	12–100	Comparison with a reservoir system; tests on 46 patients with CCP and CNCP; 91% of patients preferred the matrix system	[43]
	12–100	Comparison between Fentadur®, a reservoir system, and Durogesic Dtrans®, a matrix system, in terms of adhesion and dissolution properties; both were acceptable for skin attachment, although higher adhesion forces were expressed by the matrix system; from the point of view of the dissolution study, the release of the reservoir system was lower than that of the matrix system.	[50]
	12.5	Comparison with Durotep®, a commercial reservoir system; tests on 87 patients of both sexes with a confirmed diagnosis of cancer; successful replacement of morphine, oral oxycodone, or fentanyl injection formulations	[51]

Table 1. *Cont.*

Type of Patch	Amount of Drug [μg/h]	Main Results	Reference
	25	Comparison with Duragesic® commercial patches to evaluate the effect of elevated temperatures and compromised skin on drug release and skin permeation; the patches are comparable when applied to intact skin at normal skin temperature, whereas, in the case of compromised skin, the matrix patch has a more significant permeation enhancement effect than the reservoir system	[52]
	50	Comparison with Durogesic®, a commercial reservoir system; adherence and skin irritability results were similar, but the matrix formulation is more convenient because of the smaller size	[53]
	50	Comparison with a reservoir system; tests on 24 male patients; enhanced adherence compared to the reservoir formulation; constant fentanyl delivery rate over three days	[54]
	12.5–50	Use of Durotep® MT patch, Durogesic® Dtrans®, and Durogesic® SMAT commercial patches; comparison with a reservoir system; patients preferred the transdermal matrix patch over the transdermal reservoir patch	[55]
	12–300	Equivalent to a reservoir system in terms of TDD and analgesic efficacy in patients with CCP	[56]

In the reservoir systems, patches containing 1.25–10 mg fentanyl are available to provide 12.5–100 μg/h of fentanyl, respectively, for up to three days (72 h). It is, therefore, evident that more than 25% of fentanyl present in the patch is not released. For example, considering the 100 μg/h patch, 7.2 mg of the 10 mg present will have been released after three days [57].

Observing the data inserted in Table 1, it is clear that the matrix systems are the ones mainly used at this point in time. The transition from the first type of patches to the second occurred due to better skin compatibility, adhesive properties, wearability/comfort, and patients' general satisfaction, which improved significantly with the new matrix technology. Another reason why the second type of patch has supplanted the first is that, in reservoir patches, a significant amount of opioids remain in the patch after removal and, therefore, the patches must be discarded according to opioid disposal guidelines to prevent misuse [58]; in addition, it has been experienced that defects in such patches can cause the patch reservoir to leak onto patients skin and result in overdose [59] and that the gel can be eliminated from the patch and the content unintentionally taken by the patient, leading to a fatal outcome [60]. Conversely, the matrix systems distinctly reduce the risk of drug leakage and abuse. From the point of view of the dimensions of the patch, the matrix patches are thinner (because no reservoir is present) and smaller (for equal strengths, the matrix patch is at least one-third smaller). Considering the pharmacokinetic profiles of reservoir and matrix patches, they are very similar; in Figure 3, the results obtained by Marier et al. indicate that the mean serum concentrations are superimposable [54].

Many papers are devoted to comparing two types of systems from different perspectives. For example, Moore et al. [61] studied the effects of a heating pad on serum fentanyl concentrations with the two types of patches; for both systems, an increase in systemic exposure to fentanyl was observed when a direct external heat source was applied for 10 h; the authors concluded that heat-associated devices, such as saunas, hot tubs, tanning beds, and electric blankets should be avoided during treatment with transdermal fentanyl systems.

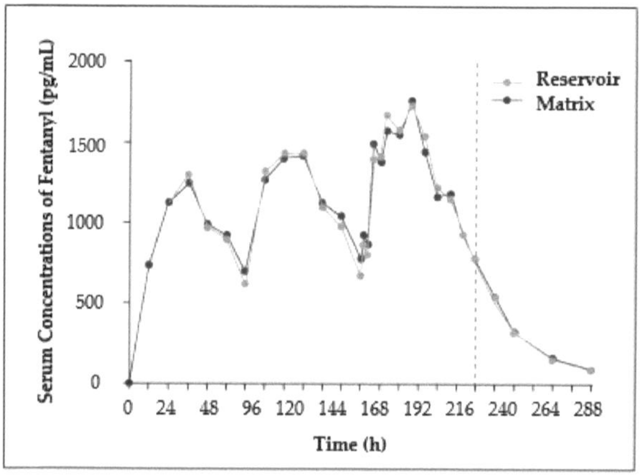

Figure 3. Mean serum concentration profiles of fentanyl after 72 h application of the reservoir and novel matrix systems (50 μg/h). Adapted with permission from [54]. Copyright © 2006 John Wiley and Sons.

Other comparative studies have been published considering different kinds of matrix systems. Zecca et al. [62] proposed a bilayer matrix fentanyl patch compared to a standard monolayer patch to improve the efficiency and reliability of delivery systems. The novel patch provides a constant fentanyl delivery over 72 h.

Considering the results in terms of fentanyl release, it is possible to observe that:

- In the case of reservoir formulations, the drug release rate from the reservoir layer is dependent on the choice of polymer and membrane layer thickness;
- In drug-in-adhesive systems and matrix patches, the control of the drug administered is diffusive, and the amount dispensed is directly proportional to the surface of the patch in contact with the skin.

A recent study by Defraeye et al. [63] aimed to verify the effect of the patient's age and the site of application of the patch on the amount of the active ingredient absorbed. The adsorption of the drug is strongly related to the different body locations because of different thicknesses of the stratum corneum or the presence of hair follicles. Indeed, the difference in fentanyl uptake between applying the patch on the shoulder and the forearm was 36% after 72 h. To evaluate the impact of the patient's age on fentanyl uptake kinetics, the authors divided the volunteers by age into six different groups. They observed that the total amount of a drug taken after 72 h differed by 26% between an 18-year-old and a 70-year-old patient, as reported in Figure 4. Furthermore, the time necessary to reach the drug's minimum effective concentration (MEC) depends on the patient's age. Considering the 18- and 70-year-old patients, the MEC was reached after 23 h in the former case and after 32 h in the latter.

In any case, fentanyl-based patches must be used with particular caution. For example, the patch has is to be used in association with any heat exposure or in cases of fever [64]. Furthermore, the patches must be applied to intact skin and must not be damaged. Unfortunately, there have been many cases of death in patients taking fentanyl. In some cases, the causes of death have been associated with using the active substance and in others not. Indeed, it is possible that manufacturing defects present in the reservoir patches favor the leakage of the reservoir on the patient's skin, resulting in an overdose [65].

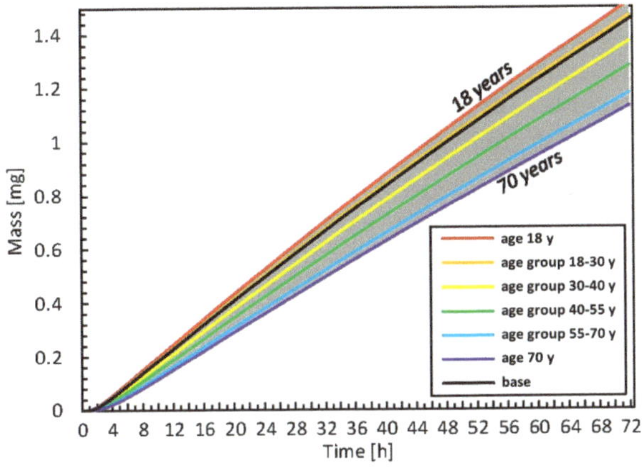

Figure 4. Cumulative amount of fentanyl taken up in the blood as a function of time for different age categories. Adapted from [63]. Copyright © 2020 Frontiers in Pharmacology.

2.2. Buprenorphine

Buprenorphine is a semisynthetic, low molecular weight, lipophilic opioid analgesic with peculiar pharmacodynamic and pharmacokinetic properties mainly used in patients who require joint administration of various drugs, the elderly, and patients who have renal or hepatic insufficiency [66,67]. It has demonstrated promising efficacy cases of moderate to severe CCP and severe CNCP unresponsive to nonopioid analgesics (for example, in the case of patients with osteoarthritis, low back pain, and other permanent pain syndromes) [68,69]. In some cases, it is preferred to other opioids because the intensity of several classic opioid-related side effects, such as respiratory depression and abuse liability, is reduced [70]. Its therapeutic effect may be correlated to the "binding affinity", which is the ability of an active principle to bind to a receptor; more molecular interactions between the buprenorphine molecule and μ-opioid receptor are present, leading to an excellent binding affinity (corresponding to a low equilibrium dissociation constant, K_i value) in comparison with that of other opioids [66]. Figure 5 compares the different binding affinities of some of the most prescribed opioids.

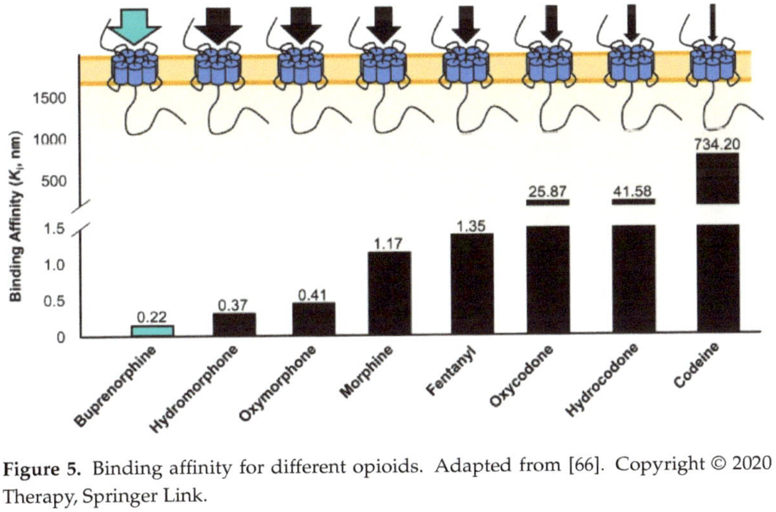

Figure 5. Binding affinity for different opioids. Adapted from [66]. Copyright © 2020 Pain and Therapy, Springer Link.

As in the case of fentanyl, the use of reservoir systems appears to be overcome. Those systems were applied until the end of the last century; ethanol- or aqueous-based reservoirs produced sustained plasma levels of buprenorphine [71]. Buprenorphine transdermal patches (BTPs) are preferably applied to the upper outer arm, upper back, sub clavicular region, or chest; they are, in general, better tolerated than transdermal fentanyl, as evidenced by a reduced need for dosage increases [72]. Different from fentanyl transdermal patches that are always worn for 72 h, BTPs can be applied (a) for 3–4 days, considering a dosage equal to 35, 52.5, and 70 μg/h [73] or (b) for more days at lower doses (5–20 μg/h) [74]. For both cases, a rotation of the application site is recommended, and a new patch must be applied to a different skin site for the following 3 or 4 weeks. Likar et al. [75] compared two durations of buprenorphine patch wearing in terms of analgesic efficacy and skin tolerability, a transdermal patch containing a buprenorphine dose of 50 μg/h in a 4-day regimen vs. a standard 3-day regimen. The tests were performed on 49 white patients of both sexes, which asserted that the quality of treatment did not differ significantly between regimens.

Patches containing a low dose of buprenorphine are commonly used in postoperative pain management or in the case of clinically diagnosed osteoarthritis or rheumatoid arthritis or for less invasive pathologies such as low back pain and joint/muscle pain. Mythili et al. [76] proposed using 10 μg/h patches in the management of postoperative pain in oral and maxillofacial surgery, including in the study of 100 people of both sexes in the age group of 18–60 years. The patches were effective because buprenorphine showed high analgesic potential and was well tolerated by the patients. Breivik et al. [77] tested low-dose (5–20 μg/h) BTPs in elderly patients with pain due to osteoarthritis. The selected patients had opioid-sensitive pain and were unable to undergo surgery. In addition, the administration of paracetamol was ineffective, and NSAIDs were not recommended. Tests were performed with 200 patients of both sexes: 100 of them were treated with 7-day BTPs and 100 with placebo patches. The patients' global impression in terms of pain reduction after a 6-month treatment period was significantly better in patients treated with BTPs than in the patients treated with a placebo. Yoon et al. [78] evaluated the effectiveness and tolerability of 5–20 μg/h BTPs in Asian patients with moderate to severe chronic musculoskeletal non-malignant pain. Treatment with BTPs resulted in effective and sustained pain relief and improved daily functioning and quality of life.

Apart from transdermal patches, Buprenorphine is sometimes administered via buccal films or tablets. Compared with the oral route, the transdermal administration avoids losses in bioavailability due to first-pass hepatic metabolism [79]. James et al. [80] compared the efficacy and tolerability of 7-day low-dose buprenorphine transdermal patches (BTPs) with buprenorphine sublingual tablets to treat osteoarthritis pain. Tests were performed on 110 patients, and the authors concluded that low-dose BTPs are as effective as sublingual buprenorphine, with a better tolerability profile. Low-dose BTPs were compared to tablets containing other opioids, such as tramadol. For example, Karlsson et al. [81] compared 5, 10, and 20 μg/h 7-day BTPs with prolonged-release of 75, 100, 150, and 200 mg tramadol tablets in patients affected by chronic osteoarthritis pain (at a level from moderate to severe). The efficacy of BTPs was non-inferior to that of tramadol tablets, and the incidence of adverse effects was similar in the two therapy groups. Consequently, most patients reported that, in case of future treatment, they would prefer the 7-day transdermal patch to the twice-daily tablet. Similar conclusions were drawn when comparing the efficacy of the same type of BTP (7-day 5–20 μg/h) with higher doses (150–300 mg) of prolonged-release tramadol tablets for the treatment of postoperative pain after single-level spinal fusion surgery [82]. Plosker and Lyseng-Williamson [83], in a review paper, analyzed the analgesic efficacy of low-dose BTPs, observing that they are equivalent to sublingual buprenorphine, non-inferior to prolonged-release tramadol, and generally superior to a matching transdermal placebo patch.

3. Conclusions and Perspectives

This review focused on using transdermal patches containing opioids (fentanyl and buprenorphine) to treat chronic cancer and non-cancer pain. The different kinds of patches can be classified into two main categories: reservoir and matrix systems. The second one has replaced the first kind of patch, as it has better skin compatibility, adhesive properties, and wearability/comfort, coupled with the identical pharmacokinetics of the active ingredient. Furthermore, in the matrix system patches, the active ingredient is almost wholly released so that no abuse phenomena have been observed in using the device; instead, abuse phenomena have been frequently noted using the reservoir systems.

This review also focused on comparing patches containing fentanyl and those containing buprenorphine, underscoring for both the priority areas of use, advantages, and disadvantages. Without a doubt, the challenges to be faced in the field of drug administration are many. The challenges include having to improve the bioavailability of poorly water-soluble drugs to having to control the release of some active ingredients. Regarding transdermal administration, research should focus on developing patches to be applied as infrequently as possible. In this way, the patient could have an important psychological advantage with the same pain-relieving effect. An interesting way to pursue this could be to couple the opioid with another active ingredient, such as a different type of analgesic or a drug that assists the therapy the patient is following.

Funding: This research received no external funding.

Data Availability Statement: Not applicable.

Conflicts of Interest: The author declares no conflict of interest.

References

1. Raja, S.N.; Carr, D.B.; Cohen, M.; Finnerup, N.B.; Flor, H.; Gibson, S.; Keefe, F.J.; Mogil, J.S.; Ringkamp, M.; Sluka, K.A.; et al. The revised International Association for the Study of Pain definition of pain: Concepts, challenges, and compromises. *Pain* **2020**, *161*, 1976–1982. [CrossRef] [PubMed]
2. Leppert, W.; Malec-Milewska, M.; Zajaczkowska, R.; Wordliczek, J. Transdermal and Topical Drug Administration in the Treatment of Pain. *Molecules* **2018**, *23*, 681. [PubMed]
3. Bussières, A.E.; Stewart, G.; Al-Zoubi, F.; Decina, P.; Descarreaux, M.; Hayden, J.; Hendrickson, B.; Hincapié, C.; Pagé, I.; Passmore, S.; et al. The Treatment of Neck Pain-Associated Disorders and Whiplash-Associated Disorders: A Clinical Practice Guideline. *J. Manip. Physiol. Ther.* **2016**, *39*, 523–564.e527. [CrossRef]
4. Banerjee, S.; Argáez, C. Multidisciplinary Treatment Programs for Patients with Acute or Subacute Pain: A Review of Clinical Effectiveness, Cost-Effectiveness, and Guidelines. Ottawa (ON): Canadian Agency for Drugs and Technologies in Health. Available online: https://www.ncbi.nlm.nih.gov/books/NBK546002/ (accessed on 7 May 2023).
5. Fitzcharles, M.A.; Cohen, S.P.; Clauw, D.J.; Littlejohn, G.; Usui, C.; Häuser, W. Nociplastic pain: Towards an understanding of prevalent pain conditions. *Lancet* **2021**, *397*, 2098–2110. [CrossRef] [PubMed]
6. Devers, A.; Galer, B.S. Topical Lidocaine Patch Relieves a Variety of Neuropathic Pain Conditions: An Open-Label Study. *Clin. J. Pain* **2000**, *16*, 205–208. [CrossRef]
7. Cohen, S.P.; Vase, L.; Hooten, W.M. Chronic pain: An update on burden, best practices, and new advances. *Lancet* **2021**, *397*, 2082–2097. [CrossRef]
8. Ripamonti, C.I.; Bandieri, E.; Roila, F. Management of cancer pain: ESMO clinical practice guidelines. *Ann. Oncol.* **2011**, *22*, vi69–vi77. [CrossRef]
9. Ventafridda, V.; Saita, L.; Ripamonti, C.; De Conno, F. WHO guidelines for the use of analgesics in cancer pain. *Int. J. Tissue React.* **1985**, *7*, 93–96.
10. Carr, D.B.; Goudas, L.C. Acute pain. *Lancet* **1999**, *353*, 2051–2058. [CrossRef]
11. Franco, P.; De Marco, I. Supercritical CO_2 adsorption of non-steroidal anti-inflammatory drugs into biopolymer aerogels. *J. CO2 Util.* **2020**, *36*, 40–53. [CrossRef]
12. Upton, R.N.; Semple, T.J.; Macintyre, P.E. Pharmacokinetic Optimisation of Opioid Treatment in Acute Pain Therapy. *Clin. Pharmacokinet.* **1997**, *33*, 225–244. [CrossRef] [PubMed]
13. Gordon, D.B.; Dahl, J.; Phillips, P.; Frandsen, J.; Cowley, C.; Foster, R.L.; Fine, P.G.; Miaskowski, C.; Fishman, S.; Finley, R.S. The use of "as-needed" range orders for opioid analgesics in the management of acute pain: A consensus statement of the American Society for Pain Management Nursing and the American Pain Society. *Pain Manag. Nurs.* **2004**, *5*, 53–58. [CrossRef] [PubMed]
14. Walder, B.; Schafer, M.; Henzi, I.; Tramer, M. Efficacy and safety of patient-controlled opioid analgesia for acute postoperative pain: A quantitative systematic review. *Acta Anaesthesiol. Scand.* **2001**, *45*, 795–804. [CrossRef] [PubMed]

15. Furlan, A.D.; Sandoval, J.A.; Mailis-Gagnon, A.; Tunks, E. Opioids for chronic noncancer pain: A meta-analysis of effectiveness and side effects. *CMAJ* **2006**, *174*, 1589–1594. [CrossRef]
16. Sarzi-Puttini, P.; Vellucci, R.; Zuccaro, S.M.; Cherubino, P.; Labianca, R.; Fornasari, D. The appropriate treatment of chronic pain. *Clin. Drug Investig.* **2012**, *32*, 21–33. [CrossRef]
17. McCleane, G.J. *Pain Management: Expanding the Pharmacological Options*; Wiley-Blackwell: Hoboken, NJ, USA, 2009; pp. 1–192. ISBN 978140517823. [CrossRef]
18. Manchikanti, L.; Singh, V.; Caraway, D.L.; Benyamin, R.M. Breakthrough pain in chronic non-cancer pain: Fact, fiction, or abuse. *Pain Phys.* **2011**, *14*, E103–E117. [CrossRef]
19. Leonardi, C.; Vellucci, R.; Mammucari, M.; Fanelli, G. Opioid risk addiction in the management of chronic pain in primary care: The addition risk questionnaire. *Eur. Rev. Med. Pharmacol. Sci.* **2015**, *19*, 4898–4905.
20. Fallon, M.; Giusti, R.; Aielli, F.; Hoskin, P.; Rolke, R.; Sharma, M.; Ripamonti, C.I. Management of cancer pain in adult patients: ESMO Clinical Practice Guidelines. *Ann. Oncol.* **2018**, *29*, iv166–iv191. [CrossRef]
21. Volkow, N.; Benveniste, H.; McLellan, A.T. Use and Misuse of Opioids in Chronic Pain. In *Annual Review of Medicine*; Annual Reviews Inc.: San Mateo, CA, USA, 2018; Volume 69, pp. 451–465.
22. Latina, R.; Varrassi, G.; Di Biagio, E.; Giannarelli, D.; Gravante, F.; Paladini, A.; D'Angelo, D.; Iacorossi, L.; Martella, C.; Alvaro, R.; et al. Chronic Non-cancer Pain Management in a Tertiary Pain Clinic Network: A Retrospective Study. *Pain Ther.* **2023**, *12*, 151–164. [CrossRef]
23. Alkilani, A.Z.; McCrudden, M.T.; Donnelly, R.F. Transdermal Drug Delivery: Innovative Pharmaceutical Developments Based on Disruption of the Barrier Properties of the stratum corneum. *Pharmaceutics* **2015**, *7*, 438–470. [CrossRef]
24. Schoellhammer, C.M.; Blankschtein, D.; Langer, R. Skin permeabilization for transdermal drug delivery: Recent advances and future prospects. *Expert Opin. Drug Deliv.* **2014**, *11*, 393–407. [CrossRef] [PubMed]
25. Mathews, L.M.; Roy, A. Management of pain using transdermal patches—A review. *Asian J. Pharm. Clin. Res.* **2016**, *9*, 32–35. [CrossRef]
26. Ciurba, A.; Antonoaea, P.; Todoran, N.; Rédai, E.; Vlad, R.A.; Tătaru, A.; Muntean, D.L.; Bîrsan, M. Polymeric films containing tenoxicam as prospective transdermal drug delivery systems: Design and characterization. *Processes* **2021**, *9*, 136. [CrossRef]
27. Tsai, J.H.; Liu, I.T.; Su, P.F.; Huang, Y.T.; Chiu, G.L.; Chen, Y.Y.; Lai, W.S.; Lin, P.C. Lidocaine transdermal patches reduced pain intensity in neuropathic cancer patients already receiving opioid treatment. *BMC Palliat. Care* **2023**, *22*, 4. [CrossRef] [PubMed]
28. Franco, P.; Pessolano, E.; Belvedere, R.; Petrella, A.; De Marco, I. Supercritical impregnation of mesoglycan into calcium alginate aerogel for wound healing. *J. Supercrit. Fluids* **2020**, *157*, 104711. [CrossRef]
29. Belvedere, R.; Novizio, N.; Morello, S.; Petrella, A. The combination of mesoglycan and VEGF promotes skin wound repair by enhancing the activation of endothelial cells and fibroblasts and their cross-talk. *Sci. Rep.* **2022**, *12*, 11041. [CrossRef]
30. Khandaker, M.; Alkadhem, N.; Progri, H.; Nikfarjam, S.; Jeon, J.; Kotturi, H.; Vaughan, M.B. Glutathione Immobilized Polycaprolactone Nanofiber Mesh as a Dermal Drug Delivery Mechanism for Wound Healing in a Diabetic Patient. *Processes* **2022**, *10*, 512. [CrossRef]
31. Tyeb, S.; Verma, V.; Kumar, N. Polysaccharide based transdermal patches for chronic wound healing: Recent advances and clinical perspective. *Carbohydr. Polym.* **2023**, *316*, 121038. [CrossRef]
32. Jeong, W.Y.; Kwon, M.; Choi, H.E.; Kim, K.S. Recent advances in transdermal drug delivery systems: A review. *Biomater. Res.* **2021**, *25*, 24. [CrossRef]
33. Kaestli, L.Z.; Wasilewski-Rasca, A.F.; Bonnabry, P.; Vogt-Ferrier, N. Use of transdermal drug formulations in the elderly. *Drugs Aging* **2008**, *25*, 269–280. [CrossRef]
34. Tanner, T.; Marks, R. Delivering drugs by the transdermal route: Review and comment. *Ski Res. Technol.* **2008**, *14*, 249–260. [CrossRef] [PubMed]
35. Bird, D.; Ravindra, N.M. Transdermal drug delivery and patches—An overview. *Med. Devices Sens.* **2020**, *3*, e10069. [CrossRef]
36. Sozio, P.; Cerasa, L.S.; Marinelli, L.; Di Stefano, A. Transdermal donepezil on the treatment of Alzheimer's disease. *Neuropsychiatr. Dis. Treat.* **2012**, *8*, 361–368. [CrossRef] [PubMed]
37. He, J.; Zhang, Y.; Yu, X.; Xu, C. Wearable patches for transdermal drug delivery. *Acta Pharm. Sin. B* **2023**, *13*, 2298–2309. [CrossRef] [PubMed]
38. McKenna, P.E.; Abbate, M.T.A.; Vora, L.K.; Sabri, A.H.; Peng, K.; Volpe-Zanutto, F.; Tekko, I.A.; Permana, A.D.; Maguire, C.; Dineen, D.; et al. Polymeric Microarray Patches for Enhanced Transdermal Delivery of the Poorly Soluble Drug Olanzapine. *ACS Appl. Mater. Interfaces* **2023**, *15*, 31300–31319. [CrossRef]
39. Naser, Y.A.; Tekko, I.A.; Vora, L.K.; Peng, K.; Anjani, Q.K.; Greer, B.; Elliott, C.; McCarthy, H.O.; Donnelly, R.F. Hydrogel-forming microarray patches with solid dispersion reservoirs for transdermal long-acting microdepot delivery of a hydrophobic drug. *J. Control. Release* **2023**, *356*, 416–433. [CrossRef]
40. Wermeling, D.P.; Banks, S.L.; Hudson, D.A.; Gill, H.S.; Gupta, J.; Prausnitz, M.R.; Stinchcomb, A.L. Microneedles permit transdermal delivery of a skin-impermeant medication to humans. *Proc. Natl. Acad. Sci. USA* **2008**, *105*, 2058–2063. [CrossRef]
41. Kochhar, J.S.; Lim, W.X.S.; Zou, S.; Foo, W.Y.; Pan, J.; Kang, L. Microneedle integrated transdermal patch for fast onset and sustained delivery of lidocaine. *Mol. Pharm.* **2013**, *10*, 4272–4280. [CrossRef]
42. Gupta, J.; Felner, E.I.; Prausnitz, M.R. Rapid pharmacokinetics of intradermal insulin administered using microneedles in type 1 diabetes subjects. *Diabetes Technol. Ther.* **2011**, *13*, 451–456. [CrossRef]

43. Freynhagen, R.; Von Giesen, H.J.; Busche, P.; Sabatowski, R.; Konrad, C.; Grond, S. Switching from reservoir to matrix systems for the transdermal delivery of fentanyl: A prospective, multicenter pilot study in outpatients with chronic pain. *J. Pain Symptom Manag.* **2005**, *30*, 289–297. [CrossRef]
44. Grape, S.; Schug, S.A.; Lauer, S.; Schug, B.S. Formulations of fentanyl for the management of pain. *Drugs* **2010**, *70*, 57–72. [CrossRef] [PubMed]
45. Jeal, W.; Benfield, P. Transdermal fentanyl—A review of its pharmacological properties and therapeutic efficacy in pain control. *Drugs* **1997**, *53*, 109–138. [CrossRef]
46. Roy, S.D.; Flynn, G.L. Transdermal Delivery of Narcotic Analgesics: pH, Anatomical, and Subject Influences on Cutaneous Permeability of Fentanyl and Sufentanil. *Pharm. Res. Off. J. Am. Assoc. Pharm. Sci.* **1990**, *7*, 842–847. [CrossRef]
47. Gupta, S.K.; Southam, M.; Gale, R.; Hwang, S.S. System functionality and physicochemical model of fentanyl transdermal system. *J. Pain Symptom Manag.* **1992**, *7*, S17–S26. [CrossRef] [PubMed]
48. Takashina, Y.; Naito, T.; Mino, Y.; Yagi, T.; Ohnishi, K.; Kawakami, J. Impact of CYP3A5 and ABCB1 gene polymorphisms on fentanyl pharmacokinetics and clinical responses in cancer patients undergoing conversion to a transdermal system. *Drug Metab. Pharmacokinet.* **2012**, *27*, 414–421. [CrossRef]
49. Prodduturi, S.; Smith, G.J.; Wokovich, A.M.; Doub, W.H.; Westenberger, B.J.; Buhse, L. Reservoir based fentanyl transdermal drug delivery systems: Effect of patch age on drug release and skin permeation. *Pharm. Res.* **2009**, *26*, 1344–1352. [CrossRef]
50. Kim, T.S.; Jin, S.E.; Sun, B.; Kim, M.S.; Hwang, S.J. Comparison of adhesion and dissolution of fentanyl patches: Fentadur® and Durogesic DTrans®. *J. Pharm. Investig.* **2015**, *45*, 475–480. [CrossRef]
51. Miyazaki, T.; Hanaoka, K.; Namiki, A.; Ogawa, S.; Kitajima, T.; Hosokawa, T.; Ishida, T.; Nogami, S.; Mashimo, S. Efficacy, safety and pharmacokinetic study of a novel fentanyl-containing matrix transdermal patch system in Japanese patients with cancer pain. *Clin. Drug Investig.* **2008**, *28*, 313–325. [CrossRef]
52. Prodduturi, S.; Sadrieh, N.; Wokovich, A.M.; Doub, W.H.; Westenberger, B.J.; Buhse, L. Transdermal delivery of fentanyl from matrix and reservoir systems: Effect of heat and compromised skin. *J. Pharm. Sci.* **2010**, *99*, 2357–2366. [CrossRef]
53. Marier, J.F.; Lor, M.; Morin, J.; Roux, L.; Di Marco, M.; Morelli, G.; Sædder, E.A. Comparative bioequivalence study between a novel matrix transdermal delivery system of fentanyl and a commercially available reservoir formulation. *Br. J. Clin. Pharmacol.* **2007**, *63*, 121–124. [CrossRef]
54. Marier, J.F.; Lor, M.; Potvin, D.; DiMarco, M.; Morelli, G.; Sædder, E.A. Pharmacokinetics, tolerability, and performance of a novel matrix transdermal delivery system of fentanyl relative to the commercially available reservoir formulation in healthy subjects. *J. Clin. Pharmacol.* **2006**, *46*, 642–653. [CrossRef] [PubMed]
55. Hoy, S.M.; Keating, G.M. Fentanyl Transdermal Matrix Patch (Durotep® MT Patch; Durogesic® DTrans®; Durogesic® SMAT). *Drugs* **2008**, *68*, 1711–1721. [CrossRef] [PubMed]
56. Hair, P.I.; Keating, G.M.; McKeage, K. Transdermal matrix fentanyl membrane patch (Matrifen®): In severe cancer-related chronic pain. *Drugs* **2008**, *68*, 2001–2009. [CrossRef] [PubMed]
57. Lane, M.E. The transdermal delivery of fentanyl. *Eur. J. Pharm. Biopharm.* **2013**, *84*, 449–455. [CrossRef]
58. Marquardt, K.A.; Tharratt, R.S.; Musallam, N.A. Fentanyl remaining in a transdermal system following three days of continuous use. *Ann. Pharmacother.* **1995**, *29*, 969–971. [CrossRef]
59. Oliveira, G.; Hadgraft, J.; Lane, M.E. Toxicological implications of the delivery of fentanyl from gel extracted from a commercial transdermal reservoir patch. *Toxicol. In Vitro* **2012**, *26*, 645–648. [CrossRef]
60. Kuhlman, J.J., Jr.; McCaulley, R.; Valouch, T.J.; Behonick, G.S. Fentanyl use, misuse, and abuse: A summary of 23 postmortem cases. *J. Anal. Toxicol.* **2003**, *27*, 499–504. [CrossRef] [PubMed]
61. Moore, K.T.; Sathyan, G.; Richarz, U.; Natarajan, J.; Vandenbossche, J. Randomized 5-treatment crossover study to assess the effects of external heat on serum fentanyl concentrations during treatment with transdermal fentanyl systems. *J. Clin. Pharmacol.* **2012**, *52*, 1174–1185. [CrossRef]
62. Zecca, E.; Manzoni, A.; Centurioni, F.; Farina, A.; Bonizzoni, E.; Seiler, D.; Perrone, T.; Caraceni, A. Pharmacokinetic study between a bilayer matrix fentalyl patch and a monolayer matrix fentanyl patch: Single dose administration in healthy volunteers. *Br. J. Clin. Pharmacol.* **2015**, *80*, 110–115. [CrossRef]
63. Defraeye, T.; Bahrami, F.; Ding, L.; Malini, R.I.; Terrier, A.; Rossi, R.M. Predicting Transdermal Fentanyl Delivery Using Mechanistic Simulations for Tailored Therapy. *Front. Pharmacol.* **2020**, *11*, 585393. [CrossRef]
64. Shomaker, T.S.; Zhang, J.; Ashburn, M.A. Assessing the impact of heat on the systemic delivery of fentanyl through the transdermal fentanyl delivery system. *Pain Med.* **2000**, *1*, 225–230. [CrossRef]
65. Luckenbill, K.; Thompson, J.; Middleton, O.; Kloss, J.; Apple, F. Fentanyl postmortem redistribution: Preliminary findings regarding the relationship among femoral blood and liver and heart tissue concentrations. *J. Anal. Toxicol.* **2008**, *32*, 639–643. [CrossRef] [PubMed]
66. Gudin, J.; Fudin, J. A Narrative Pharmacological Review of Buprenorphine: A Unique Opioid for the Treatment of Chronic Pain. *Pain Ther.* **2020**, *9*, 41–54. [CrossRef] [PubMed]
67. Al-Tawil, N.; Odar-Cederlöf, I.; Berggren, A.C.; Johnson, H.E.; Persson, J. Pharmacokinetics of transdermal buprenorphine patch in the elderly. *Eur. J. Clin. Pharmacol.* **2013**, *69*, 143–149. [CrossRef]
68. Plosker, G.L. Buprenorphine 5, 10 and 20 μg/h Transdermal Patch. *Drugs* **2011**, *71*, 2491–2509. [CrossRef]

69. Böhme, K. Buprenorphine in a transdermal therapeutic system—A new option. *Clin. Rheumatol.* **2002**, *21*, S13–S16. [CrossRef] [PubMed]
70. Ehrlich, A.T.; Darcq, E. Recommending buprenorphine for pain management. *Pain Manag.* **2019**, *9*, 13–16. [CrossRef]
71. Wilding, I.R.; Davis, S.S.; Rimoy, G.H.; Rubin, P.; Kurihara-Bergstrom, T.; Tipnis, V.; Berner, B.; Nightingale, J. Pharmacokinetic evaluation of transdermal buprenorphine in man. *Int. J. Pharm.* **1996**, *132*, 81–87. [CrossRef]
72. Sittl, R.; Nuijten, M.; Nautrup, B.P. Changes in the prescribed daily doses of transdermal fentanyl and transdermal buprenorphine during treatment of patients with cancer and noncancer pain in Germany: Results of a retrospective cohort study. *Clin. Ther.* **2005**, *27*, 1022–1031. [CrossRef]
73. Ahn, J.S.; Lin, J.; Ogawa, S.; Yuan, C.; O'Brien, T.; Le, B.H.; Bothwell, A.M.; Moon, H.; Hadjiat, Y.; Ganapathi, A. Transdermal buprenorphine and fentanyl patches in cancer pain: A network systematic review. *J. Pain Res.* **2017**, *10*, 1963–1972. [CrossRef]
74. Evans, H.C.; Easthope, S.E. Transdermal buprenorphine. *Drugs* **2003**, *63*, 1999–2010. [CrossRef] [PubMed]
75. Likar, R.; Lorenz, V.; Korak-Leiter, M.; Kager, I.; Sittl, R. Transdermal Buprenorphine Patches Applied in a 4-Day Regimen Versus a 3-Day Regimen: A Single-Site, Phase III, Randomized, Open-Label, Crossover Comparison. *Clin. Ther.* **2007**, *29*, 1591–1606. [CrossRef] [PubMed]
76. Mythili, N.; Navaneetham, R.; Sham, M.E.; Menon, S.; Kumar, V.; Archana, S. Efficacy of Transdermal Buprenorphine Patch in Postoperative Pain Management in Oral and Maxillofacial Surgery. *J. Maxillofac. Oral Surg.* **2022**, 1–6. [CrossRef]
77. Breivik, H.; Ljosaa, T.M.; Stengaard-Pedersen, K.; Persson, J.; Aro, H.; Villumsen, J.; Tvinnemose, D. A 6-months, randomised, placebo-controlled evaluation of efficacy and tolerability of a low-dose 7-day buprenorphine transdermal patch in osteoarthritis patients naïve to potent opioids. *Scand. J. Pain* **2010**, *1*, 122–141. [CrossRef] [PubMed]
78. Yoon, D.H.; Bin, S.-I.; Chan, S.K.-C.; Chung, C.K.; In, Y.; Kim, H.; Lichauco, J.J.; Mok, C.C.; Moon, Y.-W.; Ng, T.K.-T.; et al. Effectiveness and tolerability of transdermal buprenorphine patches: A multicenter, prospective, open-label study in Asian patients with moderate to severe chronic musculoskeletal pain. *BMC Musculoskelet. Disord.* **2017**, *18*, 337. [CrossRef]
79. Brown, L.; Langer, R. Transdermal delivery of drugs. *Annu. Rev. Med.* **1988**, *39*, 221–229. [CrossRef]
80. James, I.G.; O'Brien, C.M.; McDonald, C.J. A randomized, double-blind, double-dummy comparison of the efficacy and tolerability of low-dose transdermal buprenorphine (BuTrans seven-day patches) with buprenorphine sublingual tablets (Temgesic) in patients with osteoarthritis pain. *J. Pain Symptom Manag.* **2010**, *40*, 266–278. [CrossRef]
81. Karlsson, M.; Berggren, A.C. Efficacy and safety of low-dose transdermal buprenorphine patches (5, 10, and 20 microg/h) versus prolonged-release tramadol tablets (75, 100, 150, and 200 mg) in patients with chronic osteoarthritis pain: A 12-week, randomized, open-label, controlled, parallel-group noninferiority study. *Clin. Ther.* **2009**, *31*, 503–513. [CrossRef]
82. Kim, H.J.; Ahn, H.S.; Nam, Y.; Chang, B.S.; Lee, C.K.; Yeom, J.S. Comparative study of the efficacy of transdermal buprenorphine patches and prolonged-release tramadol tablets for postoperative pain control after spinal fusion surgery: A prospective, randomized controlled non-inferiority trial. *Eur. Spine J.* **2017**, *26*, 2961–2968. [CrossRef]
83. Plosker, G.L.; Lyseng-Williamson, K.A. Buprenorphine 5, 10 and 20 μg/h transdermal patch: A guide to its use in chronic non-malignant pain. *CNS Drugs* **2012**, *26*, 367–373. [CrossRef]

processes

MDPI

Review

Progress of Research on Antioxidants and Carriers for Skin Wound Repair

Xiaoxuan Zhu [1,†], Wenlu Yuan [2,†], Zhuozheng Li [2,†], Yin Lin [2], Wenfang Li [2], Lei Ji [3], Dan Wang [4,*,‡], Huafang Zhang [5,*,‡] and Yingshuai Wang [2,*,‡]

1 School of Clinical Medicine, Weifang Medical University, Weifang 261053, China; 17806274923@163.com
2 Department of Bioscience and Technology, Weifang Medical University, 7166# Baotong West Street, Weifang 261053, China; wenluyuan2002@163.com (W.Y.); 17861282087@163.com (Z.L.); linyin1014@163.com (Y.L.); liwenfang@wfmc.edu.cn (W.L.)
3 Library, Weifang Medical University, 7166# Baotong West Street, Weifang 261053, China; npujeff@hotmail.com
4 Department of Physical Education, School of Foundation Medical, Weifang Medical University, Weifang 261053, China
5 The Department of Health Care/The Third Department of Geriatrics, Weifang People's Hospital, No. 151, Guangwen Street, Kuiwen District, Weifang 261041, China
* Correspondence: wangdan198021@163.com (D.W.); huafangzhang@sina.com (H.Z.); yingshuaiwang1987@163.com (Y.W.)
† These authors contributed equally to this work.
‡ These authors contributed equally to this work.

Abstract: The skin is the first protective barrier of the human body, and oxidative damage is one of the main mechanisms of skin injury. Effective antioxidant therapy plays an important role in skin healing. Therefore, exploring antioxidants and suitable drug delivery methods that can be used for skin injury repair is of great value in regulating skin repair and regeneration and promoting wound healing. Based on this, this paper presents a review of the progress of research on (1) antioxidants and (2) antioxidant carriers for skin repair in order to summarize the research results and provide reference for the subsequent development of new drug-carrier structures and new skin repair strategies.

Keywords: skin damage repair; oxidative damage; antioxidants; antioxidant carriers

Citation: Zhu, X.; Yuan, W.; Li, Z.; Lin, Y.; Li, W.; Ji, L.; Wang, D.; Zhang, H.; Wang, Y. Progress of Research on Antioxidants and Carriers for Skin Wound Repair. *Processes* **2023**, *11*, 2069. https://doi.org/10.3390/pr11072069

Academic Editor: Paolo Trucillo

Received: 16 June 2023
Revised: 7 July 2023
Accepted: 8 July 2023
Published: 11 July 2023

1. Introduction

The skin, as the largest organ of the human body, can resist microbial invasion, maintain body fluid and water balance, regulate body temperature and is an important physical protective barrier against external pressure and injury [1]. After a skin injury, the process of wound healing involves inflammation, epithelial remodeling and tissue remodeling [2]. The inflammatory stage of skin tissue regeneration is the key stage of normal wound healing [3]. However, free radicals and reactive oxygen species (ROS) are released from the wound at this stage [4]. The accumulation of these products increases the level of oxidative stress in the cells, destroys the balance between antioxidants and pro-oxidants, affects redox signal transduction, leads to the cell and tissue damage [5] and causes a series of diseases, such as infection [6], delayed wound healing [7], chronic diseases and cancer [8]. Oxidative stress is caused by the excessive accumulation of ROS or metabolites, and when ROS are excessive, a series of free radical chain reactions will lead to oxidative damage in the cells and tissues, seriously affecting cell activity and wound healing [9]. Therefore, the occurrence of a skin injury is closely related to oxidative stress (Figure 1).

Antioxidants are a kind of substances that can help to capture and neutralize free radicals and delay or prevent the oxidation of other chemical substances so as to eliminate the damage of free radicals to the human body, and different antioxidants have different abilities to scavenge free radicals [10]. Antioxidants can not only prevent the related adverse reactions of oxidation but also are one of the important factors in maintaining and balancing the physiological environment in the body. Using antioxidants to reduce ROS damage is an important research direction in the field of skin damage repair (Figure 2). However, the administration of antioxidants via oral and parenteral routes often requires high doses, repeated doses and long-term treatment, causing adverse effects on human health [2]. Traditional wound dressings are not sufficient to promote hemostasis and adhesion and keep the wound moist [11]. Therefore, in recent years, the development of new drug carriers such as nanofibers and hydrogels has attracted extensive attention. Drug carriers can improve the utilization rate, safety and timeliness of drugs, reduce the frequency of administration, improve the bad odor of drugs and enhance the accuracy of dose administration and release of drugs to targeted tissues and organs [12,13]. Therefore, the selection of suitable antioxidants and their carriers is crucial for the development of novel skin repair materials.

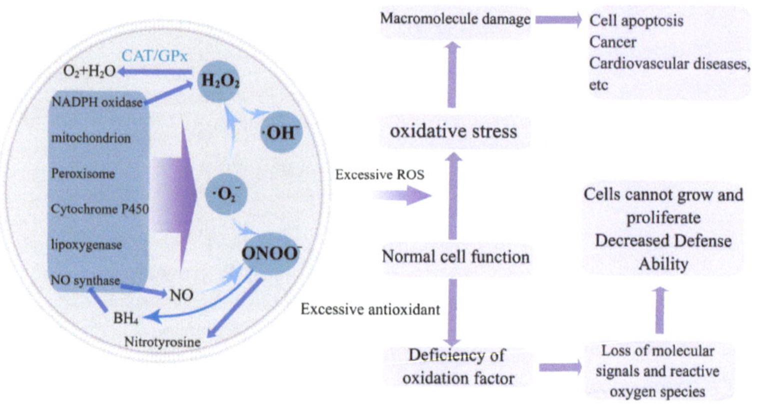

Figure 1. The relationship between the amount of intracellular ROS and cell activity. By Figdraw (www.figdraw.com, accessed on 11 June 2023).

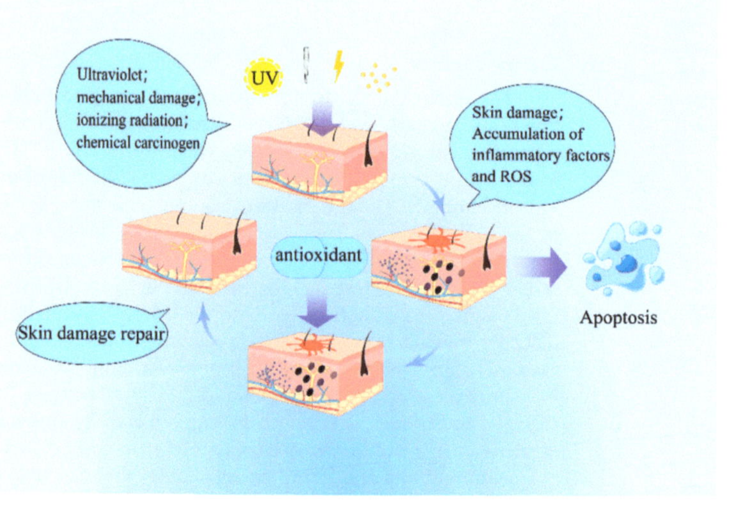

Figure 2. Schematic diagram of antioxidation and skin damage repair. By Figdraw (www.figdraw.com, accessed on 11 June 2023).

Based on the above, this paper discusses the progress of research on antioxidants and their carriers for skin wound repair, which is mainly divided into two parts: (1) antioxidants and (2) antioxidant carriers. In this paper, the sources of antioxidants are classified, and the carriers that can release drugs are initially discussed in order to provide guidance for the development of novel nanoantioxidants and the study of new skin repair systems.

2. Antioxidants

Antioxidants can be roughly divided into two categories: natural antioxidants and artificial antioxidants. Among them, natural antioxidants can be divided into plant antioxidants, animal antioxidants and microbial antioxidants. Artificial antioxidants can be divided into four categories: antioxidants purified by artificial extraction, synthetic antioxidants, nanoantioxidants and gaseous antioxidants. Different types of antioxidants have their own advantages and disadvantages in terms of cost, function and biosafety. Artificial antioxidants are developing in the direction of continuously reducing their side effects and exerting their most appropriate and more effective functions.

2.1. Natural Antioxidants

Natural antioxidants come from a wide range of sources mainly divided into three categories: plants, animals and microorganisms. Among them, the plant source is the most extensive, and the peptides from animals have better activity, while the microbial source may bring new inspiration for future treatments.

2.1.1. Plant Antioxidants

In plants, anthocyanins, vitamins, flavonoids, ascorbic acid, polyphenols, etc., are well known as good antioxidants [14–17]. Fruits, vegetables, spices and herbs are generally rich in carotenoids, phenols, polysaccharides, vitamins, trace elements and other antioxidants, which are safe and effective, natural and nontoxic. They inhibit microbial growth and have antioxidative, anti-inflammatory and anti-aging properties [10] (Table 1).

Table 1. Some plant metabolites with antioxidant activity [15].

Secondary Metabolites	Common Dietary Sources	References
Ascorbic acid	Peppers, strawberries, kiwifruit and citrus fruits	[16]
Polyphenols	Fruits, vegetables, coffee, tea and cereals	[16,18]
Anthocyanins	Strawberries, black rice, berries, cherries, etc.	[16,19]
Flavones	Blueberries, blood orange juice	[17]
Flavonols	Cherries, chokeberries, elderberries, goji berries (wolfberries)	[17]
Resveratrol	Purple wine, peanuts	[20,21]
Theaflavins	Black tea	[19,22]
Carotenoids	Carrots, tomatoes, pumpkins, peppers, among others	[16,23]
Lycopene	Tomatoes, watermelons, red peppers, papayas, apricots, pink grapefruit	[24,25]

Perilla frutescens is rich in antioxidant substances such as polyphenols, which can repair the DNA damage and keratinocytes in the skin exposed to UVB light, and has the effects of antioxidation, anti-inflammation and anti-allergy [26]. Cinnamaldehyde (CIN), the main component of cinnamon, also has strong antioxidant activity [27]. Korean ginseng plays an important role in anti-photodamaging, anti-radiation, anti-inflammatory, antioxidative, anti-aging, anti-melanogenic and wound healing activities [28]. One study [29] found that Curcuma species (Curcuma longa and Curcuma aeruginosa) and curcumin not only influenced the antioxidative and anti-inflammatory process in the production of hyaluronic acid, increased skin moisture and reduced axillary hair growth but also promoted wound healing and prevented chronic UVB injuries. Moreover, Curcuma was also found to improve the symptoms of psoriasis and lesions of radiation dermatitis.

Spirulina contains antioxidants such as chlorophyll a and carotenoids, which play an important role in its antioxidative system [30]. Microalgae contain a variety of antioxidants such as carotenoids, ascorbic acid and glutathione, and these antioxidants play an important role in maintaining the normal life activities of microalgae [31].

Fruits and vegetables that we often eat in our daily life also contain a lot of antioxidants. For example, blueberries are rich in bioactive anthocyanins and have significant antioxidative and ROS-scavenging, anti-inflammatory and antibacterial activities [32–34]. Jujube (*Ziziphus jujuba* Mill.) fruit contains a variety of bioactive substances, such as polysaccharides, polyphenols, amino acids, nucleotides, triterpenoid acids, alkaloids and other nutrients, and has antioxidant and other physiological functions [35]. Cabbage contains glucosinolates, which not only has good antioxidant capacity but also plays an important role in cancer prevention [36]. Lotus root contains phenolic compounds, which have strong antioxidant and scavenging capacities against oxidative factors, so it will become a potential material for extracting antioxidants [37].

2.1.2. Animal Antioxidants

In nature, antioxidants are also common in animals. They typically include antioxidant peptides, astaxanthin and so on.

Antioxidant peptides are typical antioxidants of animal origin. It has been found that a new peptide named OM-LV20, identified in the skin secretion of odorous frog Odorrana margaretae, exhibits antioxidant and strong wound-healing activities in a mouse model of full-thickness skin wounds [38]. Tyrosine-OA1 is also a new type of antioxidant peptide found in amphibians. While having a good antioxidant capacity, it can also promote the healing of injuries and accelerate the recovery of wounded and formation of new epithelial tissues [10]. Astaxanthin is a kind of natural ketocarotenoid, which is widely found in shellfish such as shrimp, crabs and oysters. It has good antioxidant properties and also shows anti-inflammatory, immunoregulatory, DNA-repairing and anti-photodamaging properties [39].

One study [40] showed that subcutaneous injection of hucMSC-ex (exocrine bodies derived from the mesenchymal stem cells of the human umbilical cord) can have an antioxidative and anti-inflammatory effect on ultraviolet-radiation-induced apoptosis and DNA damage and promote skin regeneration and repair. Another study [41] showed that laminin from the cells of the mouse dorsal skin can reduce the concentration of peroxide ions in the cells while exerting the antioxidant power to repair skin damage.

2.1.3. Microbial Antioxidants

Microorganisms are also important sources of antioxidants, such as actinomycetes, bacteria, cyanobacteria, fungi and lichens. They contain various antioxidants in their own complete antioxidant system, which plays an important role in maintaining their normal life activities [10]. In SKH-1 hairless mice, topical application of the supernatant of milk fermented with Lactobacillus helveticus NS8 (NS8-FS) alleviated UVB-induced skin photodamage, including the improvement of the appearance of epidermal thickness, transepidermal water loss and lipid peroxidation levels [42]. Oral administration of Bifidobacterium breve in hairless mice can inhibit the increase in UVB-induced hydrogen peroxide formation, protein oxidation and xanthine oxidase activity in the skin [43]. These studies suggest that exploring the effects of microbiological therapy on skin healing and other antioxidant treatments may lead to new strategies.

2.2. Artificial Antioxidants

2.2.1. Antioxidants Purified by Artificial Extraction

The content of antioxidants in animals and plants is often low. In order to improve the antioxidant effects, antioxidants from animals and plants are often extracted. The extracted product has the characteristics of high safety and strong antioxidant capacity, is easy to be prepared in large quantities and so on. Extraction is the most commonly used method. For example, the extract of watermelon rind is rich in antioxidant substances, which can effectively increase the stability of soybean oil, thereby prolonging its shelf life oil [44].

Ultrasound-assisted extraction (UAE) can be used to extract polyphenolic antioxidants from carotene. Compared with traditional extraction technology, UAE provides higher antioxidant yield and shorter processing time [45].

2.2.2. Synthetic Antioxidants

Artificial synthesis is also one of the common ways to obtain antioxidants. At present, there are mainly butylated hydroxyanisole (BHA), butylated hydroxytoluene (BHT), propyl gallate (PG) and tert-butyl hydroquinone (TBHQ).

TBHQ is a food-grade phenolic antioxidant. It has been found that TBHQ can antagonize a decrease in arsenic-induced cell viability, the production of ROS and lipid peroxidation, as well as a decrease in superoxide-dismutase (SOD) and catalase (CAT) activities, and improve arsenic-induced intracellular damage and apoptosis [46]. BHA is also a phenolic antioxidant, which can reduce DNA damage and lipid peroxidation induced by mechlorethamine (HN2) and reduce the toxicity of HN2 to A-431 skin cells [47]. N-acetylcysteine, as an antioxidant, can also promote wound healing and has been used as an adjuvant to treat a variety of skin conditions, as well as protect against radiation-induced skin damage, including photoaging, photocarcinogenesis and radiation dermatitis [48].

However, compared with natural antioxidants, synthetic antioxidants have many safety risks, such as excessive use can cause teratogenesis, cancer and some chronic diseases. TBHQ may cause skin irritation in people who are allergic to Nigella sativa oil [49] and even in people who are allergic to resins [50]. Adding the antioxidant BHA to the diet of F344 rats resulted in a high incidence of anterior gastric papilloma and squamous cell carcinoma [51]. Propyl gallate is a very effective sunscreen; however, Kahn and Curry [52] reported that it was a strong contact sensitizer. Therefore, the safety evaluation of synthetic antioxidants is very important for their development.

2.2.3. Nanoantioxidants

In recent years, nanoantioxidants have gradually become a new trend. Nanomaterials with antioxidant properties have attracted much attention because of their stable structure, renewable activity and antibacterial and antioxidant activity.

The application of nanoparticles in oxidative stress and inflammation can improve the activities of antioxidant enzymes such as SOD and glutathione peroxidase in organisms. At the same time, they can also remove harmful peroxide ions and free radicals in biological cells [53], showing high medical potential in many aspects [54]. There are a variety of nanoparticles. The nanocomposite films prepared on the basis of chitosan and gelatin show good co-persistence and absorption in wound dressings; they not only have good antioxidant capacity but also the capacity to promote wound recovery [55]. A biodegradable hybrid film composed of green synthesized zinc oxide nanoparticles (ZnONPs) and a chitosan (CS) matrix exhibited remarkable antibacterial and antioxidant properties with strong visible emission and UV-blocking activities in the 480 nm region [56]. Nano-enzyme is a kind of nanomaterial with enzyme-mimicking activity, which has relatively high physical and chemical stability in a harsh environment, higher durability and lower cost compared to natural enzymes [57]. Li et al. [58] prepared $CuCo_2S_4$ nanoenzyme, using solvothermal synthesis, which showed excellent peroxidase-like activity under neutral conditions and was proved to have the capacity to accelerate the healing of pH-neutral burn wounds. Cerium oxide nanoparticles (nanoceria) (Figure 3) [59] are a kind of multifunctional enzymes with biocompatibility and a unique biomimetic activity, which can shield UV radiation, protect tissues from UV-oxidation damage and open the way to a smarter and safer reduction in skin damage and cancer caused by UV rays.

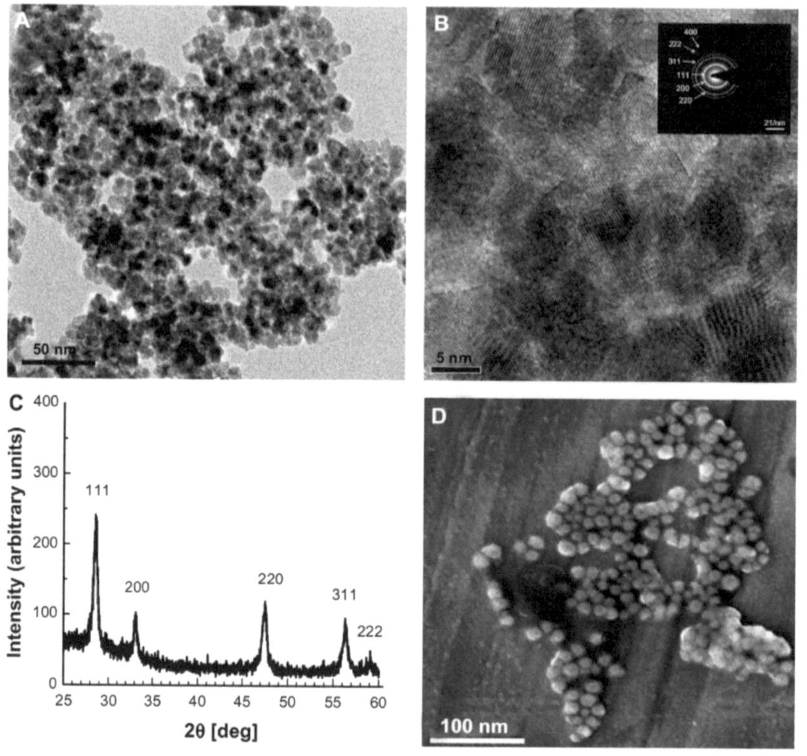

Figure 3. Characterization of nanoceria. (**A**) High-resolution transmission electron microscopy (HR-TEM) photomicrograph. (**B**) HR-TEM micrograph of nanoceria. (**C**) X-ray diffraction of nanoceria showing the characteristic peaks of the crystal. (**D**) Scanning electron microscopy analysis showing the spherical shape and homogenous surface topology of nanoparticles [60].

Silver nanoparticles (AgNPs) has attracted much attention in the field of biomedicine because of their anti-inflammatory, bacteriostatic, antioxidant and wound-healing capacities. One study [61] found that the biocompatible AgNPs synthesized from antioxidant-rich aqueous extract of Aerva javanica had good antioxidant potential, promoted rapid wound healing after adding it to a hydrogel and showed low toxicity in vitro and in vivo. As one of the important metal oxidation nanomaterials, zinc oxide nanoparticles (ZnONPs) are widely used in the biomedical field because of their ultraviolet-absorbing, antibacterial, anti-inflammatory and other properties. The green synthesis of ZnONPs from the aqueous extract of sea lavender showed effective antibacterial and antifungal activities, as well as considerable antioxidant potential [62].

However, many experiments have shown that metal nanoparticles can pass through the skin and enter the body. After the external application of nano-silver dressing on a burn wound, silver nanoparticles can enter the blood through the damaged skin and accumulate in the body to cause liver and kidney toxicity [63]. Moreover, ZnONPs and TiO_2NPs can enhance DNA oxidative damage by increasing the production of ROS, causing concerns about the safety of ZnONPs and TiO_2NPs [64]. Therefore, exploring how to reduce the toxicity of metal nanomaterials and increase their biosafety has become a new development of and strategy for the application of metal nanomaterials in the field of biomedicine.

2.2.4. Gaseous Antioxidants

As an antioxidant gas, H2S can resist oxidative substances such as reactive oxygen species (ROS) and reactive nitrogen species (RNS) and protect skin from oxidative stress. [65]. It has been found that NaHS can promote the migration of skin fibroblasts and keratinocytes by alleviating ROS and increasing mitochondrial membrane potential, indicating that H2S can promote skin repair through antioxidation [66].

Hydrogen has been widely researched in recent years, and more and more studies find that hydrogen has the effects of antioxidation, anti-inflammation, anti-apoptosis and so on. As a small molecular substance, hydrogen can easily pass through membrane structures, quickly reach the damage site, react with the highly oxidizing ROS and scavenge oxidation products such as free radicals [67]. Hydrogen can also enhance the antioxidant capacity of peroxidase, thus reducing the damage of oxidative ions such as ROS [68]. A Japanese study [69] showed that for hospitalized elderly patients with severe stress ulcers, the intake of hydrogen-rich water through tube feeding can achieve wound size reduction and early recovery, which might be related to the construction of type I collagen in skin fibroblasts or the enhancement of mitochondrial reduction and ROS inhibition in epidermal keratinocytes. Another study [70] showed that inhaling H_2 can significantly reduce wound area, 8-oxo-dG level (oxidative DNA damage) and apoptosis rate in skin injuries. Zhao et al. [71] found that hydrogen-rich water can significantly improve the survival of a skin flap after ischemia and reperfusion and reduce the ischemia–reperfusion injury, leukocyte infiltration and the production of lipid peroxides and inflammatory cytokines. Hydrogen has been widely studied and explored, and it plays a variety of physiological roles in skin injury repair.

3. Antioxidant Carriers

Drug carriers can change the way drugs enter and are distributed in the body, control the speed of drug release and deliver drugs to the targeted organ system so as to improve the utilization, safety and effectiveness of drugs. The main carriers of antioxidants used for skin repair are electrospun nanofibers, nanoemulsions, microemulsions, nanoliposomes, hydrogels and so on.

3.1. Electrospun Nanofibers

Among different nanomaterials, nanofibers prepared by electrospinning are widely used in the best finishing materials because of their excellent physical and mechanical properties. By adjusting the arrangement of nanofibers, the controlled release of loaded drugs and superior mechanical properties adapted to limb movements can be realized [72]. Porous nanofiber mats readily produced by the electrospinning process offer a promising solution in the management of wounds [73].

Various polymers can be selected for electrospinning materials according to the specific requirements of damaged skin [72]. Brahatheeswaran et al. [74] prepared fiber membranes by electrospinning zein fibers in TFE solutions containing curcumin at a constant flow rate of 1.0 mL h^{-1} under the electrostatic field strength of 2.0 kV cm^{-1}, which had good cell adhesion and proliferation, strong antioxidation and good sustained-release effects and can be used in wound dressing and as drug carriers. Dan et al. [75] loaded curcumin (Cur) into a polycaprolactone (PCL) core and broad-spectrum antibacterial tetracycline hydrochloride (TH) into a gelatin (GEL) shell and prepared a PCL–Cur/GEL-TH core–shell nanofiber membrane using coaxial electrospinning. The membrane showed good antioxidant activity, excellent antibacterial activity, good water absorption capacity and hydrophilic and mechanical properties and had great potential to promote wound repair (Figure 4). Li et al. [76] successfully created a wound dressing core–shell nanofiber membrane (PGEC: poly(L-Lactic-co-caprolactone) (PLCL), gelatin and epigallocatechin-3-O-gallate (EGCG)) using a coaxial electrospinning technology, which improved the low bioavailability of EGCG in vivo, had excellent biocompatibility, good antibacterial and antioxidant capacities, can slow the release of drugs and promote wound regeneration and healing.

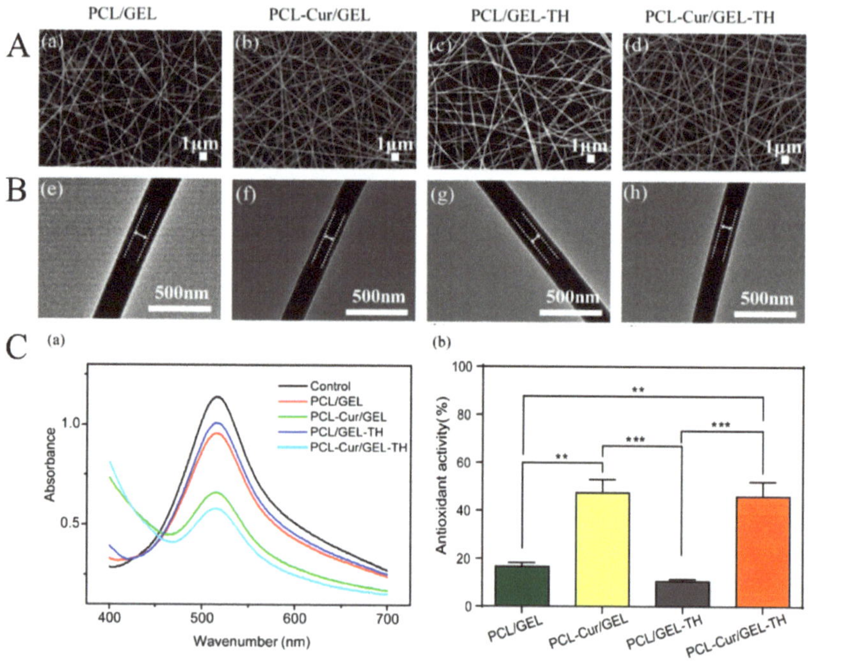

Figure 4. (**A**) (**a–d**) SEM images. (**B**) (**e–h**) TEM images. The white dash line in (**B**) (**e–h**) indicates the border between the PCL core and the GEL shell. (**C**) (**a**) Absorbance of DPPH solution with different nanofibrous membranes; (**C**) (**b**) DPPH-scavenging activities of different nanofiber membranes; ** $p < 0.01$, *** $p < 0.001$ [75].

In summary, based on electrospinning loading technology, nanoparticles and bioactive substances can interact with the electrospun film components and improve the performance of drug delivery.

3.2. Microemulsions and Nanoemulsions

Microemulsions are usually composed of an oil phase, a surfactant, a cosurfactant and an aqueous phase and have optical transparency, thermodynamic stability and low interfacial tension, which can enable the effective dissolution and higher bioavailability of the drug [77]. Resveratrol is a particularly effective antioxidant with anti-inflammatory and anti-proliferative properties [78]. It has been found that a microemulsion composed of sucrose oleate (SO), ethanol, isopropyl myristate (IPM) and water (MESO-E) shows a significant increase in the amount of resveratrol incorporated into the skin, making it an effective delivery carrier of resveratrol for the skin [79]. However, because microemulsions contain a large amounts of surfactants and cosurfactants, they may be irritating to the skin and have not been widely used in skin repair.

Nanoemulsions are colloidal systems formed by mixing oil, emulsifier and water, which are dynamically stable and thermodynamically unstable and are divided into oil-in-water (O/W) and water-in-oil (W/O) systems, of which oil-in-water nanoemulsions are the most commonly used [80]. As traditional drug carriers, nanoemulsions are not only simple to prepare but also can cooperate with drugs to achieve a safer, more stable and more significant drug-use effect. It was found that an optimized curcumin-encapsulated α-tocopherol nanoemulsion system, obtained by adjusting the content of ingredients in the formulation and using a high-speed homogenization technique, can significantly reduce oxidative stress, enhance collagen deposition, prevent wound bacterial contamination and accelerate the process of skin tissue regeneration [81]. Sunflower oil contains vitamin E which can be used as a natural sunscreen to absorb the UVB light. One study found that sunflower oil nanoemulsions prepared by a self-emulsifying method with F3 (Tween 180%, sorbitol 38%), F22 (Tween 280%, sorbitol 36%) and F24 (Tween 380%, sorbitol 34%) and 26% of sunflower oil as sunscreen substances had a higher SPF value and can be considered more effective when used in sunscreen cosmetics [82].

3.3. Nanoliposomes

Nanoliposomes are liposome structures with a particle size smaller than 100 nm. They have the special effect of nanoparticles in terms of stability, absorption and distribution in vivo. They can carry drugs directly to the target tissue to play a pharmacological role. The seeds of moringa tree (*Moringa oleifera* Lam.) are rich in isothiocyanates (MITC), which have excellent antibacterial, anti-inflammatory, antioxidant and antitumor activities [83]. It was found that amphiphilic hyaluronic acid (HA) coupled with ceramides (CEs) was used to modify MITC nanoliposomes by a thin-film hydration method to prepare HACE/MITC nanoparticles (NPs), which can significantly increase the activity of antioxidant enzymes, eliminate ROS induced by UVB and reduce the expression of MMP-1, MMP-3 and MMP-9 caused by radiation-induced photoaging [84]. Anthocyanins are easily degraded under the conditions of high pH value, light, heat and oxygen. A large number of studies have shown that nanoliposomes improve the stability, bioavailability and biological activity of anthocyanins [85]. One study [86] found that a combination of propylene glycol liposomes and silver nanoparticles from grape extract prepared using a new optimized one-step green preparation method with castor seed as a raw material can effectively eliminate several pathogenic microorganisms and dangerous free radicals and protect fibroblasts and keratinocytes against antioxidant stress, thus providing a suitable formula for local treatment of skin injuries.

3.4. Hydrogels

Hydrogels are a new type of dressing to prevent secondary wound injury and infection. Hydrogels attract more and more attention because of their adjustable chemical, physical and biological properties and three-dimensional cross-linked polymer networks that can absorb and retain large amounts of water [87]. One study [88] found that a hydrogel prepared using MnO2 nanosheets (EM) coated with ε-polylysine (EPL) and insulin-loaded self-assembled aldehyde Pluronic F127 (FCHO) micelles showed strong antioxidant activity in the application in chronic wounds. Another study [89] found that a AgNP-PADM hydrogel synthesized by embedding AgNPs into a PADM hydrogel can slowly release AgNPs, which showed sufficient antibacterial and antioxidant properties, negligible toxicity and can promote angiogenesis and cure skin defects. Silver nanoparticles containing chitosan–polyethylene glycol (PEG) pre-polymer solution were synthesized by reducing silver nitrate with a polyethylene glycol and chitosan solution to convert silver ions into silver nanoparticles. A chitosan–PEG hydrogel impregnated with silver nanoparticles, which were formed by crosslinking the obtained pre-polymer solution with glutaraldehyde, also improved the antibacterial and antioxidant properties and promoted the healing of diabetic wounds [90]. Yikun Ren et al. [91] prepared and characterized tyramine-grafted and hyaluronic-acid/gallic-acid-grafted quaternized chitosan (HT/QGA) hydrogels with injection and antioxidant properties, which showed excellent injectability, good antioxidant activity and biocompatibility with great potential application prospects in wound healing (Figure 5).

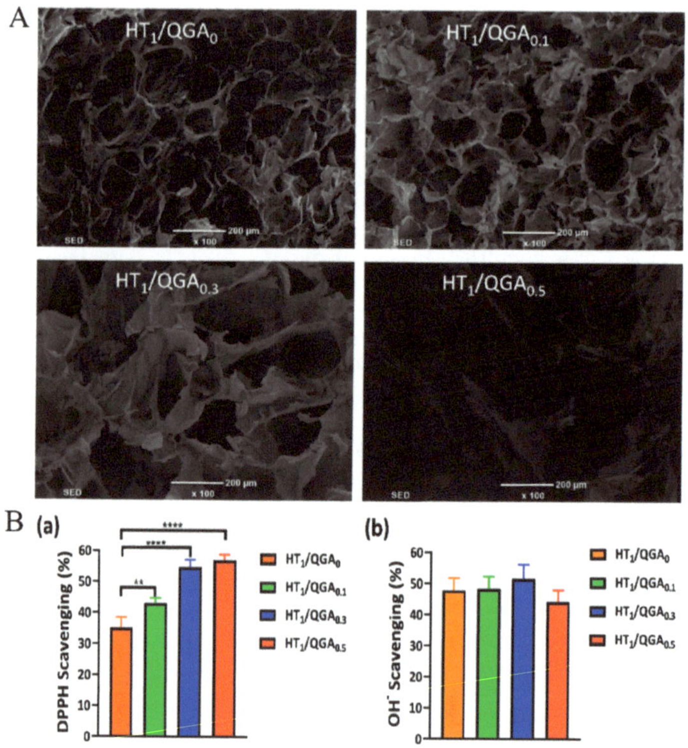

Figure 5. (**A**) The morphology of HT/QGA measured by SEM. (**B**) Antioxidant efficiency of HT/QGA hydrogel: (**a**) DPPH-radical- and (**b**) hydroxyl-radical-scavenging activities in hydrogel; ** $p < 0.01$, **** $p < 0.0001$, mean ± SD, $n = 3$ [91].

4. Conclusions

Many chronic diseases of human body are closely related to the imbalance of the reduction–oxidation reaction. Rational utilization and development of new antioxidants play an important role in human health. In this paper, the sources of antioxidants and the carriers that can realize drug release are summarized and analyzed in order to provide guidance for the development of new nanoantioxidants and research on new skin repair systems.

Natural antioxidants, a natural resource, have been proved to have strong antioxidant effects. Artificial antioxidants are also widely used because of their strong antioxidant capacity and easy mass preparation, especially the nanomaterial with antioxidant properties. They have attracted much attention because of their stable structure and renewable activity and gradually become a new trend. Gas antioxidants have also been found to play a variety of physiological roles in the repair of skin injuries. Drug carriers such as electrospun nanofibers, nanoemulsions, microemulsions, nanoliposomes and hydrogels have also been explored to improve the utilization, safety and effectiveness of antioxidants. However, there are still some deficiencies in the exploration of antioxidants, and more safety data are still needed in order to provide the experimental basis and confidence for the subsequent study of new strategies for skin damage repair and the promotion and application of new drug-carrier structures. Future studies should strive to find the most suitable antioxidants and their carriers in various fields and maximize their advantages in promoting the repair of skin damage and maintaining human health.

Author Contributions: Conceptualization, writing—original draft preparation, X.Z., W.Y. and Z.L.; investigation, resources, Y.L., W.L. and L.J.; writing—review and editing, supervision, D.W., H.Z. and Y.W.; funding acquisition, Y.W. All authors have read and agreed to the published version of the manuscript.

Funding: Natural Science Foundation of Shandong Province (ZR2022QC087) and National Natural Science Foundation of China (82202317).

Data Availability Statement: Data are contained within the article.

Conflicts of Interest: The authors declare no conflict of interest.

References

1. Wells, A.; Nuschke, A.; Yates, C.C. Skin tissue repair: Matrix microenvironmental influences. *Matrix Biol. J. Int. Soc. Matrix Biol.* **2016**, *49*, 25–36. [CrossRef]
2. Franco, P.; Pessolano, E.; Belvedere, R.; Petrella, A.; De Marco, I. Supercritical impregnation of mesoglycan into calcium alginate aerogel for wound healing. *J. Supercrit. Fluids* **2020**, *157*, 104711. [CrossRef]
3. Ross, R.; Benditt, E.P. Wound healing and collagen formation. II. Fine structure in experimental scurvy. *J. Cell Biol.* **1962**, *12*, 533–551. [CrossRef] [PubMed]
4. Yan, T.; Kong, S.; Ouyang, Q.; Li, C.; Hou, T.; Chen, Y.; Li, S. Chitosan-Gentamicin Conjugate Hydrogel Promoting Skin Scald Repair. *Mar. Drugs* **2020**, *18*, 233. [CrossRef] [PubMed]
5. de Jager, T.L.; Cockrell, A.E.; Du Plessis, S.S. Ultraviolet Light Induced Generation of Reactive Oxygen Species. *Adv. Exp. Med. Biol.* **2017**, *996*, 15–23. [CrossRef] [PubMed]
6. Rashaan, Z.M.; Krijnen, P.; Kwa, K.A.A.; van der Vlies, C.H.; Schipper, I.B.; Breederveld, R.S. Flaminal® versus Flamazine® in the treatment of partial thickness burns: A randomized controlled trial on clinical effectiveness and scar quality (FLAM study). *Wound Repair Regen.* **2019**, *27*, 257–267. [CrossRef] [PubMed]
7. Patel, S.; Srivastava, S.; Singh, M.R.; Singh, D. Preparation and optimization of chitosan-gelatin films for sustained delivery of lupeol for wound healing. *Int. J. Biol. Macromol.* **2018**, *107*, 1888–1897. [CrossRef] [PubMed]
8. da Silva, C.R.; Oliveira, M.B.; Motta, E.S.; de Almeida, G.S.; Varanda, L.L.; de Pádula, M.; Leitão, A.C.; Caldeira-de-Araújo, A. Genotoxic and Cytotoxic Safety Evaluation of Papain (*Carica papaya* L.) Using In Vitro Assays. *J. Biomed. Biotechnol.* **2010**, *2010*, 197898. [CrossRef]
9. Riley, P.A. Free radicals in biology: Oxidative stress and the effects of ionizing radiation. *Int. J. Radiat. Biol.* **1994**, *65*, 27–33. [CrossRef]
10. Flieger, J.; Flieger, W.; Baj, J.; Maciejewski, R. Antioxidants: Classification, Natural Sources, Activity/Capacity Measurements, and Usefulness for the Synthesis of Nanoparticles. *Materials* **2021**, *14*, 4135. [CrossRef]
11. Pilehvar-Soltanahmadi, Y.; Dadashpour, M.; Mohajeri, A.; Fattahi, A.; Sheervalilou, R.; Zarghami, N. An Overview on Application of Natural Substances Incorporated with Electrospun Nanofibrous Scaffolds to Development of Innovative Wound Dressings. *Mini Rev. Med. Chem.* **2018**, *18*, 414–427. [CrossRef] [PubMed]
12. Enzhong, Z.; Tao, W.; Cuixia, L.; Weixiang, S.; Xinxing, L.; Zhen, T. Robust and thermo-response graphene–PNIPAm hybrid hydrogels reinforced by hectorite clay. *Carbon* **2013**, *62*, 117–126. [CrossRef]
13. Wei, X.; Lv, X.; Zhao, Q.; Qiu, L. Thermosensitive beta-cyclodextrin modified poly(epsilon-caprolactone)-poly(ethylene glycol)-poly(epsilon-caprolactone) micelles prolong the anti-inflammatory effect of indomethacin following local injection. *Acta Biomater* **2013**, *9*, 6953–6963. [CrossRef]
14. Rao, D.; Hu, Y.; Zhao, R.; Li, H.; Chun, Z.; Zheng, S. Quantitative Identification of Antioxidant Basis for Dendrobium Nobile Flower by High Performance Liquid Chromatography-Tandem Mass Spectrometry. *Int. J. Anal. Chem.* **2022**, *2022*, 9510598. [CrossRef] [PubMed]
15. Forni, C.; Facchiano, F.; Bartoli, M.; Pieretti, S.; Facchiano, A.; D'Arcangelo, D.; Norelli, S.; Valle, G.; Nisini, R., Beninati, S.; et al. Beneficial Role of Phytochemicals on Oxidative Stress and Age-Related Diseases. *BioMed Res. Int.* **2019**, *2019*, 8748253. [CrossRef] [PubMed]
16. Manganaris, G.A.; Goulas, V.; Mellidou, I.; Drogoudi, P. Antioxidant Phytochemicals in Fresh Produce: Exploitation of Genotype Variation and Advancements in Analytical Protocols. *Front. Chem.* **2017**, *5*, 95. [CrossRef]
17. Bhagwat, S.; Haytowitz, D.B.; Holden, J.M.U.S.; Department of Agriculture Agricultural Research Service. USDA Database for the Flavonoid Content of Selected Foods. Release 3.1.; 2014; pp. 1–173. Available online: http://www.ars.usda.gov/nutrientdata (accessed on 15 June 2023).
18. Neto, C.C. Cranberry and blueberry: Evidence for protective effects against cancer and vascular diseases. *Mol. Nutr. Food Res.* **2007**, *51*, 652–664. [CrossRef]
19. Peng, C.; Wang, X.; Chen, J.; Jiao, R.; Wang, L.; Li, Y.M.; Zuo, Y.; Liu, Y.; Lei, L.; Ma, K.Y.; et al. Biology of ageing and role of dietary antioxidants. *BioMed Res. Int.* **2014**, *2014*, 831841. [CrossRef]
20. Santamaria, A.R.; Mulinacci, N.; Valletta, A.; Innocenti, M.; Pasqua, G. Effects of elicitors on the production of resveratrol and viniferins in cell cultures of Vitis vinifera L. cv Italia. *J. Agric. Food Chem.* **2011**, *59*, 9094–9101. [CrossRef]

21. Pervaiz, S.; Holme, A.L. Resveratrol: Its biologic targets and functional activity. *Antioxid. Redox Signal.* **2009**, *11*, 2851–2897. [CrossRef]
22. Takemoto, M.; Takemoto, H. Synthesis of Theaflavins and Their Functions. *Molecules* **2018**, *23*, 918. [CrossRef]
23. Ranard, K.M.; Jeon, S.; Mohn, E.S.; Griffiths, J.C.; Johnson, E.J.; Erdman, J.W., Jr. Dietary guidance for lutein: Consideration for intake recommendations is scientifically supported. *Eur. J. Nutr.* **2017**, *56*, 37–42. [CrossRef]
24. Wood, L.G.; Garg, M.L.; Smart, J.M.; Scott, H.A.; Barker, D.; Gibson, P.G. Manipulating antioxidant intake in asthma: A randomized controlled trial. *Am. J. Clin. Nutr.* **2012**, *96*, 534–543. [CrossRef]
25. Liu, Z.; Ren, Z.; Zhang, J.; Chuang, C.C.; Kandaswamy, E.; Zhou, T.; Zuo, L. Role of ROS and Nutritional Antioxidants in Human Diseases. *Front. Physiol.* **2018**, *9*, 477. [CrossRef] [PubMed]
26. Lee, H.; Park, E. Perilla frutescens Extracts Enhance DNA Repair Response in UVB Damaged HaCaT Cells. *Nutrients* **2021**, *13*, 1263. [CrossRef]
27. Tanaka, Y.; Uchi, H.; Furue, M. Antioxidant cinnamaldehyde attenuates UVB-induced photoaging. *J. Dermatol. Sci.* **2019**, *96*, 151–158. [CrossRef] [PubMed]
28. You, L.; Cho, J.Y. The regulatory role of Korean ginseng in skin cells. *J. Ginseng Res.* **2021**, *45*, 363–370. [CrossRef] [PubMed]
29. Barbalho, S.M.; de Sousa Gonzaga, H.F.; de Souza, G.A.; de Alvares Goulart, R.; de Sousa Gonzaga, M.L.; de Alvarez Rezende, B. Dermatological effects of Curcuma species: A systematic review. *Clin. Exp. Dermatol.* **2021**, *46*, 825–833. [CrossRef]
30. Zhou, J.; Wang, M.; Carrillo, C.; Zhu, Z.; Brncic, M.; Berrada, H.; Barba, F.J. Impact of Pressurized Liquid Extraction and pH on Protein Yield, Changes in Molecular Size Distribution and Antioxidant Compounds Recovery from Spirulina. *Foods* **2021**, *10*, 2153. [CrossRef]
31. Tamaki, S.; Mochida, K.; Suzuki, K. Diverse Biosynthetic Pathways and Protective Functions against Environmental Stress of Antioxidants in Microalgae. *Plants* **2021**, *10*, 1250. [CrossRef]
32. He, J.; Giusti, M.M. Anthocyanins: Natural colorants with health-promoting properties. *Annu. Rev. Food Sci. Technol.* **2010**, *1*, 163–187. [CrossRef] [PubMed]
33. Sroka, Z.; Cisowski, W. Hydrogen peroxide scavenging, antioxidant and anti-radical activity of some phenolic acids. *Food Chem. Toxicol. Int. J. Publ. Br. Ind. Biol. Res. Assoc.* **2003**, *41*, 753–758. [CrossRef] [PubMed]
34. Mazza, G.; Kay, C.D.; Cottrell, T.; Holub, B.J. Absorption of anthocyanins from blueberries and serum antioxidant status in human subjects. *J. Agric. Food Chem.* **2002**, *50*, 7731–7737. [CrossRef] [PubMed]
35. Lu, Y.; Bao, T.; Mo, J.; Ni, J.; Chen, W. Research advances in bioactive components and health benefits of jujube (Ziziphus jujuba Mill.) fruit. *J. Zhejiang Univ. Sci. B* **2021**, *22*, 431–449. [CrossRef]
36. Kapusta-Duch, J.; Kusznierewicz, B. Young Shoots of White and Red Headed Cabbages Like Novel Sources of Glucosinolates as Well as Antioxidative Substances. *Antioxidants* **2021**, *10*, 1277. [CrossRef]
37. Min, H.; Leif, H.S. Antioxidative capacity of rhizome extract and rhizome knot extract of edible lotus (Nelumbo nuficera). *Food Chem.* **2002**, *76*, 327–333. [CrossRef]
38. Li, X.; Wang, Y.; Zou, Z.; Yang, M.; Wu, C.; Su, Y.; Tang, J.; Yang, X. OM-LV20, a novel peptide from odorous frog skin, accelerates wound healing in vitro and in vivo. *Chem. Biol. Drug Des.* **2018**, *91*, 126–136. [CrossRef]
39. Ng, Q.X.; De Deyn, M.; Loke, W.; Foo, N.X.; Chan, H.W.; Yeo, W.S. Effects of Astaxanthin Supplementation on Skin Health: A Systematic Review of Clinical Studies. *J. Diet. Suppl.* **2021**, *18*, 169–182. [CrossRef]
40. Wu, P.; Zhang, B.; Han, X.; Sun, Y.; Sun, Z.; Li, L.; Zhou, X.; Jin, Q.; Fu, P.; Xu, W.; et al. HucMSC exosome-delivered 14-3-3ζ alleviates ultraviolet radiation-induced photodamage via SIRT1 pathway modulation. *Aging* **2021**, *13*, 11542–11563. [CrossRef]
41. Ahn, J.H.; Kim, D.W.; Park, C.W.; Kim, B.; Sim, H.; Kim, H.S.; Lee, T.K.; Lee, J.C.; Yang, G.E.; Her, Y.; et al. Laminarin Attenuates Ultraviolet-Induced Skin Damage by Reducing Superoxide Anion Levels and Increasing Endogenous Antioxidants in the Dorsal Skin of Mice. *Mar. Drugs* **2020**, *18*, 345. [CrossRef] [PubMed]
42. Rong, J.; Shan, C.; Liu, S.; Zheng, H.; Liu, C.; Liu, M.; Jin, F.; Wang, L. Skin resistance to UVB-induced oxidative stress and hyperpigmentation by the topical use of Lactobacillus helveticus NS8-fermented milk supernatant. *J. Appl. Microbiol.* **2017**, *123*, 511–523. [CrossRef] [PubMed]
43. Ishii, Y.; Sugimoto, S.; Izawa, N.; Sone, T.; Chiba, K.; Miyazaki, K. Oral administration of Bifidobacterium breve attenuates UV-induced barrier perturbation and oxidative stress in hairless mice skin. *Arch. Dermatol. Res.* **2014**, *306*, 467–473. [CrossRef] [PubMed]
44. Farooq, M.; Azadfar, E.; Trif, M.; Jabaleh, R.A.; Rusu, A.; Bahrami, Z.; Sharifi, M.; Bangar, S.P.; Ilyas, N.; Ştefănescu, B.E.; et al. Soybean Oil Enriched with Antioxidants Extracted from Watermelon (Citrullus colocynthis) Skin Sap and Coated in Hydrogel Beads via Ionotropic Gelation. *Coatings* **2021**, *11*, 1370. [CrossRef]
45. Christou, A.; Stavrou, I.J.; Kapnissi-Christodoulou, C.P. Continuous and pulsed ultrasound-assisted extraction of carob's antioxidants: Processing parameters optimization and identification of polyphenolic composition. *Ultrason. Sonochem.* **2021**, *76*, 105630. [CrossRef]
46. Duan, X.; Li, J.; Li, W.; Xing, X.; Zhang, Y.; Li, W.; Zhao, L.; Sun, G.; Gao, X.H.; Li, B. Antioxidant tert-butylhydroquinone ameliorates arsenic-induced intracellular damages and apoptosis through induction of Nrf2-dependent antioxidant responses as well as stabilization of anti-apoptotic factor Bcl-2 in human keratinocytes. *Free Radic. Biol. Med.* **2016**, *94*, 74–87. [CrossRef]
47. Pino, M.A.; Billack, B. Reduction of vesicant toxicity by butylated hydroxyanisole in A-431 skin cells. *Cutan. Ocul. Toxicol.* **2008**, *27*, 161–172. [CrossRef]

48. Adil, M.; Amin, S.S.; Mohtashim, M. N-acetylcysteine in dermatology. *Indian J. Dermatol. Venereol. Leprol.* **2018**, *84*, 652–659. [CrossRef]
49. Seiller, H.; Kurihara, F.; Chasset, F.; Soria, A.; Barbaud, A. Tert-butylhydroquinone is a marker for sensitivity to Nigella sativa oil allergy: Five new cases. *Contact Dermat.* **2021**, *84*, 447–449. [CrossRef]
50. Zimerson, E.; Bruze, M.; Goossens, A. Simultaneous p-tert-butylphenol-formaldehyde resin and p-tert-butylcatechol contact allergies in man and sensitizing capacities of p-tert-butylphenol and p-tert-butylcatechol in guinea pigs. *J. Occup. Environ. Med.* **1999**, *41*, 23–28. [CrossRef]
51. Ito, N.; Fukushima, S.; Tsuda, H. Carcinogenicity and modification of the carcinogenic response by BHA, BHT, and other antioxidants. *Crit. Rev. Toxicol.* **1985**, *15*, 109–150. [CrossRef]
52. Kahn, G.; Curry, M.C. Ultraviolet light protection by several new compounds. *Arch. Dermatol.* **1974**, *109*, 510–517. [CrossRef] [PubMed]
53. Zhang, W.; Chen, L.; Xiong, Y.; Panayi, A.C.; Abududilibaier, A.; Hu, Y.; Yu, C.; Zhou, W.; Sun, Y.; Liu, M.; et al. Antioxidant Therapy and Antioxidant-Related Bionanomaterials in Diabetic Wound Healing. *Front. Bioeng. Biotechnol.* **2021**, *9*, 707479. [CrossRef] [PubMed]
54. Singh, T.A.; Sharma, A.; Tejwan, N.; Ghosh, N.; Das, J.; Sil, P.C. A state of the art review on the synthesis, antibacterial, antioxidant, antidiabetic and tissue regeneration activities of zinc oxide nanoparticles. *Adv. Colloid Interface Sci.* **2021**, *295*, 102495. [CrossRef]
55. Hajji, S.; Ktari, N.; Ben Salah, R.; Boufi, S.; Debeaufort, F.; Nasri, M. Development of Nanocomposite Films Based on Chitosan and Gelatin Loaded with Chitosan-Tripolyphosphate Nanoparticles: Antioxidant Potentials and Applications in Wound Healing. *J. Polym. Environ.* **2021**, *30*, 833–854. [CrossRef]
56. Alamdari, S.; Mirzaee, O.; Nasiri Jahroodi, F.; Tafreshi, M.J.; Ghamsari, M.S.; Shik, S.S.; Ara, M.H.M.; Lee, K.Y.; Park, H.H. Green synthesis of multifunctional ZnO/chitosan nanocomposite film using wild Mentha pulegium extract for packaging applications. *Surf. Interfaces* **2022**, *34*, 102349. [CrossRef] [PubMed]
57. Ren, X.; Chen, D.; Wang, Y.; Li, H.; Zhang, Y.; Chen, H.; Li, X.; Huo, M. Nanozymes-recent development and biomedical applications. *J. Nanobiotechnol.* **2022**, *20*, 92. [CrossRef]
58. Li, D.; Guo, Q.; Ding, L.; Zhang, W.; Cheng, L.; Wang, Y.; Xu, Z.; Wang, H.; Gao, L. Bimetallic CuCo(2) S(4) Nanozymes with Enhanced Peroxidase Activity at Neutral pH for Combating Burn Infections. *ChemBioChem* **2020**, *21*, 2620–2627. [CrossRef]
59. Corsi, F.; Di Meo, E.; Lulli, D.; Deidda Tarquini, G.; Capradossi, F.; Bruni, E.; Pelliccia, A.; Traversa, E.; Dellambra, E.; Failla, C.M.; et al. Safe-Shields: Basal and Anti-UV Protection of Human Keratinocytes by Redox-Active Cerium Oxide Nanoparticles Prevents UVB-Induced Mutagenesis. *Antioxidants* **2023**, *12*, 757. [CrossRef]
60. Arya, A.; Sethy, N.K.; Singh, S.K.; Das, M.; Bhargava, K. Cerium oxide nanoparticles protect rodent lungs from hypobaric hypoxia-induced oxidative stress and inflammation. *Int. J. Nanomed.* **2013**, *8*, 4507–4520. [CrossRef]
61. Muhammad Uzair, H.; Faria, K.; Nauman, K.; Asad Abdullah, S.; Aqib, J.; Tehseen, A.; Nasir, J.; Muhammad Qasim, H.; Shah Rukh, A.; Hussnain Ahmed, J. Hydrogels incorporated with silver nanocolloids prepared from antioxidant rich Aerva javanica as disruptive agents against burn wound infections. *Colloids Surf. A Physicochem. Eng. Asp.* **2017**, *529*, 475–486. [CrossRef]
62. Naiel, B.; Fawzy, M.; Halmy, M.W.A.; Mahmoud, A.E.D. Green synthesis of zinc oxide nanoparticles using Sea Lavender (*Limonium pruinosum* L. Chaz.) extract: Characterization, evaluation of anti-skin cancer, antimicrobial and antioxidant potentials. *Sci. Rep.* **2022**, *12*, 20370. [CrossRef] [PubMed]
63. Holsapple, M.P.; Farland, W.H.; Landry, T.D.; Monteiro-Riviere, N.A.; Carter, J.M.; Walker, N.J.; Thomas, K.V. Research strategies for safety evaluation of nanomaterials, part II: Toxicological and safety evaluation of nanomaterials, current challenges and data needs. *Toxicol. Sci.* **2005**, *88*, 12–17. [CrossRef] [PubMed]
64. Tyagi, N.; Srivastava, S.K.; Arora, S.; Omar, Y.; Ijaz, Z.M.; Al-Ghadhban, A.; Deshmukh, S.K.; Carter, J.E.; Singh, A.P.; Singh, S. Comparative analysis of the relative potential of silver, Zinc-oxide and titanium-dioxide nanoparticles against UVB-induced DNA damage for the prevention of skin carcinogenesis. *Cancer Lett.* **2016**, *383*, 53–61. [CrossRef] [PubMed]
65. Xiao, Q.; Xiong, L.; Tang, J.; Li, L.; Li, L. Hydrogen Sulfide in Skin Diseases: A Novel Mediator and Therapeutic Target. *Oxidative Med. Cell. Longev.* **2021**, *2021*, 6652086. [CrossRef] [PubMed]
66. Xu, M.; Hua, Y.; Qi, Y.; Meng, G.; Yang, S. Exogenous hydrogen sulphide supplement accelerates skin wound healing via oxidative stress inhibition and vascular endothelial growth factor enhancement. *Exp. Dermatol.* **2019**, *28*, 776–785. [CrossRef]
67. Ohsawa, I.; Ishikawa, M.; Takahashi, K.; Watanabe, M.; Nishimaki, K.; Yamagata, K.; Katsura, K.; Katayama, Y.; Asoh, S.; Ohta, S. Hydrogen acts as a therapeutic antioxidant by selectively reducing cytotoxic oxygen radicals. *Nat. Med.* **2007**, *13*, 688–694. [CrossRef]
68. Keith, J.M.; Hall, M.B. Potential hydrogen bottleneck in nickel-iron hydrogenase. *Inorg. Chem.* **2010**, *49*, 6378–6380. [CrossRef]
69. Li, Q.; Kato, S.; Matsuoka, D.; Tanaka, H.; Miwa, N. Hydrogen water intake via tube-feeding for patients with pressure ulcer and its reconstructive effects on normal human skin cells in vitro. *Med. Gas Res.* **2013**, *3*, 20. [CrossRef]
70. Fang, W.; Wang, G.; Tang, L.; Su, H.; Chen, H.; Liao, W.; Xu, J. Hydrogen gas inhalation protects against cutaneous ischaemia/reperfusion injury in a mouse model of pressure ulcer. *J. Cell. Mol. Med.* **2018**, *22*, 4243–4252. [CrossRef]
71. Zhao, L.; Wang, Y.B.; Qin, S.R.; Ma, X.M.; Sun, X.J.; Wang, M.L.; Zhong, R.G. Protective effect of hydrogen-rich saline on ischemia/reperfusion injury in rat skin flap. *J. Zhejiang University. Science. B* **2013**, *14*, 382–391. [CrossRef]
72. Guo, S.; Wang, P.; Song, P.; Li, N. Electrospinning of botanicals for skin wound healing. *Front. Bioeng. Biotechnol.* **2022**, *10*, 1006129. [CrossRef]

73. Rieger, K.A.; Birch, N.P.; Schiffman, J.D. Designing electrospun nanofiber mats to promote wound healing—A review. *J. Mater. Chem. B* **2013**, *1*, 4531–4541. [CrossRef]
74. Brahatheeswaran, D.; Mathew, A.; Aswathy, R.G.; Nagaoka, Y.; Venugopal, K.; Yoshida, Y.; Maekawa, T.; Sakthikumar, D. Hybrid fluorescent curcumin loaded zein electrospun nanofibrous scaffold for biomedical applications. *Biomed. Mater. Bristol Engl.* **2012**, *7*, 045001. [CrossRef]
75. Xia, D.; Liu, Y.; Cao, W.; Gao, J.; Wang, D.; Lin, M.; Liang, C.; Li, N.; Xu, R. Dual-Functional Nanofibrous Patches for Accelerating Wound Healing. *Int. J. Mol. Sci.* **2022**, *23*, 10983. [CrossRef] [PubMed]
76. Li, A.; Li, L.; Zhao, B.; Li, X.; Liang, W.; Lang, M.; Cheng, B.; Li, J. Antibacterial, antioxidant and anti-inflammatory PLCL/gelatin nanofiber membranes to promote wound healing. *Int. J. Biol. Macromol.* **2022**, *194*, 914–923. [CrossRef] [PubMed]
77. Ait-Touchente, Z.; Zine, N.; Jaffrezic-Renault, N.; Errachid, A.; Lebaz, N.; Fessi, H.; Elaissari, A. Exploring the Versatility of Microemulsions in Cutaneous Drug Delivery: Opportunities and Challenges. *Nanomaterials* **2023**, *13*, 1688. [CrossRef] [PubMed]
78. Udenigwe, C.C.; Ramprasath, V.R.; Aluko, R.E.; Jones, P.J. Potential of resveratrol in anticancer and anti-inflammatory therapy. *Nutr. Rev.* **2008**, *66*, 445–454. [CrossRef] [PubMed]
79. Yutani, R.; Komori, Y.; Takeuchi, A.; Teraoka, R.; Kitagawa, S. Prominent efficiency in skin delivery of resveratrol by novel sucrose oleate microemulsion. *J. Pharm. Pharmacol.* **2016**, *68*, 46–55. [CrossRef] [PubMed]
80. McClements, D.J. Emulsion design to improve the delivery of functional lipophilic components. *Annu. Rev. Food Sci. Technol.* **2010**, *1*, 241–269. [CrossRef]
81. Ali, M.; Khan, N.R.; Subhan, Z.; Mehmood, S.; Amin, A.; Rabbani, I.; Rehman, F.U.; Basit, H.M.; Syed, H.K.; Khan, I.U.; et al. Novel Curcumin-Encapsulated α-Tocopherol Nanoemulsion System and Its Potential Application for Wound Healing in Diabetic Animals. *BioMed Res. Int.* **2022**, *2022*, 7669255. [CrossRef]
82. Arianto, A.; Cindy, C. Preparation and Evaluation of Sunflower Oil Nanoemulsion as a Sunscreen. *Open Access Maced. J. Med. Sci.* **2019**, *7*, 3757–3761. [CrossRef] [PubMed]
83. Tumer, T.B.; Rojas-Silva, P.; Poulev, A.; Raskin, I.; Waterman, C. Direct and indirect antioxidant activity of polyphenol- and isothiocyanate-enriched fractions from Moringa oleifera. *J. Agric. Food Chem.* **2015**, *63*, 1505–1513. [CrossRef]
84. Wang, Y.; Ouyang, Q.; Chang, X.; Yang, M.; He, J.; Tian, Y.; Sheng, J. Anti-photoaging effects of flexible nanoliposomes encapsulated Moringa oleifera Lam. isothiocyanate in UVB-induced cell damage in HaCaT cells. *Drug Deliv.* **2022**, *29*, 871–881. [CrossRef] [PubMed]
85. Chen, B.H.; Stephen Inbaraj, B. Nanoemulsion and Nanoliposome Based Strategies for Improving Anthocyanin Stability and Bioavailability. *Nutrients* **2019**, *11*, 1052. [CrossRef] [PubMed]
86. Castangia, I.; Marongiu, F.; Manca, M.L.; Pompei, R.; Angius, F.; Ardu, A.; Fadda, A.M.; Manconi, M.; Ennas, G. Combination of grape extract-silver nanoparticles and liposomes: A totally green approach. *Eur. J. Pharm. Sci.* **2017**, *97*, 62–69. [CrossRef] [PubMed]
87. Zhang, A.; Liu, Y.; Qin, D.; Sun, M.; Wang, T.; Chen, X. Research status of self-healing hydrogel for wound management: A review. *Int. J. Biol. Macromol.* **2020**, *164*, 2108–2123. [CrossRef]
88. Wang, S.; Zheng, H.; Zhou, L.; Cheng, F.; Liu, Z.; Zhang, H.; Wang, L.; Zhang, Q. Nanoenzyme-Reinforced Injectable Hydrogel for Healing Diabetic Wounds Infected with Multidrug Resistant Bacteria. *Nano Lett.* **2020**, *20*, 5149–5158. [CrossRef]
89. Dong, Q.; Zu, D.; Kong, L.; Chen, S.; Yao, J.; Lin, J.; Lu, L.; Wu, B.; Fang, B. Construction of antibacterial nano-silver embedded bioactive hydrogel to repair infectious skin defects. *Biomater. Res.* **2022**, *26*, 36. [CrossRef]
90. Masood, N.; Ahmed, R.; Tariq, M.; Ahmed, Z.; Masoud, M.S.; Ali, I.; Asghar, R.; Andleeb, A.; Hasan, A. Silver nanoparticle impregnated chitosan-PEG hydrogel enhances wound healing in diabetes induced rabbits. *Int. J. Pharm.* **2019**, *559*, 23–36. [CrossRef]
91. Ren, Y.; Zhang, D.; He, Y.; Chang, R.; Guo, S.; Ma, S.; Yao, M.; Guan, F. Injectable and Antioxidative HT/QGA Hydrogel for Potential Application in Wound Healing. *Gels* **2021**, *7*, 204. [CrossRef]

processes

MDPI

Review

Drug Carriers: A Review on the Most Used Mathematical Models for Drug Release

Paolo Trucillo

Department of Chemical, Material and Industrial Production Engineering, University of Naples Federico II, Piazzale Tecchio, 80, 80125 Naples, Italy; paolo.trucillo@unina.it; Tel.: +39-329-6566043

Abstract: Carriers are protective transporters of drugs to target cells, facilitating therapy under each points of view, such as fast healing, reducing infective phenomena, and curing illnesses while avoiding side effects. Over the last 60 years, several scientists have studied drug carrier properties, trying to adapt them to the release environment. Drug/Carrier interaction phenomena have been deeply studied, and the release kinetics have been modeled according to the occurring phenomena involved in the system. It is not easy to define models' advantages and disadvantages, since each of them may fit in a specific situation, considering material interactions, diffusion and erosion phenomena, and, no less important, the behavior of receiving medium. This work represents a critical review on main mathematical models concerning their dependency on physical, chemical, empirical, or semi-empirical variables. A quantitative representation of release profiles has been shown for the most representative models. A final critical comment on the applicability of these models has been presented at the end. A mathematical approach to this topic may help students and researchers approach the wide panorama of models that exist in literature and have been optimized over time. This models list could be of practical inspiration for the development of researchers' own new models or for the application of proper modifications, with the introduction of new variable dependency.

Keywords: drug carriers; drug release; mathematical models; drug-polymer interactions

Citation: Trucillo, P. Drug Carriers: A Review on the Most Used Mathematical Models for Drug Release. *Processes* **2022**, *10*, 1094. https://doi.org/10.3390/pr10061094

Academic Editor: Carla Silva

Received: 11 May 2022
Accepted: 27 May 2022
Published: 31 May 2022

1. Introduction

The role of drug transporters raised much interest in the pharmaceutical [1], medical [2], and engineering fields [3], especially to improve drug bioavailability and efficacy for target cells [4]. Drug Carriers (DC) have always been considered as transporters of molecules, improving and preserving their properties during administration [5–8]. Nowadays, very complex drug delivery systems have been developed in order to reduce or avoid side effects, according to drugs' therapeutic windows.

Today, the well-consolidated theory of drug carriers has been included in an emerging, more complex concept, which goes under the definition of Drug Delivery Systems (DDS) [9]. These intelligent (and often self-powered [10]) transporters [11] can be either artificial or natural objects [12], and they are generally used for the simultaneous transportation and protection of active molecules from the external environment to specific target cells or living tissues [13].

The intrinsic nature of these carriers could be very different [14] according to several fabrication factors [15], such as production process [16], materials employed [17], type of drug entrapped [18], and the eventual addition of peptides or specific ligands affecting release [19].

According to their shape, Drug Carriers can be classified [20] in spherical, cylindrical, disc, or thin film complexes [21,22], which are responsible for their way and time of releasing their drug content. However, another important parameter of distinction is related to their mean dimensions (at nanometric or micrometric level), depending on the application required by the system [23].

One of the main reasons for the use of drug carriers is related to native drug degradation tendency. Therefore, drug can be in contact with external stimuli [24], such as heat [25], temperature [26], pH variations [27], or vibrational phenomena, due to near infrared activation [28]. In addition, drug loss and degradation are often due to endogenous effects, such as the human Immune System (IS), acting against external administered molecules, which could be identified as "enemies" and subsequently eliminated. Based on the Cell-Penetrating-Peptide approach (CPP), the main goal of drug carriers stands in the preparation of specific biocompatible objects that can act as Trojan horses, releasing the beneficial effects of drugs without being blocked by the effect of the Immune System [29].

Drug carriers have been designed, developed, and used in order to avoid degradation phenomena, which are often responsible for the inefficacy of drug after administration [30]. This protection is often guaranteed to drugs incorporated in DCs, since carriers' material behave as protective barrier during the overall release time necessary from administration to cellular uptake. Another important role played by protective carriers is related to the necessity of controlling drug concentration in the blood after delivery [31]; in other words, drug concentration must be included in the therapeutic window, i.e., a region delimited by a minimum level C_{min} of efficacy and a maximum level C_{max} of toxicity. The main problem of native administered drugs is that each drug may have a very different behavior, in terms of efficacy and toxic levels, respectively. Therefore, it is actually not easy to determine those levels with great precision and in a unique manner [32,33]. Without the use of a carrier, drug molecules can easily reach the toxic level, due to the side effects that an initial burst could cause; however, diffusion through drug release guarantees a delayed or controlled release of the drug, thus reaching optimal blood concentration levels within the therapeutic window. In specific situations, the use of drug carriers can significantly extend the area of the therapeutic window [34], but this operation must be studied, according to native drug pharmacokinetics, compared to the same molecule entrapped into DDS.

As defined by Williams in 2008 [35], biocompatibility is the capability of a specific material to behave with an appropriate host response in a specific application, i.e., it performs the desired function without significant side effects or deleterious systemic phenomena. Therefore, the development of biocompatible drug carriers has been characterized by great advances in the last 10 years, especially for the technologies developed to continuously balance the drug concentration in the blood [36], maintaining it at the desired values for as long as possible. It is the case of Micro-electro-mechanical Systems (MEMS) [37] (first proposed by the end of 20th century [38] for automatized diabetes control and management [39]) that are characterized by a precise closed loop algorithm in order to take therapeutic decisions without a direct human intervention [40].

The increase in complexity in the development of drug carriers resulted in a serious necessity to start a dialogue and an efficient collaboration among biotechnologists, pharmacists, medical doctors, and engineers [41]. The knowledge obtained from scientists working in different scientific fields created a full knowledge of the problems that needed to be solved. Therefore, several kinds of drug delivery systems have been developed over years [42]; each of them has its own history and can be classified according to their shape, geometry, methods of production, and final applications (pharmaceutical, nutraceutical, cosmetic, etc.). Nowadays, one of the most employed drug carrier systems in pharmaceutical applications is characterized by a spherical shape, and it can be constituted by lipids, cholesterol, or other surfactants. This DC subset goes under the name of spherical vesicle-based carriers [43] and is mainly characterized by liposomes [44,45], niosomes [46,47], solid lipid nanoparticles [48], polymersomes [49], nanospheres [50], and microcapsules [51]. On the other hand, there is another subset of DC that is stable in the form of homogeneous and macroscopic matrices, i.e., finite elements that it is possible to see with the naked eye, if compared with colloidal nanometric suspensions. These carriers are characterized by a macroscopic and a microscopic world. Of course, this means that they are constituted by micro or nanostructures or pores [52]. Foams [53], aerogels [54], and hydrogels [55] have several characteristics in common, and they deserve to be included in the same subset [56].

In order to determine a successful production of drug carriers, different techniques have been developed with the aim of protecting the entrapped drugs after administration [57]. For example, in the case of liposomes (spherical vesicles made by a double layer of phospholipids surrounding an aqueous nucleus), hydrophilic drugs can be entrapped in the inner core, while hydrophobic can be encapsulated among the double lipidic layers. In these cases, the transportation in these two different compartments of hydrophilic and lipophilic compounds can also occur simultaneously. Spherical DC are also niosomes, which are essentially made of surfactants belonging to the family of Tween and Span (polysorbates); they can be used to entrap drugs of the same nature. Aerogels [58] and foams [59–62], as well as more complex structures, such as cubosomes, are thought to be a sort of homogeneous box containing the drug.

The entrapment of drugs into carriers has given these objects a higher chance to increase the drug's bioavailability [63], meaning their efficacy during release, soon after administration. Moreover, according to these systems, drug administration takes place more gradually, since there is a natural barrier given by the carrier itself; this contributes to delaying the loss of the active principle. The drug can diffuse from the inner part of the carrier to the external environment while being delayed by the carrier's material resistance and external viscosity. This delay time is not considered a disadvantage; on the contrary, it is a precise manner to control drug release time, tuning the exact amount of drug that the cells need to absorb per time unit and avoiding the side effects caused by native drug administration overdose. To control and learn how to tune drug loss from carriers per time unit, it is particularly important to study drug release profiles, i.e., the kinetics of the drugs diffusing from carriers to the external receiving medium. These conditions can be reproduced easily in vitro.

The theory regarding drug release gained much importance in the last decades, and it has found great help with the use of computationally aided observation techniques. In particular, the controlled release's main purpose is to maintain specific drug concentration levels in blood vessels for as long as possible, controlling the release rate and its duration in target tissues.

Drug release profiles often follow different phenomena, such as simple diffusion, erosion, degradation, and absorption uptake [64]. Scientists have proposed several mathematical models and used them as tools for the design of pharmaceutical formulations [65]. As a definition, models could be intended as mathematical interpretations of the phenomena that are observed by scientists among their experimental findings and common experience [66]. The necessity to fit the enormous amount of raw data registered, experimentally, from release profiles was also moved by the overlapping of more than one competing phenomenon, observed or hypothesized, in the release environment. Mathematical models used to describe raw data release profiles could be empirical, semi empirical, or built on the physical concepts of their variables. Often, some of these models are derived from well-known equations reported in the literature. However, only a precise combination of the experimental observations with the proposed models can result in a full understanding of the release mechanisms, thereby enabling a correct prediction of release kinetics and material interactions.

Drug carriers differ in shape, dimension, drug dissolution profile, and functionality [67], thus resulting in the impossibility to develop a unique model for predicting release profile in a universal manner. Therefore, it is crucial to choose the best model that fits correctly in each case of study [68].

After a careful reading of the literature, it appears to be of fundamental importance to study the physical value of each variable included in the model used in order to understand the effect of different factors affecting the drug dissolution rate and how this can influence the efficacy of the drug, according to the therapy requested for patients.

Once defined, the main goal of this review work, an overall comment on the most used drug release modeling profiles, is proposed. Each reported fitting model will be analyzed from the mathematical and physical point of view, indicating the type of function that is described and its related application. An average fitting quality evaluation will follow this analysis for the same models, according to parameter significance, and study the effect of the trends on modeling.

2. Description of Models for Drug Release

The interest of researchers in drug delivery is abundantly reported in literature. To have a concrete idea on the attention of scientists and researchers to this field, the certified source Academia.edu has been consulted (accessed on the 11 May 2022) in order to get information on the approximate number of papers that mention, in their full text some specific keywords.

In particular, by searching "drug delivery" as a very generic expression in this academic search engine, more than 873 thousand papers appeared, citing these two words, specifically. Among these, 164 thousand report a more specific "drug delivery modeling" in their full text. This means that around 18.8% of papers related to drug carriers are effectively sensible and interested in the problem of modeling raw data obtained from drug delivery experiments or explanations. A similar result has been obtained by searching "drug delivery systems" and "drug delivery systems modeling", obtaining, in the first case, about 579 thousand mentions and, in the second case, about 137 thousand, with a 23% ratio. As a third confirmation, a final search has been performed with "drug carriers" and with "drug carriers modeling", obtaining a ratio of 22.3%. On average, it appears that one paper over five is referring to models, denoting that only 20% of the works related to drug delivery are effectively described by a mathematical prediction of the phenomena.

In Table 1 and Figure 1, the research has been performed by going deeply into details of the kind of mathematical models used in drug delivery field. This research regards the 10 most used models of drug delivery developed or modified over the years. The searching criteria consisted of writing the name of the author's proposing model, followed by "drug delivery model" as a full expression.

Scientists have studied raw data obtained during experimentations; drug release tests have been reproduced in laboratories, trying to describe the phenomena involving drug specific behavior. For example, the resistances offered by materials and media represent a barrier among drugs and target tissues. These molecules need to overcome those barriers during their diffusion from the inner part of the carrier to the external receiving liquid bulk.

Several mathematical models have been reported in the literature over the years, and they were employed by a large number of scientists to fit their drug release raw data [69]. Moreover, drug release previously consisted only in simple observations; then, the advancement of computer simulation and data analysis interpretation resulted in an improved pharmaceutical technology. The use of well-known mathematical models and the development of new ones contributed to correctly designing new drug formulations and dosages. Nowadays, the main challenge that scientists are facing is the development of mathematical models that simultaneously combine the theories describing drug release phenomena and the effects on drugs during transportation to the human body. In other words, it is not simple to define models that take into account both the diffusion phenomena and the uptake response of cells, simultaneously. However, this field of study is in continuous evolution, especially from the industrial point of view, due to its enormous potential.

The kind of drug, its hydrophobicity or lipophilicity, the drug amount incorporated in carriers, the eventual excipients, and the bulk delivery conditions could affect the drug release. Some of these features influencing release are listed as: polymer degradation, polymer or drug dissolution in the external medium, possible degradation of excipient or stabilizers, creation of hydrostatic pressure inside the system, coalescence of drug carriers among them, variation of pH or temperature induced by not programmed stimuli, or environmental phenomena. However, it is practically impossible to simultaneously predict all of these phenomena, especially considering the drug transport coupled with cellular response.

Table 1. Diffusion of specific authors' and scientists' models for drug release.

Authors and Scientists	Nr. of Full Text Mentions	Use Frequency, %
Crank and Baker	2748	5
Peppas	16,117	27
Higuchi	19,243	33
Hixson-Crowell	2094	3
Peppas and Sahlin	640	1
Bake & Lonsdale	1716	3
Hopfenberg	403	1
Corrigan	10,297	18
Weibull	4238	7
Sivak	1047	2

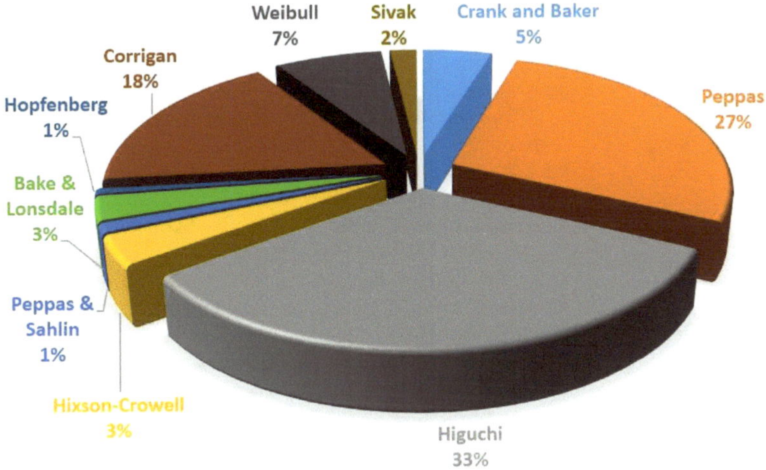

Figure 1. Frequency use distribution of the most famous drug release models.

Mathematical solutions are generally characterized by complex partial differential equations, with dependencies on time and tridimensional space. The use of numerical approximations, limiting the introduced error as much as possible, resulted in a reduced level of complexity in drug release models. In this section, the most famous models for drug release will be described one by one, describing the physics behind the involved variables and providing application purposes.

In order to correctly describe the models for drug delivery systems, it is worth adding the concept of drug encapsulation efficiency into carriers. This means that the effective amount of a drug entrapped in the carrier could not be equal to the amount of drug dissolved in the preparation solutions before production of the carriers themselves. For example, when preparing the precursor solutions, an operator will dissolve M_0 (theoretical initial mass of drug) into a defined initial volume V. This will result in a drug concentration (defined as C_0). However, after the processing of drugs and materials for carriers, a partial loss of the drug could occur, defining a final M_f value of the drug incorporated in its final volume V_f, thus resulting in a final concentration C_f. This theory can be described by the following equation:

$$EE, \% = \frac{C_0 - C_f}{C_f} 100 \tag{1}$$

In case the volume of the system is constant, this equation can easily become

$$EE, \% = \frac{M_0 - M_f}{M_f} 100 \quad (2)$$

Generally, these measurements are performed using a UV-Vis (UltraViolet-VISible) spectrophotometer at the specific wavelength of the incorporated drug (reported in tables or databases for each molecule). The ultraviolet region is generally included among 100 nm and 400 nm, while the visible region is among 400 and 700 nm. This instrument works, with a good approximation, with the Lambert–Beer equation (Equation (3)), which linearly correlates the concentration of drug dissolved in the bulk solution to the absorbance of the drug in that medium:

$$A = K\,C \quad (3)$$

where A is the absorbance of that drug (measured after setting the wavelength), C is the concentration, and K is the slope obtained by the calibration line at known drug concentrations. The slope and the concentration should also be multiplied by the optical path value, which is often equal to one.

The definition of the encapsulation efficiency is extremely important, since it is strictly correlated with the drug release study. In case of representation of the cumulative drug release percentage, it is correct to multiply the following modeled drug concentration by the encapsulation efficiency obtained for the prepared drug carriers. Otherwise, the drug release pharmacokinetics could be overestimated.

A tool to identify the efficacy of a model in a specific system is equally necessary, and it is worth citing it in this section. In a certain manner, the following Equation (4) represents a model-for-the-models; this means that this mathematical correlation can be used to compare the other models among them and evaluate their predictive efficacy. Before defining that a model fits his/her necessities, the acute researchers or students have the chance to minimize the value coming out from the following mathematical correlation. This equation goes under the classical definition of Summed Square Error (SSE):

$$\text{SSE} = \sum_{i=0}^{n} (Value_{model} - Value_{data})^2 \quad (4)$$

where the subtraction, reported in parenthesis, is among the value calculated using the model and the value observed or registered by experimentation.

2.1. Fick's Classic Model

Drug release could be described as simple diffusion controlled by a constant coefficient. This is the specific case of drug depots, which consist of reservoirs of drugs surrounded by a homogeneous polymer-based membrane. Another example is characterized by monolithic systems, also known as one-block delivery systems. The important feature is that there should be no separation between the drug reservoir and the controlling barrier. In other words, the simplicity of the system could be easily approximated by a generic Fick's law [70–72]; in case of complex or multiple barriers controlling the release rate, it is not possible to use this model.

According to the classification proposed by Siepmann [73–75], in case drug permeability through the polymeric barrier is constant, the release kinetic could be approximated as the first order:

$$\frac{dM_{(t)}}{dt} = \frac{A\,D\,K\left(M_0 - M_{(t)}\right)}{L\,V} \quad (5)$$

where $M_{(t)}$ represents the amount of the drug released with time dependency, M_0 is the initial mass of the drug, V is the total volume of the drug depot, A is the total surface area of the device, and L is the thickness of the polymeric membrane, which separates the drug depot from the external environment. Obviously, D is the diffusion coefficient of the drug among the membrane and in the external bulk. As it is possible to see, in these systems, the release rate is considered constant, as well as the diffusion rate and the A, K, L, and V coefficients; that is the reason why this partial differential equation has a linear solution in the time variable and could be an easy approximation. Vice versa, in case the initial drug concentration is larger than the drug solubility in the reservoir liquid bulk, drug crystals will be created in the depot, resulting in saturated conditions at the interphase internal/external of the drug carrier. In this case, the differential variation of drug concentration with time is constant and directly proportional to A, D, K, and C_s, with this last value being the solubility of the drug in the depot. Therefore, the drug release system will consist of a zero-order release rate, as long as there will be an excess of drugs in the drug depot.

In the case of monolithic solutions, the geometry of the system will significantly affect the drug release. For example, for thin films, Crank and Baker proposed a cumulative approach model, as follows:

$$\frac{M_{(t)}}{M_0} = 4\sqrt{\frac{D\,t}{\pi\,L^2}} \tag{6}$$

According to this model, $M_{(t)}/M_0$ ratio should be a value between 0 and 0.6. In particular, $M_{(t)}$ is the cumulative amount released at time t, M_0 is the initial drug amount, D is the drug diffusion coefficient in the considered matrix, and L is the thickness of the film. Similar equations were adapted by Crank in 1975 for spherical and cylindrical matrices, which simply differ by some geometrical modifications, but are essentially similar.

It is also worth adding that M_0 is not always equal to M_{inf}. Indeed, the initial mass of the drug and the mass of the drug released at time infinite are equal only in the case that all the drug content is released from the carrier, thus indicating $M_{(t)}/M_0$ as the fraction released at time t. A simulation of the Crank and Baker model has been ideally reported in Figure 2, where the model evolution over time is discretized in black dots. It is quite important to note that the varying parameters have been chosen in order to obtain a clear visualization of the trends on physical effects. However, this choice has been taken according to the physical significance of these parameters.

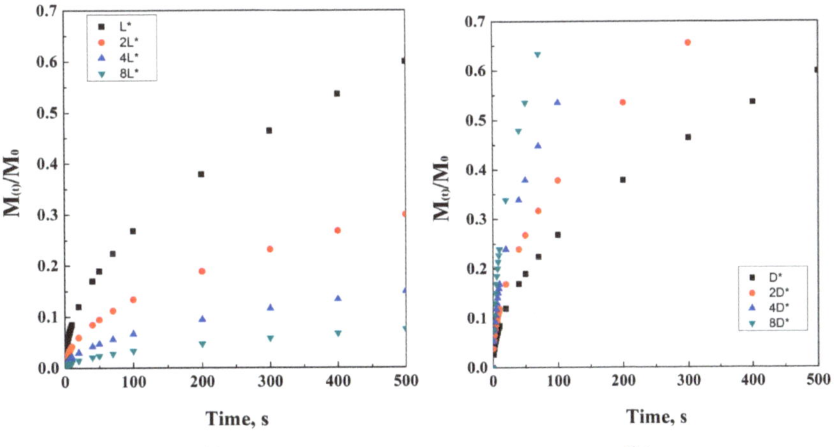

(a) (b)

Figure 2. Crank and Baker model at different thickness values (**a**) and diffusion coefficients (**b**).

In case the diffusion controls totally release the system, it could be also independent from the shape or geometry of the carrier. In that situation, the release model could be simplified to a classical power law, also proposed by Peppas, as described in the following sections.

In case of osmotically controlled systems [76–79], the following equation could describe the system, as follows:

$$M_t = K\,\frac{A\,L_p\,\sigma\,C\,\pi_s}{\delta}\,t \tag{7}$$

where L_p is a coefficient of permeability (unit of measure of length over time), C is the concentration of the drug, δ is the thickness of the carrier matrix, π_s is the osmotic pressure of water, A is the cross-section area of the carrier, σ is a reflection coefficient, and K is an empirical coefficient.

2.2. Higuchi's Model

Drug dissolution from matrices requires a different mathematical approach [80]; this model is well-known in the literature, since it was firstly introduced in the 1960s by Takeru Higuchi, who derived it while studying drug release from ointments. However, this famous scientist was also an expert in release phenomena from semi-solid and solid matrix systems. The first studied systems were characterized by planar mixtures releasing a lipophilic compound. According to this model, the concentration of drugs liberated increases with the square root of time. In this case, the mathematical model used was:

$$\frac{M(t)}{M_\infty} = \sqrt{D(2A - C_s)\,C_s t} \tag{8}$$

where $M_{(t)}/M_\infty$ represents the drug fraction released per unit surface area. This fraction has dimensions of $[\text{Length}]^{-2}$ and is a function of the square root of time. C_s is the solubility of the drug in the external matrix medium, D is the diffusivity in the matrix medium, and A is the loading of solute, in case it exceeds its solubility in the matrix. A theoretical simulation of Higuchi's model has been proposed in Figure 3.

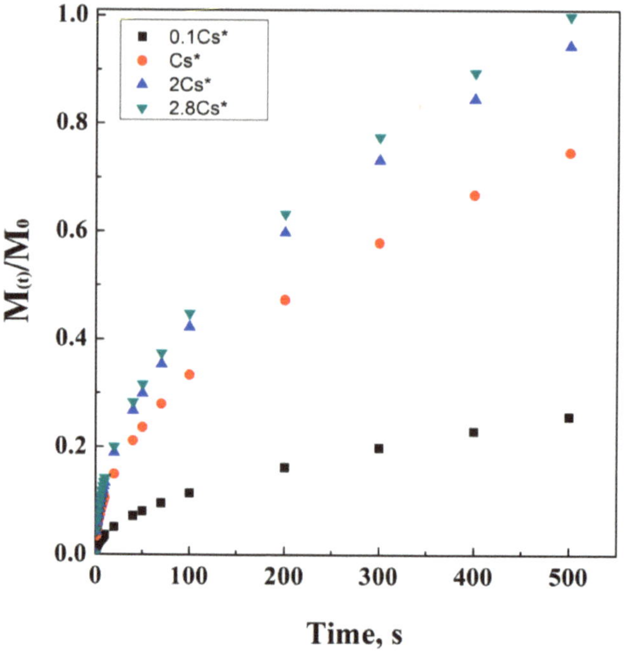

Figure 3. Higuchi's model at different drug solubility concentrations.

This model, of course, is particularly applicable in a saturated system, and it has also been proposed for other kinds of shapes and geometries, such as spherical. Available in homogeneous and not homogeneous systems, the equation can be adapted to porous systems, as follows:

$$\frac{M(t)}{M_\infty} = \sqrt{\frac{D\,\varepsilon}{\tau}(2A - \varepsilon\, C_s)\, C_s t} \tag{9}$$

where ε is the porosity of the matrix and τ is its tortuosity factor. Figure 4 reports an ideal simulation of Higuchi's modified equation, including tortuosity and porosity dependencies. It is interesting noting how the direct proportionality is related to porosity, while the indirect one is related to tortuosity.

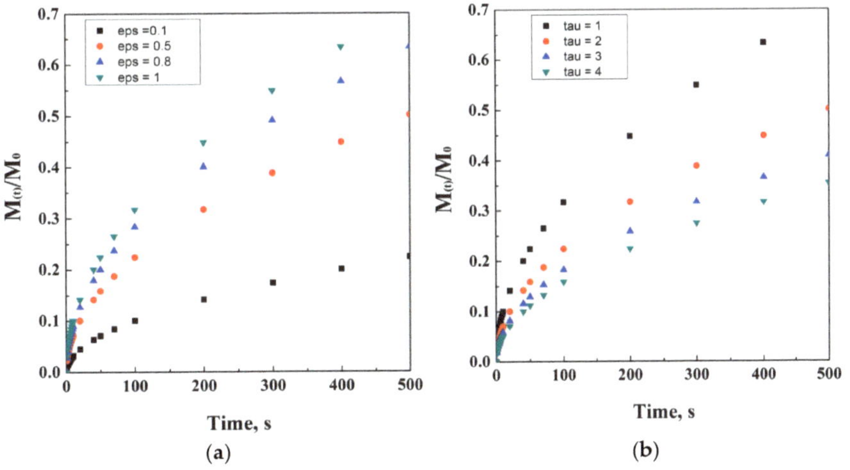

Figure 4. Higuchi's model at different porosity (**a**) and tortuosity (**b**) values.

However, another modification was proposed for this equation, in case of a matrix already saturated with a drug, that excluded C_s and introduced C_0, intended as the concentration of the diffusing drug in a porous matrix:

$$\frac{M(t)}{M_\infty} = \sqrt{\frac{2\,C_0\,D\,\varepsilon\,t}{\tau\,\pi}} = K_H\,\sqrt{t} \tag{10}$$

This equation establishes that the fraction of the drug released is proportional to the square root of release time; according to this last equation, all the concentration, diffusion, tortuosity and porosity coefficients are assumed constant and are indicated as a unique constant K_H, named Higuchi's constant. The possibility to use this model is linked to conditions; among those, the initial drug concentration must be higher than saturation conditions of the drug in the defined matrix; moreover, diffusion must be considered unidirectional and constant; last, but not least, dissolution of the matrix and border effects are negligible.

2.3. Hixson-Crowell's Model

The researchers Hixson and Crowell [81–83] proposed a model that correlates initial drug mass amount fed to the system to the amount of remaining drug (i.e., not released) at time t. However, in this case, the cubic roots of these two mass values are linearly correlated with time. Therefore, it is possible to define the following Equation (11):

$$\sqrt[3]{M_i(t)} = \sqrt[3]{M_0} - K_{HC}t \tag{11}$$

where M_0 is the initial amount of drug, $M_i(t)$ is a function of time, describing the amount of drug that still remains confined at each time lapse, K_{HC} is defined as a constant of incorporation, which is, respectively, a function of surface and volume of the drug carrier. Generally, this model is used for specific DC geometries, such as tablets or parallel planes.

The value of this constant could be additionally defined as follows: by introducing the number of particles N, the diffusivity D, the solubility at saturated conditions and fixed temperature C_s, and being no less important, the thickness of the thin layer δ.

$$K_{HC} = \frac{K' \sqrt[3]{N}\, D\, C_s\, t}{\delta} \tag{12}$$

This equation introduces a new constant, K', that refers to the average density of the carrier. Of course, it is in correlation with not only the number of particles but also with its mass concentration in the bulk solution. Moreover, the use of this model is strictly correlated to the literal possibility to count particles in colloidal suspensions; this could be performed using proper instruments or technologies, such as Nanoparticle Tracking Analysis (NTA). However, this technique is not sufficient by itself, since it gives only information about particles number and their velocity, following their Brownian motion. However, the main characteristics of these particles, pores and shape, could be observed and critically commented on by only using a scanning electron microscope, which also provides the possibility to measure pores dimensions. If the dissolution of the drug is constant, especially from spherical particles, K' becomes constant and determined experimentally.

2.4. Peppas' Models

The model proposed by Peppas [84–86] in 1983 consisted of a simple power law; this necessity resulted from the fact that, in general, it is not easy to reach a zero order kinetic with a linear correlation among the drug fraction released over time t. Indeed, the laboratory experience of Peppas and his co-workers induced the development of a new semi-empirical model, which described drug release as an exponential function of time. Indicating $M(t)$ as the amount of the drug released at time t and M_∞ as the final amount of drug at the equilibrium, the following simple model was proposed:

$$\frac{M(t)}{M_\infty} = K\, t^n \tag{13}$$

where n is the exponent of time t and it is related to the drug release mechanisms, which can be also obtained experimentally. In particular, K is the constant that depends on the shape and geometry of the drug carrier vesicles or matrix. Indeed, this equation can be easily re-written as

$$M(t) = K\, M_\infty\, t^n \tag{14}$$

where M_∞ is a constant and could be englobed in K as:

$$K' = K\, M_\infty \tag{15}$$

It is worth clarifying that the encapsulation efficiency plays an important role in defining the value of M_∞; in fact, if the encapsulation efficiency is equal to the total initial drug amount fed to the system before the preparation of the drug carrier, the value M_∞ could be assimilated into the initial amount of the drug dissolved into the system. On the contrary, if the Encapsulation Efficiency (EE) is not equal to 100%, it is necessary to multiply M_∞ by the EE percentage. An ideal representation of Peppa's model is reported in Figure 5.

However, it is possible to modify this equation by the necessity to consider the latency time, i.e., the lag phase in which the drug carrier is overcoming the resistance to diffusion, offered by the external layers of the drug carrier and/or by the external initial bulk conditions. This power law model has been ideally simulated in Figure 6. In Equation (16), M_{i-l} represents the initial mass, considering that the administration starts at the end of the lag phase.

$$M_{i-l} = K\, M_\infty\, (t - t_l)^n \tag{16}$$

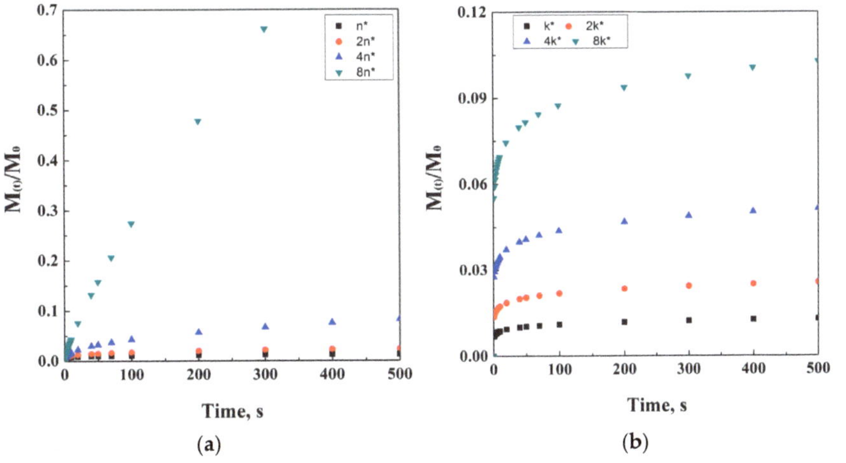

Figure 5. Power law at different exponent values (**a**) and k coefficients (**b**).

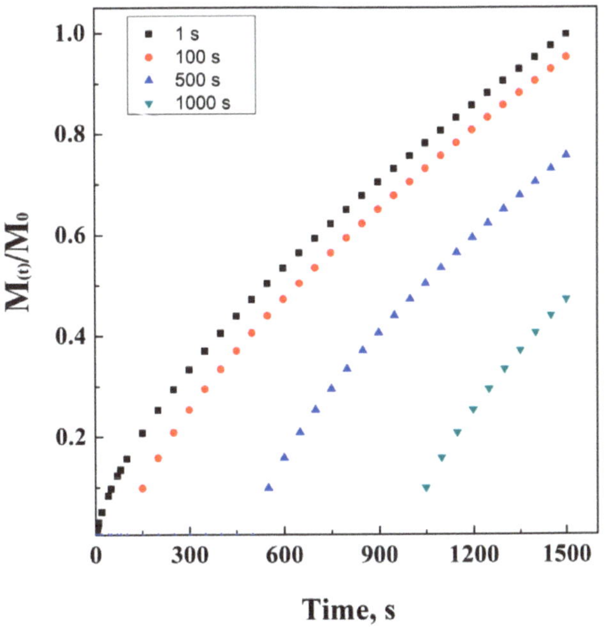

Figure 6. Power law at different lag times.

Furthermore, it is possible to correct this model in case an initial burst of drug occurs, as follows:

$$\frac{M(t)}{M_\infty} = K\,t^n + b \tag{17}$$

where b describes the initial concentration of the drug rapidly released to the system soon after the time-zero from the beginning of the phenomenon.

However, Peppas is the author of several papers reporting different models developed in collaboration with his co-workers. For example, the description of uncommon simultaneous release phenomena (diffusion and relaxation of polymer chains) was included in this model as a sum of two additive contribution of power laws. This new equation by Peppas and Sahlin [87,88] is described as follows:

$$\frac{M(t)}{M_\infty} = K_1\ t^n + K_2\ t^{2\,n} \tag{18}$$

where three constants have been introduced: K_1, K_2, and n. While K_1 represents the Fick's diffusional contribution from matrices of any shape, K_2 is related to the relaxation contribution.

The relaxation time is correlated to the diffusion time, calculating the following Deborah number:

$$De = \frac{\lambda}{\theta} \tag{19}$$

where λ is the relaxation time of polymers chains, and θ the diffusion time. In the case that De is much lower than 1, diffusion controls the release process; on the contrary, when it is much larger than 1, relaxation phenomenon controls the process. In case of a value near to 1, the two contributions need to be evaluated together.

2.5. Baker's and Lonsdale's Model

These two scientists derived a model from Higuchi's equation [89–91], adapting it for drug release from spherical matrices. This model was recombined as follows:

$$\left(1 - \frac{M_t}{M_\infty}\right)^{2/3} + \frac{2}{3}\frac{M_t}{M_\infty} = K + \frac{2}{3}\,t \tag{20}$$

In this model, again, M_t is a function of time. Describing the amount of a drug released over time, M_∞ is the maximum amount released considering an infinite time, t is the independent variable, and K is determined as follows:

$$K = 1 - \frac{2\,D\,S}{r_0^2\,C_0} \tag{21}$$

where D represents the diffusion coefficient of the drug, S is the solubility of the drug in the carrier, r_0 is the initial radius of the carrier, and C_0 is the initial concentration of the drug. In case the solubility and the diffusion coefficient in the carrier matrix are constant over time, the value of K could be considered a constant. K values could vary according to the kind of the entrapped drug and the type of carrier matrix. Of course, the values of D and S are related to a matrix that is just releasing drug, reducing its diameter and mass without degradation phenomena. In the case of foams of other porous matrices, the K value becomes:

$$K = 1 - \frac{2\,D\,S\,\varepsilon}{r_0^2\,C_0\,\tau} \tag{22}$$

where ε and τ represent the porosity of the carrier matrix and the tortuosity during drug diffusion, respectively.

In this case, a graphical simulation of the model is not reported, since it requires a numerical solution by successive attempts.

2.6. Hopfenberg's Model

Hopfenberg [92,93] proposed a model to describe drug release from polymers that could be degraded or eroded during drug loss, independently from their shape or dimensions.

The main representing equation for this model is conditioned by the erosion grade constant, named k_0, and by the initial concentration of drug entrapped in the matrix, C_0:

$$\frac{M_{(t)}}{M_\infty} = 1 - \left[1 - \frac{k_0\,t}{C_0\,a_0}\right]^n \tag{23}$$

where, as always, $M_{(t)}$ and M_{∞} are the mass of drug released at time t and at plateau level, respectively; t is the independent variable of time, and a_0 is the initial length of the carrier matrix. Of course, this last variable could be a function of time; however, in this case, it can be approximated to the initial value. Indeed, a_0 only depends on the shape of the matrix (radius in case of a sphere or cylinder, or half thickness in case of a thin layer). The exponent "n" is set to 1 in case of a thin film shape, to 2 in case of a cylinder, and 3 in case of a sphere.

As it is possible to see, in case of thin layer, the equation becomes:

$$\frac{M_{(t)}}{M_{\infty}} = \frac{k_0\ t}{C_0\ a_0} \tag{24}$$

where the three constants could be included under a unique k_1 value, as follows:

$$\frac{M_{(t)}}{M_{\infty}} = k_1\ t \tag{25}$$

In case of a not negligible latency time (t_L) from the beginning of release experiments or administration, a correction could be performed on the following equation:

$$\frac{M_{(t)}}{M_{\infty}} = 1 - \left[1 - \frac{k_0\ t\ (t - t_L)}{C_0\ a_0}\right]^n \tag{26}$$

This equation represents an evolution of the exponential model. In this case, there is no graphical simulation, since this model is easily linkable to one of the previously mentioned equations.

2.7. Corrigan's Model

Dr. Owen I. Corrigan [94–96] proposed the use of poly(lactic-co-glycolic) acids (PLGA), for the design of innovative drug delivery systems, for the entrapment of several kinds of active principles, such as steroids, anti-inflammatory drugs, antigens and, in general, therapeutic agents. These scientists demonstrated that copolymers could be characterized by complex simultaneous release phenomena, probably also due to drug-polymer interactions and to drug molecular weight. In the case of large molecules, such as Bovine Serum Albumin (BSA), the phenomenon of release from polymeric particles was characterized by a fast initial burst, followed by a subsequent slower, and continuous release. Corrigan attributed this behavior not only to degradation of polymer but also to the affinity between BSA and PLGA.

For specific compounds, even three steps were detected during drug delivery observation: an initial burst due to diffusion, a subsequent lag phase, and a final completion of drug release due to polymer mass loss. In particular, the kinetics of polymer mass loss, for pure microparticles of PLGA, are described by the following equation.

$$\ln\left(\frac{x}{1-x}\right) = k\,t - k\,t_{max} = k\,(t - t_{max}) \tag{27}$$

where x represents the polymer mass loss, t is the time lapse since the beginning of the observation, while k and t_{max} are, respectively, time and rate constants directly related to the bulk degradation of the polymer.

Drug release from copolymers such as PLGA/PLA (see Abbreviation List) were described using a two phase model, according to which there was a first controlling step, where the drug diffuses among the interface of the polymer, followed by a second release of the drug entrapped into the polymer structure, with this last step starting its own degradation. The initial diffusion step is described by the following equation.

$$F_B = F_{BIN}\left(1 - e^{-k_b\,t}\right) \tag{28}$$

where F_B is the released amount due to the diffusion, F_{BIN} is the burst fraction considering an infinite time of observation, and k_b represents the first order constant rate associated with diffusion step. In particular, k_b is also directly dependent on the diffusion coefficient (D), solubility of the drug (C_S), and surface area of drug that takes part in the dissolution. The second degradative step describes the release due to degradation phenomena of the polymer; this contribution is described by the following equation:

$$F_{Deg} = (1 - F_{BIN}) \left(\frac{e^{k\,(t-t_{max})}}{1 + e^{k\,(t-t_{max})}} \right) \quad (29)$$

Combining the two contributes, the Corrigan overall formula becomes the following:

$$F_{TOT} = F_B + F_{Deg} = F_{BIN}\left(1 - e^{-k_b}\right) + (1 - F_{BIN}) \left(\frac{e^{k\,(t-t_{max})}}{1 + e^{k\,(t-t_{max})}} \right) \quad (30)$$

Moreover, the quantification of the active drug released (A_t) is given by the product of the fraction F_{TOT} and the amount of drug loaded. An example that came out from the studies of Corrigan and co-workers describes the production of ketoprofen-loaded PLGA nanoparticles, which resulted in an encapsulation efficiency between 40 and 65%. Drug release studies showed a first plateau at 5 days due to diffusion, followed by a second plateau reached at about 20 days, due to polymer degradation. An idealistic representation of Corrigan's model is proposed in Figure 7. In details, in Figure 7b,c, lines represent guides for the eyes.

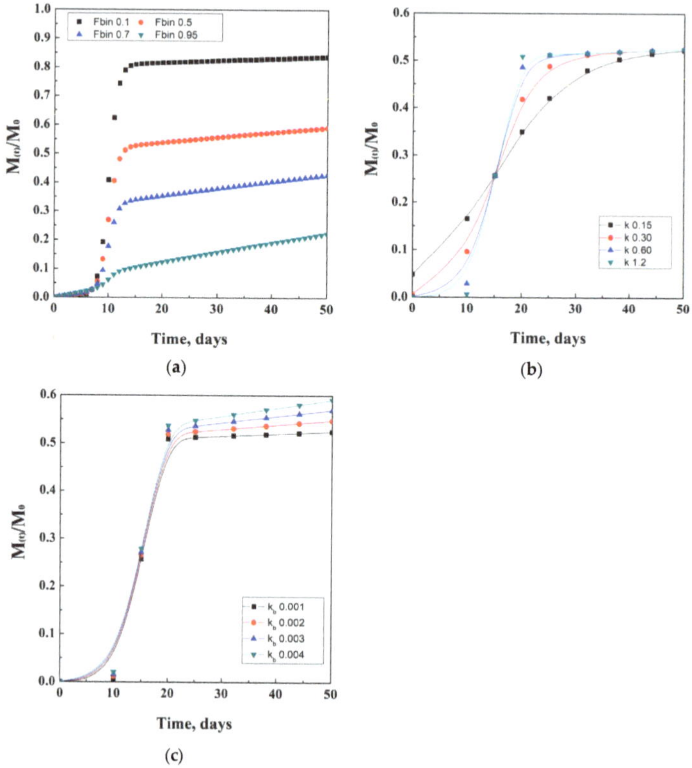

Figure 7. Corrigan model depending on the F_{bin} coefficient (**a**), k degradation coefficient (**b**), and k_b diffusion coefficient (**c**).

2.8. Weibull's Model

Weibull [97–99] proposed a model to describe the phenomena of drug release during a finite time of observation.

This mathematical model is characterized by three newly introduced parameters. The first one is $k = 1/a$, where "a" is the time-scale parameter, and it is obtained empirically; t is the time elapsed from administration; t_c is the central value of time, related to the latency time from the beginning of the release. This value can be, of course, equal to zero in case of negligible latency time of the release system; the last parameter is the exponent d, strictly correlated to the type of modeling curve. In particular, d is equal to 1 in case of a simple exponential equation, while it is more than 1 in case of a sigmoid and less than 1 in case of parabolic modeling function.

$$\frac{M_t}{M_\infty} = 1 - e^{-(k\,(t-t_c))^d} \quad (31)$$

This model is generally used as a comparative reference to check the accuracy of different models among them, describing drug release from polymeric matrices. Another reason linked to its use is that it does not depend on the simultaneous effect of different phenomena occurring during release. An ideal mathematical simulation of this model is proposed in Figure 8, where for t_c a unit of time has been used.

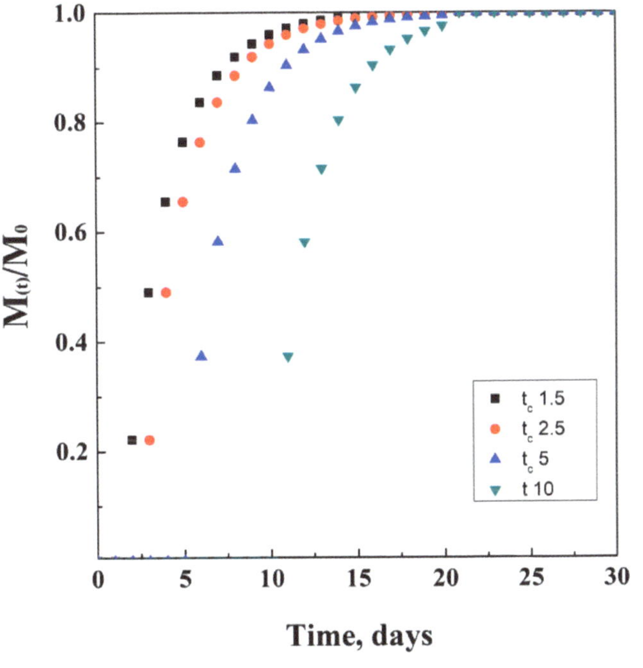

Figure 8. Weibul model at different t_c values.

2.9. Sivak Model

Sivak, in 2009, proposed a model [100] for the simultaneous release of two anti-cancer compounds (DB-67 and doxorubicin [101]) from polyurethane foams [102]. In particular, the two indicated compounds are characterized by several functional groups, which can act as points to create chemical bonds with polyurethane networks. These tests were performed in a phosphate-buffered saline (PBS) medium at controlled temperatures of 4 °C, 22 °C, 37 °C, and 70 °C, for a maximum period of 10 weeks.

The drug-released amount was detected using fluorescence spectroscopy; raw data were used to develop this model; in particular, the amount of drug diffused into the external medium is expressed by the following equation, in terms of molar ratio:

$$mol_{drug} = \frac{1}{MW}\left[3\ \sum_i^j \left(\frac{m}{V}\right)_i + \left(\frac{m}{V}\right)_j (V_T - 3\ V_j)\right] \tag{32}$$

where m and V are the mass and the volume of each collected sample at the i-th time interval, V_T is the total volume of the system in which the release experiment is performed, and j represents the last time interval. Authors showed that, working at 37 °C, an initial rapid burst is detected, especially, for DB-67, but then, drug concentration stabilizes after about 10 days. Instead, working at 70 °C, a rapid increase of the curve slope was detected, over time, without obtaining a plateau after more than 70 days.

2.10. Pulsatile Drug Delivery

Pulsed delivery [103–105] consists of the artificial intervention during delivery and after administration. For example, let us imagine that a spherical drug carrier has been delivered to a blood vessel; it will start releasing drug content only after receiving a stimulus from an external agent that is activated by an operator. Then, after a defined therapy time, the operator interrupts this stimulus, stopping the release of the drug from the carrier. The release function suddenly equals a drug concentration of a fraction equal to zero.

In detail, each of the previously described models are eligible for pulsed drug delivery. A generic system of this kind can be characterized as follows:

$$\begin{cases} C(0 < t < k) = f(t) \\ C(k < t < m) = 0 \end{cases} \tag{33}$$

where $C(t)$ is a generic function of time, depending on the release phenomena and type of carrier employed; instead, k represents a generic variable that, in this case, determines the timing value of the function. Let k be 1 h; this means that the carrier will be programmed to release its content for the entire first hour after administration, following the model indicated by the generic function $f(t)$. Soon after, the release will be stopped by the interruption of the external stimulus, and the concentration of the drug released will go to zero, until reaching time m. Of course, this alternation of "release" and "no release" could be programmed to be cyclic, as well as the associated stimulus that induces the release. This kind of mechanism could be very useful to have a fast initial burst of drug without a significant drug accumulation in the target tissues, avoiding consequent toxic side effects. A tentative representation of this model is ideally reported in Figure 9, considering $f(t)$ as the equation adapted from the Crank and Baker model, which is assumed as the example.

However, it is possible to simplify this model as follows:

$$\begin{cases} C(b < t < k) = p \\ C(k < t < m) = 0 \end{cases} \tag{34}$$

where p represents a constant value of drug released concentration, in case the system is able to lose a time-independent amount of drug; b represents the initial time where the drug's initial burst occurs: this last value could be approximated to 0 in the case of very long release times at a constant value of $C(t)$.

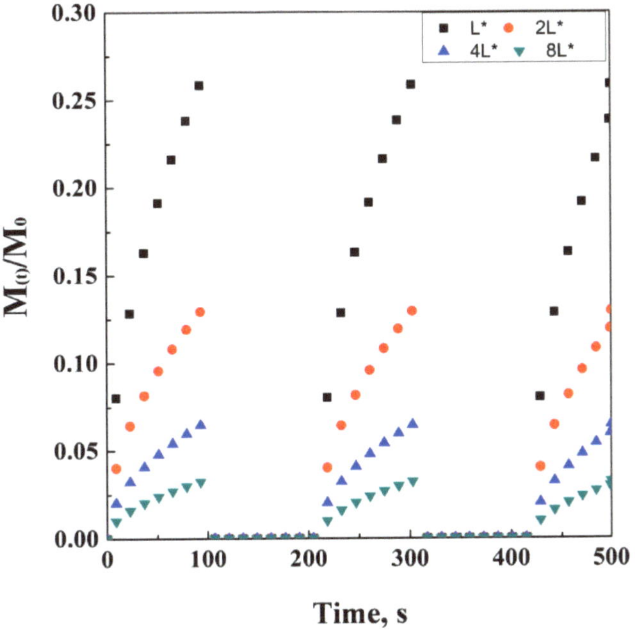

Figure 9. An example of pulsed delivery, adapted from the Crank and Baker parametric, on thickness L.

3. Discussion

According to the wide literature treating these topics, the previous section represented just a brief summary of the mathematical models for drug delivery. The difficulty that researchers and scientists may find during their experimental campaigns stands not only in the interaction that eventually occur among drug and carrier's materials but also on different phenomena that may happen during delivery, such as degradation, bio-degradation, erosion, and bio-erosion [106–109]. In order to distinguish these phenomena, it is necessary to recall their definitions, as summarized in Table 2.

Table 2. Definition of general disrupting phenomena occurring in drug delivery.

Phenomena	Definition	Ref.
Degradation	Scission of a polymer chain in oligomers or monomers	[110]
Bio-degradation	Degradation due to the action of a biological system	[111]
Erosion	Mass loss phenomenon from the polymeric interphase	[112]
Bio-erosion	Erosion due to the action of a biological system	[113]

In detail, the scission of a chain is a chemical phenomenon that causes the separation of a long polymeric chain into monomers and oligomers. This is considered a degradation phenomenon, since the polymeric barrier created using the long chain is disrupted, thus characterizing the fast and unwanted diffusion of the entrapped drug from the inner core to the external bulk. Chain scission could be caused either by artificial stimuli or by the action of microorganisms. Instead, erosion is actually mass loss, occurring at the interphase between the DC external surface and the aqueous bulk. This boundary layer phenomenon is controlled by reactions that only occur on the surface of the DC, and it may be, in this case, caused by artificial or natural stimuli, such as microorganisms' action.

Mathematical models can be considered empirical when they are purely characterized by trial and error attempts to let the variable work correctly to describe the release profiles; otherwise, the models are classified according to their physic-chemical dependencies, in case they describe the diffusive phenomena and the chemical reaction occurring in the chosen system. In these last cases, they can be classified as semi-empirical models.

In more detail, depending on the specific composition of an erodible support (i.e., on the kind of polymer), on the loading of additives, and on the geometry and shape of the matrix, many phenomena of mass transfer and chemical reactions may influence the release kinetics of a drug. This set of phenomena (listed in Table 3) should be considered as a subset of Table 1, since they go, deeply, into the effect of degradation on materials.

Table 3. Phenomena of degradation.

Phenomena	Reference
External water invades the inner core of the matrix	[114,115]
Drug degradation	[116,117]
Polymer degradation	[118,119]
Creation of aqueous pores	[120,121]
Degradative phenomena of interference drug/polymer	[122,123]
Polymer swelling	[124,125]
Drug/polymer adsorption	[126,127]

In details the invasions of the external bulk, in the inner core of DC, is due to probable interphase disruption and the creation of a pressure gradient directed to the inner compartment. Drug degradation may occur during an increase in temperature and energy of the system, resulting in total or partial loss of efficacy of a drug during administration. Polymer degradation may occur during preparation of DC or delivery, probably due to the overcoming of glass transition conditions. The creation of aqueous pores is a sort of partial invasion of the external water in the inner compartments of DC. Degradative phenomena, due to drug/polymer interference, are essentially related to reactions that may occur in the disruptive polymeric membrane, separating inner compartments from external bulk. Polymer swelling is caused when a solvent penetrates into the polymeric network, causing a sudden change in the volume. Drug adsorption on polymers is the well-known phenomenon of physical uptake of a solute (in this case, the drug) on a solid matrix (the polymer).

Depending on the complexity of physical phenomena, each release system must be preliminarily characterized; then, a modeling attempt can be performed. The information obtained after this analysis can be of fundamental importance to the use of a mathematical model, which either already exists, in which case it needs to be optimized, or is created ex novo. If there is not sufficient information on the release system, it will be difficult to find a model that properly describes the phenomena. For this reason, it is essential to study, with great attention, the property variations of the system during the release, over time.

Indeed, this work was intended as a collection of mathematical correlations, leaving, to the reader, the critical choice of the one that best fits with their research and development activities. Several scientists have successfully and brilliantly studied the deep functionalization of complex carriers, thus obtaining outstanding results; some examples are related to the use of dietary supplements, proteins, aerogel, liposomes, hydrogels, nanogels, and nanocomposites [128–136], whose data have been successfully used and applied to the existing models [137,138], taking into account the occurring phenomena naturally or artificially induced during therapy [139–144].

Purely empirical models generally do not take into consideration the overlapping of different transport phenomena effects, such as diffusion of water and the drug, swelling, or degradation phenomena that may also bring to a zero order kinetic. Instead, using a non-empirical model, physical and chemical phenomena are described precisely, with a more complex mathematical expression dedicated to them. Summarizing, empirical models have a less precise approach, but they are particularly powerful with a more simplified kinetic; non-empirical models describe all the phenomena occurring in the system but with a complex mathematical approach. Last, but not less important, semi-empirical models are a mix of previous characteristics and properties, with variable kinetic orders.

4. Conclusions

Recalling the aims of this paper, the author tried to provide a general and basic knowledge on this topic without presuming to give the proper tools to predict any kind of complex simultaneous phenomena and occurring during drug delivery, such as predicting bio-relevant media for drug dissolution experiments or providing the best carrier material for a specific molecule. The use of mathematical tools represents a modern approach for the correct exploitation of drugs and their beneficial effects on human tissues [145]. Release mechanisms can be controlled and ruled by diffusion, chemical reactions, osmotic pressure, or swelling. Moreover, the correct prediction of the physics of the systems is a strong tool also for its potential reduction in the number of experiments; for example, professor Higuchi was considered a sort of "father" of the mathematical modeling for drug carriers, since his method was recognized to be theoretically stable and easy to apply [146].

However, the preliminary understanding of these phenomena is of fundamental importance [147] before approaching the problem using a mathematical model. Indeed, the theory of drug delivery requires accurate molecular, transport, and thermodynamic evaluations of the carriers' behavior in the chosen environmental system, in order to predict the expected interactions. Of course, a physical and chemical approach to the system needs to be validated with several experimental runs, especially in terms of drug dosage optimization; one actual example is represented by the genetic material delivery, which is particularly complex. This will require much more studies and wider validations across semi-empirical models, in order to approach a rational design of modern delivery systems. To reach this goal, the most powerful and efficient high-pressure systems [148,149] and computer-aided analysis [150–152] and simulations will be significantly helpful in the decisional process for delivery description and modeling, becoming an integral and essential part of this theory.

Funding: This research received no external funding.

Conflicts of Interest: The authors declare no conflict of interest.

Abbreviations

DC	Drug Carriers
DDS	Drug Delivery Systems
IS	Immune System
CPP	Cell-Penetrating-Peptide
MEMS	Micro-electro-mechanical Systems
UV-Vis	Ultraviolet-Visible
SSE	Summed Square Error
EE	Encapsulation Efficiency
PLGA	poly(lactic-co-glycolic) Acids
BSA	Bovine Serum Albumin
PLA	Poly-lactic Acid
PBS	Phosphate-buffered saline

References

1. Chothe, P.P.; Nakakariya, M.; Rotter, C.J.; Sandoval, P.; Tohyama, K. Recent advances in drug transporter sciences: Highlights from the year 2020. *Drug Metab. Rev.* **2001**, *53*, 321–349. [CrossRef]
2. Borst, P.; Evers, R.; Kool, M.; Wijnholds, J. A family of drug transporters: The multidrug resistance-associated proteins. *J. Natl. Cancer Inst.* **2000**, *92*, 1295–1302. [CrossRef]
3. Sahoo, S.K.; Labhasetwar, V. Nanotech approaches to drug delivery and imaging. *Drug Discov. Today* **2003**, *8*, 1112–1120.
4. Zylberberg, C.; Matosevic, S. Pharmaceutical liposomal drug delivery: A review of new delivery systems and a look at the regulatory landscape. *Drug Deliv.* **2016**, *23*, 3319–3329. [CrossRef]

5. Coelho, J.; Trucillo, P.; Nobre, B.; Palavra, A.F.; Campardelli, R.; Reverchon, E. Extraction and bioprocessing with supercritical fluids. *Phys. Sci. Rev.* **2020**, *5*, 1–14. [CrossRef]
6. Balasubramanian, V.; Onaca, O.; Enea, R.; Hughes, D.W.; Palivan, C.G. Protein delivery: From conventional drug delivery carriers to polymeric nanoreactors. *Expert Opin. Drug Deliv.* **2010**, *7*, 63–78. [CrossRef]
7. D'Angelo, I.; Conte, C.; La Rotonda, M.I.; Miro, A.; Quaglia, F.; Ungaro, F. Improving the efficacy of inhaled drugs in cystic fibrosis: Challenges and emerging drug delivery strategies. *Adv. Drug Deliv. Rev.* **2014**, *75*, 92–111. [CrossRef]
8. Beloqui, A.; Solinís, M.Á.; Rodríguez-Gascón, A.; Almeida, A.J.; Préat, V. Nanostructured lipid carriers: Promising drug delivery systems for future clinics. *Nanomed. Nanotechnol. Biol. Med.* **2016**, *12*, 143–161. [CrossRef]
9. Jahromi, L.P.; Ghazali, M.; Ashrafi, H.; Azadi, A. A comparison of models for the analysis of the kinetics of drug release from PLGA-based nanoparticles. *Heliyon* **2020**, *6*, e03451. [CrossRef]
10. Patra, D.; Sengupta, S.; Duan, W.; Zhang, H.; Pavlick, R.; Sen, A. Intelligent, self-powered, drug delivery systems. *Nanoscale* **2013**, *5*, 1273–1283. [CrossRef]
11. Dutta, R.C. Drug carriers in pharmaceutical design: Promises and progress. *Curr. Pharm. Des.* **2007**, *13*, 761–769. [CrossRef]
12. Alvarez-Lorenzo, C.; Bromberg, L.; Concheiro, A. Light-sensitive intelligent drug delivery systems. *Photochem. Photobiol.* **2009**, *85*, 848–860. [CrossRef]
13. Langer, R.; Peppas, N.A. Advances in biomaterials, drug delivery, and bionanotechnology. *AIChE J.* **2003**, *49*, 2990–3006. [CrossRef]
14. Nigam, S.K. What do drug transporters really do? *Nat. Rev. Drug Discov.* **2015**, *14*, 29–44. [CrossRef]
15. Goldberg, M.; Langer, R.; Jia, X. Nanostructured materials for applications in drug delivery and tissue engineering. *J. Biomater. Sci.* **2007**, *18*, 241–268. [CrossRef]
16. Trucillo, P. Drug carriers: Classification, administration, release profiles, and industrial approach. *Processes* **2021**, *9*, 470. [CrossRef]
17. Lutz, J.F. Polymerization of oligo (ethylene glycol)(meth) acrylates: Toward new generations of smart biocompatible materials. *J. Polym. Sci. Part A Polym. Chem.* **2008**, *46*, 3459–3470. [CrossRef]
18. Langer, R. New methods of drug delivery. *Science* **1990**, *249*, 1527–1533. [CrossRef]
19. Davoodi, P.; Lee, L.Y.; Xu, Q.; Sunil, V.; Sun, Y.; Soh, S.; Wang, C.H. Drug delivery systems for programmed and on-demand release. *Adv. Drug Deliv. Rev.* **2018**, *132*, 104–138. [CrossRef]
20. Champion, J.A.; Katare, Y.K.; Mitragotri, S. Particle shape: A new design parameter for micro-and nanoscale drug delivery carriers. *J. Control. Release* **2007**, *121*, 3–9. [CrossRef]
21. Champion, J.A.; Katare, Y.K.; Mitragotri, S. Making polymeric micro-and nanoparticles of complex shapes. *Proc. Natl. Acad. Sci. USA* **2007**, *104*, 11901–11904. [CrossRef]
22. Holder, S.J.; Sommerdijk, N.A. New micellar morphologies from amphiphilic block copolymers: Disks, toroids and bicontinuous micelles. *Polym. Chem.* **2011**, *2*, 1018–1028. [CrossRef]
23. Trucillo, P.; Campardelli, R.; Scognamiglio, M.; Reverchon, E. Control of liposomes diameter at micrometric and nanometric level using a supercritical assisted technique. *J. CO2 Util.* **2019**, *32*, 119–127. [CrossRef]
24. Chen, Y.; Chen, N.; Feng, X. The role of internal and external stimuli in the rational design of skin-specific drug delivery systems. *Int. J. Pharm.* **2021**, *592*, 120081. [CrossRef]
25. Fuller, E.G.; Sun, H.; Dhavalikar, R.D.; Unni, M.; Scheutz, G.M.; Sumerlin, B.S.; Rinaldi, C. Externally triggered heat and drug release from magnetically controlled nanocarriers. *ACS Appl. Polym. Mater.* **2019**, *1*, 211–220. [CrossRef]
26. Choi, S.W.; Zhang, Y.; Xia, Y. A temperature-sensitive drug release system based on phase-change materials. *Angew. Chem. Int. Ed.* **2010**, *49*, 7904–7908. [CrossRef]
27. Wu, W.; Zhang, Q.; Wang, J.; Chen, M.; Li, S.; Lin, Z.; Li, J. Tumor-targeted aggregation of pH-sensitive nanocarriers for enhanced retention and rapid intracellular drug release. *Polym. Chem.* **2014**, *5*, 5668–5679. [CrossRef]
28. Luo, L.; Bian, Y.; Liu, Y.; Zhang, X.; Wang, M.; Xing, S.; Li, L.; Gao, D. Combined near infrared photothermal therapy and chemotherapy using gold nanoshells coated liposomes to enhance antitumor effect. *Small* **2016**, *12*, 4103–4112. [CrossRef]
29. Huang, Y.; Jiang, Y.; Wang, H.; Wang, J.; Shin, M.C.; Byun, Y.; He, H.; Liang, Y.; Yang, V.C. Curb challenges of the "Trojan Horse" approach: Smart strategies in achieving effective yet safe cell-penetrating peptide-based drug delivery. *Adv. Drug Deliv. Rev.* **2013**, *65*, 1299–1315. [CrossRef]
30. Casalini, T.; Rossi, F.; Lazzari, S.; Perale, G.; Masi, M. Mathematical modeling of PLGA microparticles: From polymer degradation to drug release. *Mol. Pharm.* **2014**, *11*, 4036–4048. [CrossRef]
31. Yun, Y.H.; Lee, B.K.; Park, K. Controlled drug delivery: Historical perspective for the next generation. *J. Control. Release* **2015**, *219*, 2–7. [CrossRef]
32. Bowers, L.D. Therapeutic monitoring for cyclosporine: Difficulties in establishing a therapeutic window. *Clin. Biochem.* **1991**, *24*, 81–87. [CrossRef]
33. Hampton Atkinson, J.; Patel, S.M.; Meyer, J.M.; Slater, M.A.; Zisook, S.; Capparelli, E. Is there a therapeutic window with some antidepressants for analgesic response? *Curr. Pain Headache Rep.* **2009**, *13*, 93–99. [CrossRef]
34. Khera, E.; Thurber, G.M. Pharmacokinetic and immunological considerations for expanding the therapeutic window of next-generation antibody–drug conjugates. *BioDrugs* **2018**, *32*, 465–480. [CrossRef]
35. Williams, D.F. On the mechanisms of biocompatibility. *Biomaterials* **2008**, *29*, 2941–2953. [CrossRef]
36. Kohane, D.S.; Langer, R. Biocompatibility and drug delivery systems. *Chem. Sci.* **2010**, *1*, 441–446. [CrossRef]

37. Tsai, N.C.; Sue, C.Y. Review of MEMS-based drug delivery and dosing systems. *Sens. Actuators A Phys.* **2007**, *134*, 555–564. [CrossRef]
38. Liepmann, D.; Pisano, A.P.; Sage, B. Microelectromechanical systems technology to deliver insulin. *Diabetes Technol. Ther.* **1999**, *1*, 469–476. [CrossRef]
39. Hultström, M.; Roxhed, N.; Nordquist, L. Intradermal insulin delivery: A promising future for diabetes management. *J. Diabetes Sci. Technol.* **2014**, *8*, 453–457. [CrossRef]
40. Trevitt, S.; Simpson, S.; Wood, A. Artificial pancreas device systems for the closed-loop control of type 1 diabetes: What systems are in development? *J. Diabetes Sci. Technol.* **2016**, *10*, 714–723. [CrossRef]
41. Abrahams, E.; Ginsburg, G.S.; Silver, M. The personalized medicine coalition. *Am. J. Pharm.* **2005**, *5*, 345–355. [CrossRef] [PubMed]
42. Patra, J.K.; Das, G.; Fraceto, L.F.; Campos, E.V.R.; Rodriguez-Torres, M.D.P.; Acosta-Torres, L.S.; Diaz-Torrez, L.A.; Grillo, R.; Swamy, K.M.K.; Sharma, S.; et al. Nano based drug delivery systems: Recent developments and future prospects. *J. Nanobiotechnol.* **2018**, *16*, 71. [CrossRef]
43. Dan, N. Vesicle-based drug carriers: Liposomes, polymersomes, and niosomes. In *Design and Development of New Nanocarriers*; William Andrew Publishing: Norwich, NY, USA, 2018; pp. 1–55.
44. Brandl, M. Liposomes as drug carriers: A technological approach. *Biotechnol. Annu. Rev.* **2001**, *7*, 59–85. [PubMed]
45. Torchilin, V.P. Liposomes as targetable drug carriers. *Crit. Rev. Ther. Drug Carr. Syst.* **1985**, *2*, 65–115.
46. Namdeo, A.; Jain, N.K. Niosomes as drug carriers. *Indian J. Pharm. Sci.* **1996**, *58*, 41–47.
47. Shahiwala, A.; Misra, A. Studies in topical application of niosomally entrapped nimesulide. *J. Pharm. Pharm. Sci.* **2002**, *5*, 220–225. [PubMed]
48. Wissing, S.A.; Kayser, O.; Müller, R.H. Solid lipid nanoparticles for parenteral drug delivery. *Adv. Drug Deliv. Rev.* **2004**, *56*, 1257–1272. [CrossRef] [PubMed]
49. Lee, J.S.; Feijen, J. Polymersomes for drug delivery: Design, formation and characterization. *J. Control. Release* **2012**, *161*, 473–483. [CrossRef]
50. Yu, C.Y.; Jia, L.H.; Yin, B.C.; Zhang, X.Z.; Cheng, S.X.; Zhuo, R.X. Fabrication of nanospheres and vesicles as drug carriers by self-assembly of alginate. *J. Phys. Chem. C* **2008**, *112*, 16774–16778. [CrossRef]
51. Zhang, J.; Li, C.; Wang, Y.; Zhuo, R.X.; Zhang, X.Z. Controllable exploding microcapsules as drug carriers. *Chem. Commun.* **2011**, *47*, 4457–4459. [CrossRef]
52. Ran, R.; Sun, Q.; Baby, T.; Wibowo, D.; Middelberg, A.P.; Zhao, C.X. Multiphase microfluidic synthesis of micro-and nanostructures for pharmaceutical applications. *Chem. Eng. Sci.* **2017**, *169*, 78–96. [CrossRef]
53. Chakravarty, P.; Famili, A.; Nagapudi, K.; Al-Sayah, M.A. Using supercritical fluid technology as a green alternative during the preparation of drug delivery systems. *Pharmaceutics* **2019**, *11*, 629. [CrossRef] [PubMed]
54. Lovskaya, D.D.; Lebedev, A.E.; Menshutina, N.V. Aerogels as drug delivery systems: In vitro and in vivo evaluations. *J. Supercrit. Fluids* **2015**, *106*, 115–121. [CrossRef]
55. Ghasemiyeh, P.; Mohammadi-Samani, S. Hydrogels as drug delivery systems; pros and cons. *Trends Pharm. Sci.* **2019**, *5*, 7–24.
56. Zhao, S.; Malfait, W.J.; Guerrero-Alburquerque, N.; Koebel, M.M.; Nyström, G. Biopolymer aerogels and foams: Chemistry, properties, and applications. *Angew. Chem. Int. Ed.* **2018**, *57*, 7580–7608. [CrossRef]
57. Yousefi, M.; Narmani, A.; Jafari, S.M. Dendrimers as efficient nanocarriers for the protection and delivery of bioactive phytochemicals. *Adv. Colloid Interface Sci.* **2020**, *278*, 102125. [CrossRef]
58. García-González, C.A.; Sosnik, A.; Kalmár, J.; De Marco, I.; Erkey, C.; Concheiro, A.; Alvarez-Lorenzo, C. Aerogels in drug delivery: From design to application. *J. Control. Release* **2021**, *332*, 40–63. [CrossRef]
59. Tamarkin, D.; Friedman, D.; Shemer, A. Emollient foam in topical drug delivery. *Expert Opin. Drug Deliv.* **2006**, *3*, 799–807. [CrossRef]
60. Trucillo, P.; Di Maio, E. Classification and Production of Polymeric Foams among the Systems for Wound Treatment. *Polymers* **2021**, *13*, 1608. [CrossRef]
61. Marrazzo, C.; Maio, E.; Iannace, S. Foaming of synthetic and natural biodegradable polymers. *J. Cell. Plast.* **2007**, *43*, 123–133. [CrossRef]
62. Marrazzo, C.; Di Maio, E.; Iannace, S.; Nicolais, L. Process-structure relationships in PCL foaming. *J. Cell. Plast.* **2008**, *44*, 37–52. [CrossRef]
63. Iqbal, M.A.; Md, S.; Sahni, J.K.; Baboota, S.; Dang, S.; Ali, J. Nanostructured lipid carriers system: Recent advances in drug delivery. *J. Drug Target.* **2012**, *20*, 813–830. [CrossRef] [PubMed]
64. Soares, J.S.; Zunino, P. A mixture model for water uptake, degradation, erosion and drug release from polydisperse polymeric networks. *Biomaterials* **2010**, *31*, 3032–3042. [PubMed]
65. Raza, S.N.; Khan, N.A. Role of mathematical modelling in controlled release drug delivery. *Int. J. Med. Res. Pharm. Sci* **2017**, *4*, 84–95.
66. Grassi, M.; Grassi, G. Mathematical modelling and controlled drug delivery: Matrix systems. *Curr. Drug Deliv.* **2005**, *2*, 97–116. [CrossRef] [PubMed]
67. Zhang, Y.; Huo, M.; Zhou, J.; Zou, A.; Li, W.; Yao, C.; Xie, S. DDSolver: An add-in program for modeling and comparison of drug dissolution profiles. *AAPS J.* **2010**, *12*, 263–271.

68. Arifin, D.Y.; Lee, L.Y.; Wang, C.H. Mathematical modeling and simulation of drug release from microspheres: Implications to drug delivery systems. *Adv. Drug Deliv. Rev.* **2006**, *58*, 1274–1325.
69. Siepmann, J.; Siepmann, F. Mathematical modeling of drug delivery. *Int. J. Pharm.* **2008**, *364*, 328–343. [CrossRef]
70. Siepmann, J.; Siegel, R.A.; Siepmann, F. Diffusion controlled drug delivery systems. In *Fundamentals and Applications of Controlled Release Drug Delivery*; Springer: Boston, MA, USA, 2012; pp. 127–152.
71. Sripetch, S.; Loftsson, T. Topical drug delivery to the posterior segment of the eye: Thermodynamic considerations. *Int. J. Pharm.* **2021**, *597*, 120332. [CrossRef]
72. Serra, L.; Doménech, J.; Peppas, N.A. Engineering design and molecular dynamics of mucoadhesive drug delivery systems as targeting agents. *Eur. J. Pharm. Biopharm.* **2009**, *71*, 519–528. [CrossRef]
73. Siepmann, J.; Peppas, N.A. Modeling of drug release from delivery systems based on hydroxypropyl methylcellulose (HPMC). *Adv. Drug Deliv. Rev.* **2012**, *64*, 163–174. [CrossRef]
74. Siepmann, J.; Peppas, N.A. Hydrophilic matrices for controlled drug delivery: An improved mathematical model to predict the resulting drug release kinetics (the "sequential layer" model). *Pharm. Res.* **2000**, *17*, 1290–1298. [CrossRef] [PubMed]
75. Muschert, S.; Siepmann, F.; Leclercq, B.; Carlin, B.; Siepmann, J. Prediction of drug release from ethylcellulose coated pellets. *J. Control. Release* **2009**, *135*, 71–79. [CrossRef]
76. Kedem, O.; Katchalsky, A. Thermodynamic analysis of the permeability of biological membranes to non-electrolytes. *Biochim. Et Biophys. Acta* **1958**, *27*, 229–246. [CrossRef]
77. Peppas, N.A.; Narasimhan, B. Mathematical models in drug delivery: How modeling has shaped the way we design new drug delivery systems. *J. Control. Release* **2014**, *190*, 75–81. [CrossRef]
78. Verma, R.K.; Mishra, B.; Garg, S. Osmotically controlled oral drug delivery. *Drug Dev. Ind. Pharm.* **2000**, *26*, 695–708. [CrossRef]
79. Gupta, B.P.; Thakur, N.; Jain, N.P.; Banweer, J.; Jain, S. Osmotically controlled drug delivery system with associated drugs. *J. Pharm. Pharm. Sci.* **2010**, *13*, 571–588. [CrossRef]
80. Paul, D.R. Elaborations on the Higuchi model for drug delivery. *Int. J. Pharm.* **2011**, *418*, 13–17. [CrossRef]
81. Ramteke, K.H.; Dighe, P.A.; Kharat, A.R.; Patil, S.V. Mathematical models of drug dissolution: A review. *Sch. Acad. J. Pharm.* **2014**, *3*, 388–396.
82. Singhvi, G.; Singh, M. In-vitro drug release characterization models. *Int. J. Pharm. Stud. Res.* **2011**, *2*, 77–84.
83. Rehman, Q.; Akash, M.S.H.; Rasool, M.F.; Rehman, K. Role of Kinetic Models in Drug Stability. In *Drug Stability and Chemical Kinetics*; Springer: Singapore, 2020; pp. 155–165.
84. Peppas, N.A. A model of dissolution-controlled solute release from porous drug delivery polymeric systems. *J. Biomed. Mater. Res.* **1983**, *17*, 1079–1087. [CrossRef] [PubMed]
85. Campelo, P.; Neves-Moreira, F.; Amorim, P.; Almada-Lobo, B. Consistent vehicle routing problem with service level agreements: A case study in the pharmaceutical distribution sector. *Eur. J. Oper. Res.* **2009**, *273*, 131–145. [CrossRef]
86. Colombo, P.; Bettini, R.; Santi, P.; Peppas, N.A. Swellable matrices for controlled drug delivery: Gel-layer behaviour, mechanisms and optimal performance. *Pharm. Sci. Technol. Today* **2000**, *3*, 198–204. [CrossRef]
87. Peppas, N.A.; Sahlin, J.J. A simple equation for the description of solute release. III. Coupling of diffusion and relaxation. *Int. J. Pharm.* **1989**, *57*, 169–172. [CrossRef]
88. Kiortsis, S.; Kachrimanis, K.; Broussali, T.; Malamataris, S. Drug release from tableted wet granulations comprising cellulosic (HPMC or HPC) and hydrophobic component. *Eur. J. Pharm. Biopharm.* **2005**, *59*, 73–83. [CrossRef]
89. Paarakh, M.P.; Jose, P.A.; Setty, C.M.; Christoper, G.P. Release kinetics–concepts and applications. *Int. J. Pharm. Res. Technol.* **2018**, *8*, 12–20.
90. Adrover, M.E.; Pedernera, M.; Bonne, M.; Lebeau, B.; Bucalá, V.; Gallo, L. Synthesis and characterization of mesoporous SBA-15 and SBA-16 as carriers to improve albendazole dissolution rate. *Saudi Pharm. J.* **2020**, *28*, 15–24. [CrossRef]
91. Bhardwaj, S.B.; Shukla, A.J.; Collins, C.C. Effect of varying drug loading on particle size distribution and drug release kinetics of verapamil hydrochloride microspheres prepared with cellulose esters. *J. Microencapsul.* **1995**, *12*, 71–81. [CrossRef]
92. Hopfenberg, H.B. Membranes. In *Polymers in Medicine and Surgery*; Springer: Boston, MA, USA, 1975; pp. 99–107.
93. Katzhendler, I.; Hoffman, A.; Goldberger, A.; Friedman, M. Modeling of drug release from erodible tablets. *J. Pharm. Sci.* **1997**, *86*, 110–115. [CrossRef]
94. Corrigan, O.I.; Li, X. Quantifying drug release from PLGA nanoparticulates. *Eur. J. Pharm. Sci.* **2009**, *37*, 477–485. [CrossRef]
95. Coughlan, D.C.; Quilty, F.P.; Corrigan, O.I. Effect of drug physicochemical properties on swelling/deswelling kinetics and pulsatile drug release from thermoresponsive poly (N-isopropylacrylamide) hydrogels. *J. Control. Release* **2004**, *98*, 97–114. [CrossRef] [PubMed]
96. Fitzgerald, J.F.; Corrigan, O.I. Mechanisms governing drug release from poly-α-hydroxy aliphatic esters: Diltiazem base release from poly-lactide-co-glycolide delivery systems. *Polym. Deliv. Syst.* **1993**, *23*, 311–326.
97. Papadopoulou, V.; Kosmidis, K.; Vlachou, M.; Macheras, P. On the use of the Weibull function for the discernment of drug release mechanisms. *Int. J. Pharm.* **2006**, *309*, 44–50. [CrossRef] [PubMed]
98. Kosmidis, K.; Argyrakis, P.; Macheras, P. A reappraisal of drug release laws using Monte Carlo simulations: The prevalence of the Weibull function. *Pharm. Res.* **2003**, *20*, 988–995. [CrossRef]
99. Ghosal, K.; Chandra, A.; Rajabalaya, R.; Chakraborty, S.; Nanda, A. Mathematical modeling of drug release profiles for modified hydrophobic HPMC based gels. *Die Pharm.-Int. J. Pharm. Sci.* **2012**, *67*, 147–155.

100. Sivak, W.N.; Zhang, J.; Petoud, S.; Beckman, E.J. Simultaneous drug release at different rates from biodegradable polyurethane foams. *Acta Biomater.* **2009**, *5*, 2398–2408. [CrossRef]
101. Sivak, W.N. Synthesis and Characterization of Novel Polyurethane Drug Delivery Systems. Ph.D. Thesis, University of Pittsburgh, Pittsburgh, PA, USA, 2007.
102. Cherng, J.Y.; Hou, T.Y.; Shih, M.F.; Talsma, H.; Hennink, W.E. Polyurethane-based drug delivery systems. *Int. J. Pharm.* **2013**, *450*, 145–162. [CrossRef]
103. De Geest, B.G.; Mehuys, E.; Laekeman, G.; Demeester, J.; De Smedt, S.C. Pulsed drug delivery. *Expert Opin. Drug Deliv.* **2006**, *3*, 459–462. [CrossRef]
104. Marin, A.; Muniruzzaman, M.; Rapoport, N. Acoustic activation of drug delivery from polymeric micelles: Effect of pulsed ultrasound. *J. Control. Release* **2001**, *71*, 239–249. [CrossRef]
105. Kikuchi, A.; Okano, T. Pulsatile drug release control using hydrogels. *Adv. Drug Deliv. Rev.* **2002**, *54*, 53–77. [CrossRef]
106. Göpferich, A. Mechanisms of polymer degradation and erosion. In *The Biomaterials: Silver Jubilee Compendium*; Elsevier: Amsrerdam, The Netherlands, 1996; pp. 117–128.
107. Zhu, X.; Braatz, R.D. A mechanistic model for drug release in PLGA biodegradable stent coatings coupled with polymer degradation and erosion. *J. Biomed. Mater. Res. Part A* **2015**, *103*, 2269–2279. [CrossRef] [PubMed]
108. Göpferich, A.; Tessmar, J. Polyanhydride degradation and erosion. *Adv. Drug Deliv. Rev.* **2002**, *54*, 911–931. [CrossRef]
109. Sevim, K.; Pan, J. A model for hydrolytic degradation and erosion of biodegradable polymers. *Acta Biomater.* **2018**, *66*, 192–199. [CrossRef] [PubMed]
110. Lao, L.L.; Peppas, N.A.; Boey, F.Y.C.; Venkatraman, S.S. Modeling of drug release from bulk-degrading polymers. *Int. J. Pharm.* **2011**, *418*, 28–41. [CrossRef] [PubMed]
111. Siepmann, J.; Faisant, N.; Akiki, J.; Richard, J.; Benoit, J.P. Effect of the size of biodegradable microparticles on drug release: Experiment and theory. *J. Control. Release* **2004**, *96*, 123–134. [CrossRef] [PubMed]
112. Batycky, R.P.; Hanes, J.; Langer, R.; Edwards, D.A. A theoretical model of erosion and macromolecular drug release from biodegrading microspheres. *J. Pharm. Sci.* **1997**, *86*, 1464–1477. [CrossRef]
113. Guo, Q.; Knight, P.T.; Mather, P.T. Tailored drug release from biodegradable stent coatings based on hybrid polyurethanes. *J. Control. Release* **2009**, *137*, 224–233. [CrossRef]
114. Siegel, R.A.; Rathbone, M.J. Overview of controlled release mechanisms. *Fundam. Appl. Control. Release Drug Deliv.* **2012**, *2*, 19–43.
115. Bajpai, A.K.; Shukla, S.K.; Bhanu, S.; Kankane, S. Responsive polymers in controlled drug delivery. *Prog. Polym. Sci.* **2008**, *33*, 1088–1118. [CrossRef]
116. Hotha, K.K.; Roychowdhury, S.; Subramanian, V. Drug-excipient interactions: Case studies and overview of drug degradation pathways. *Am. J. Anal. Chem.* **2016**, *7*, 107–140. [CrossRef]
117. von Ahn, A.; Dallegrave, A.; dos Santos, J.H.Z. Evaluation of the Cefalexin Drug Degradation Profile in Pharmaceutical Capsule Forms Based on Forced Degradation Studies. *Chromatographia* **2022**, *85*, 263–279. [CrossRef]
118. Prabhu, S.; Hossainy, S. Modeling of degradation and drug release from a biodegradable stent coating. *J. Biomed. Mater. Res. Part A* **2007**, *80*, 732–741. [CrossRef] [PubMed]
119. Hurrell, S.; Cameron, R.E. The effect of initial polymer morphology on the degradation and drug release from polyglycolide. *Biomaterials* **2002**, *23*, 2401–2409. [CrossRef]
120. Di Colo, G. Controlled drug release from implantable matrices based on hydrophobia polymers. *Biomaterials* **1992**, *13*, 850–856. [CrossRef]
121. Weaver, J.C.; Vaughan, T.E.; Chizmadzhev, Y. Theory of electrical creation of aqueous pathways across skin transport barriers. *Adv. Drug Deliv. Rev.* **1999**, *35*, 21–39. [CrossRef]
122. Wischke, C.; Behl, M.; Lendlein, A. Drug-releasing shape-memory polymers–the role of morphology, processing effects, and matrix degradation. *Expert Opin. Drug Deliv.* **2013**, *10*, 1193–1205. [CrossRef]
123. Visan, A.I.; Popescu-Pelin, G.; Socol, G. Degradation Behavior of Polymers Used as Coating Materials for Drug Delivery—A Basic Review. *Polymers* **2021**, *13*, 1272. [CrossRef]
124. Proikakis, C.S.; Mamouzelos, N.J.; Tarantili, P.A.; Andreopoulos, A.G. Swelling and hydrolytic degradation of poly (D, L-lactic acid) in aqueous solutions. *Polym. Degrad. Stab.* **2006**, *91*, 614–619. [CrossRef]
125. Kaunisto, E.; Marucci, M.; Borgquist, P.; Axelsson, A. Mechanistic modelling of drug release from polymer-coated and swelling and dissolving polymer matrix systems. *Int. J. Pharm.* **2011**, *418*, 54–77. [CrossRef]
126. Rizwan, M.I.; Damodharan, N. Mathematical modelling of dissolution kinetics in dosage forms. *Res. J. Pharm. Technol.* **2020**, *13*, 1339–1345.
127. Rossi, F.; Castiglione, F.; Ferro, M.; Moioli, M.; Mele, A.; Masi, M. The role of drug–drug interactions in hydrogel delivery systems: Experimental and model study. *ChemPhysChem* **2016**, *17*, 1615–1622. [CrossRef] [PubMed]
128. Iftime, M.M.; Dobreci, D.L.; Irimiciuc, S.A.; Agop, M.; Petrescu, T.; Doroftei, B. A theoretical mathematical model for assessing diclofenac release from chitosan-based formulations. *Drug Deliv.* **2020**, *27*, 1125–1133. [CrossRef] [PubMed]
129. Clercq, S.; Temelli, F.; Badens, E. In-Depth Study of Cyclodextrin Complexation with Carotenoids toward the Formation of Enhanced Delivery Systems. *Mol. Pharm.* **2021**, *18*, 1720–1729. [CrossRef] [PubMed]
130. Comin, L.M.; Temelli, F.; Saldaña, M.D. Barley β-glucan aerogels as a carrier for flax oil via supercritical CO_2. *J. Food Eng.* **2012**, *111*, 625–631. [CrossRef]

131. Zhao, L.; Temelli, F. Preparation of liposomes using supercritical carbon dioxide via depressurization of the supercritical phase. *J. Food Eng.* **2015**, *158*, 104–112. [CrossRef]
132. Zhao, L.; Temelli, F.; Curtis, J.M.; Chen, L. Encapsulation of lutein in liposomes using supercritical carbon dioxide. *Food Res. Int.* **2017**, *100*, 168–179. [CrossRef]
133. Li, X.; Sun, H.; Li, H.; Hu, C.; Luo, Y.; Shi, X.; Pich, A. Multi-Responsive Biodegradable Cationic Nanogels for Highly Efficient Treatment of Tumors. *Adv. Funct. Mater.* **2021**, *31*, 2100227. [CrossRef]
134. Liang, K.; Li, Z.; Luo, Y.; Zhang, Q.; Yin, F.; Xu, L.; Chwen, H.; Wang, H. Intelligent Nanocomposites with Intrinsic Blood–Brain-Barrier Crossing Ability Designed for Highly Specific MR Imaging and Sonodynamic Therapy of Glioblastoma. *Small* **2020**, *16*, 1906985. [CrossRef]
135. Trucillo, P.; Martino, M.; Reverchon, E. Supercritical Assisted Production of Lutein-Loaded Liposomes and Modelling of Drug Release. *Processes* **2021**, *9*, 1162. [CrossRef]
136. Sharmeen, S.; Rahman, A.M.; Lubna, M.M.; Salem, K.S.; Islam, R.; Khan, M.A. Polyethylene glycol functionalized carbon nanotubes/gelatin-chitosan nanocomposite: An approach for significant drug release. *Bioact. Mater.* **2018**, *3*, 236–244. [CrossRef]
137. Salgado, M.; Santos, F.; Rodríguez-Rojo, S.; Reis, R.L.; Duarte, A.R.C.; Cocero, M.J. Development of barley and yeast β-glucan aerogels for drug delivery by supercritical fluids. *J. CO2 Util.* **2017**, *22*, 262–269. [CrossRef]
138. Haghiralsadat, F.; Amoabediny, G.; Helder, M.N.; Naderinezhad, S.; Sheikhha, M.H.; Forouzanfar, T.; Zandieh-Doulabi, B. A comprehensive mathematical model of drug release kinetics from nano-liposomes, derived from optimization studies of cationic PEGylated liposomal doxorubicin formulations for drug-gene delivery. *Artif. Cells Nanomed. Biotechnol.* **2018**, *46*, 169–177. [CrossRef] [PubMed]
139. Drakoulas, G.; Kokkinos, C.; Fotiadis, D.; Kokkinos, S.; Loukas, K.; Moulas, A.N.; Semertzioglou, A. Coupled FEA Model with Continuum Damage Mechanics for the Degradation of Polymer-based coatings on Drug-Eluting Stents. In Proceedings of the 43rd Annual International Conference of the IEEE Engineering in Medicine & Biology Society, Virtual, 30 October–5 November 2021; pp. 4319–4323.
140. Yang, Z.; Luo, Y.; Hu, Y.; Liang, K.; He, G.; Chen, Q.; Wang, Q.; Chen, H. Photothermo-Promoted Nanocatalysis Combined with H2S-Mediated Respiration Inhibition for Efficient Cancer Therapy. *Adv. Funct. Mater.* **2021**, *31*, 2007991. [CrossRef]
141. Li, J.; Yu, X.; Jiang, Y.; He, S.; Zhang, Y.; Luo, Y.; Pu, K. Second near-infrared photothermal semiconducting polymer nanoadjuvant for enhanced cancer immunotherapy. *Adv. Mater.* **2021**, *33*, 2003458. [CrossRef] [PubMed]
142. Sun, H.; Yu, T.; Li, X.; Lei, Y.; Li, J.; Wang, X.; Peng, P.; Ni, D.; Wang, X.; Luo, Y. Second near-infrared photothermal-amplified immunotherapy using photoactivatable composite nanostimulators. *J. Nanobiotechnol.* **2021**, *19*, 433. [CrossRef]
143. Zhang, C.; He, S.; Zeng, Z.; Cheng, P.; Pu, K. Smart Nano-PROTACs Reprogram Tumor Microenvironment for Activatable Photo-metabolic Cancer Immunotherapy. *Angew. Chem.* **2022**, *61*, e202114957.
144. Sun, X.; Agate, S.; Salem, K.S.; Lucia, L.; Pal, L. Hydrogel-based sensor networks: Compositions, properties, and applications—A review. *ACS Appl. Bio. Mater.* **2000**, *4*, 140–162. [CrossRef]
145. Dash, S.; Murthy, P.N.; Nath, L.; Chowdhury, P. Kinetic modeling on drug release from controlled drug delivery systems. *Acta Pol. Pharm.* **2010**, *67*, 217–223.
146. Elmas, A.; Akyüz, G.; Bergal, A.; Andaç, M.; Andac, O. Mathematical modelling of drug release. *Res. Eng. Struct. Mater.* **2000**, *6*, 63–86. [CrossRef]
147. Quiño, J.; Franke, K.; Ivanova, M.; Kareth, S.; Petermann, M.; Braeuer, A.S. In situ measurement of drug transport in porous silica gel. *Microporous Mesoporous Mater.* **2018**, *260*, 17–23. [CrossRef]
148. Badens, E.; Masmoudi, Y.; Mouahid, A.; Crampon, C. Current situation and perspectives in drug formulation by using supercritical fluid technology. *J. Supercrit. Fluids* **2018**, *134*, 274–283. [CrossRef]
149. Bouledjouidja, A.; Masmoudi, Y.; Sergent, M.; Trivedi, V.; Meniai, A.; Badens, E. Drug loading of foldable commercial intraocular lenses using supercritical impregnation. *Int. J. Pharm.* **2016**, *500*, 85–99. [CrossRef] [PubMed]
150. Metwally, A.A.; Hathout, R.M. Computer-assisted drug formulation design: Novel approach in drug delivery. *Mol. Pharm.* **2015**, *12*, 2800–2810. [CrossRef] [PubMed]
151. Mizera, M.; Muratov, E.N.; Alves, V.M.; Tropsha, A.; Cielecka-Piontek, J. Computer-aided discovery of new solubility-enhancing drug delivery system. *Biomolecules* **2020**, *10*, 913. [CrossRef]
152. Piotto, S.; Di Biasi, L.; Fino, R.; Parisi, R.; Sessa, L. Yada: A novel tool for molecular docking calculations. *J. Comput.-Aided Mol. Des.* **2016**, *30*, 753–759. [CrossRef]

MDPI AG
Grosspeteranlage 5
4052 Basel
Switzerland
Tel.: +41 61 683 77 34

Processes Editorial Office
E-mail: processes@mdpi.com
www.mdpi.com/journal/processes

www.ingramcontent.com/pod-product-compliance
Lightning Source LLC
Chambersburg PA
CBHW041411210926
51850CB00013B/3241
* 9 7 8 3 7 2 5 8 3 9 8 7 2 *